Applications of Homogenization Theory to the Study of Mineralized Tissue

Monographs and Research Notes in Mathematics

Series Editors:
John A. Burns, Thomas J. Tucker, Miklos Bona, Michael Ruzhansky

About the Series

This series is designed to capture new developments and summarize what is known over the entire field of mathematics, both pure and applied. It will include a broad range of monographs and research notes on current and developing topics that will appeal to academics, graduate students, and practitioners. Interdisciplinary books appealing not only to the mathematical community, but also to engineers, physicists, and computer scientists are encouraged.

This series will maintain the highest editorial standards, publishing well-developed monographs as well as research notes on new topics that are final, but not yet refined into a formal monograph. The notes are meant to be a rapid means of publication for current material where the style of exposition reflects a developing topic.

Summable Spaces and Their Duals, Matrix Transformations and Geometric Properties
Feyzi Basar, Hemen Dutta

Spectral Geometry of Partial Differential Operators (Open Access)
Michael Ruzhansky, Makhmud Sadybekov, Durvudkhan Suragan

Linear Groups: The Accent on Infinite Dimensionality
Martyn Russel Dixon, Leonard A. Kurdachenko, Igor Yakov Subbotin

Morrey Spaces: Introduction and Applications to Integral Operators and PDE's, Volume I
Yoshihiro Sawano, Giuseppe Di Fazio, Denny Ivanal Hakim

Morrey Spaces: Introduction and Applications to Integral Operators and PDE's, Volume II
Yoshihiro Sawano, Giuseppe Di Fazio, Denny Ivanal Hakim

Tools for Infinite Dimensional Analysis
Jeremy J. Becnel

Semigroups of Bounded Operators and Second-Order Elliptic and Parabolic Partial Differential Equations
Luca Lorenzi, Abdelaziz Rhandi

For more information about this series please visit: https://www.crcpress.com/Chapman—HallCRC-Monographs-and-Research-Notes-in-Mathematics/book-series/CRCMONRESNOT

Applications of Homogenization Theory to the Study of Mineralized Tissue

Robert P. Gilbert, Ana Vasilic, Sandra Klinge,
Alex Panchenko, and Klaus Hackl

CRC Press
Taylor & Francis Group
Boca Raton London New York

CRC Press is an imprint of the
Taylor & Francis Group, an **informa** business
A CHAPMAN & HALL BOOK

First edition published 2021
by CRC Press
6000 Broken Sound Parkway NW, Suite 300, Boca Raton, FL 33487-2742

and by CRC Press
2 Park Square, Milton Park, Abingdon, Oxon, OX14 4RN

CRC Press is an imprint of Taylor & Francis Group, LLC

ISBN: 9781584887911 (hbk)
ISBN: 9780429143380 (ebk)

Contents

Preface

One application of particular interest to the authors is bone acoustics as a means of bone quality assessment and non-invasive diagnostics of osteoporosis. Osteoporosis is a systemic disease defined by low bone mass and micro-architectural deterioration, leading to a high incidence of nontraumatic fractures. Currently, the diagnosis of osteoporosis is based on a measure of bone mineral density (**BMD**) using x-ray absorptiometric techniques. Even though BMD is considered to be the best available surrogate parameter for the risk of osteoporotic fractures [231], bone strength is a multi-factor issue and is determined by a combination of bone size, shape and material properties [297]. BMD can explain a substantial portion of the effects of these factors on bone strength, but does not completely explain fracture incidence [297, 290]. For example, osteoporotic BMD explains less than 50% of vertebral fracture incidence [371]. Changes in BMD due to therapies explain only a portion of the reductions in fracture incidence only 50 to 60 % are improved using anti-resorptive treatments [101, 94, 360]. Micro-architecture deterioration of trabecular bone and variations in tissue material properties, likely due to improper bone turnover, are considered to be likely responsible for these discrepancies. Indeed, micro-architecture analysis, in conjunction with changes in tissue material properties, are likely to explain the observed variations in trabecular bone mechanical properties relative to density [104, 234, 293, 307]. There is therefore a need for assessment of bone quality, i.e. factors independent of bone quantity as measured by DXA, that will influence biomechanical competence of the bone. One particular ultrasonic technique for assessing bone mineral density (BMD) is by calcaneal broadband ultrasonic attenuation (**BUA**) and speed of sound (**SOS**), which are highly correlated with calcaneal **BMD** [396, 394], [162]. In this method, the time is measured for sound to travel, in water, the distance between the two transducers. Then the experiment is repeated with a bone sample placed between the transducers. Two time measurements are taken, first without the bone sample in place and then with the bone sample in place. From the known velocity of sound in water, and the size of the bone sample, the velocity of the compression wave through the bone sample can be calculated. Another experiment involves computing the spectra of the phase velocity, $c(\omega)$, and that of the attenuation, $\alpha(\omega)$, where ω is the frequency of the sound wave. Using the above experimental set up, the spectral decomposition of the wave may be obtained by a simple computation [328]. Many investigations report attenuation depends linearly on frequency from 200 kHz to 600 kHz and in the range of 600 kHz to 1 MHz [256, 321, 320]. In this work, attenuation $a(f)$ is a linear function of frequency, i.e. $a(f) = (BUA)f + K$, with K being a constant. The term (BUA) is the gradient in $\frac{dB}{MHz}$ evaluated by linear regression. Despite the fact that this technique,which is based on linear regression of the attenuation and does not have any physical basis, it is commonly accepted that the (BUA) gradient is relevant to the evaluation of osteoporosis [121]. In this book we wish to provide a mathematical background for understanding the use of ultrasound methodology and to determine whether it is possible to diagnose osteoporosis in cancellous bone using ultrasound simulations and to recognize the onset and the prevention of osteoporosis. A good ultrasound model will include the trabecular matrix with an interstitial blood-marrow fluid. This should lead to a bio-mechanically feasible model containing various bone parameters which quantify the bone matrix. If the model accurately describes the bone matrix, it should be possible to assess the brittleness of the bone and more generally to measure bone characteristics relevant to predicting, bone strength/bone resistance to fracture. This would be done by determining the bone parameters by ultrasound interrogation. This process is referred to as an **inverse problem**. We have made several initial investigations of this type using the Biot model of a porous media to represent the cancellous bone [49, 54, 51, 52, 53].

With better macroscopic models, including more physical information at the microscopic scale it is anticipated that we should have more success than with the Biot model. Not only would we be concerned with obtaining more realistic *in vitro* models incorporating the complicated blood-bone constitutive law and the complex microstructure, we would like to consider the *in vivo* environment by including muscle in our model.

The geometry of our problem is associated with two length scales: a macroscopic scale L and a microscopic scale l. The latter is related to the size of the microstructural periodicity cell. The ratio $\varepsilon = \ell/L$ is a small parameter of the problem. As $\varepsilon \to 0$, the sequences of solutions converge (in a sense we make precise) to a solution of the effective system. The effective equations have the typical two-velocity structure, where the effective velocity equation is coupled to the corrector equation. The latter is needed to account for the contribution of high-frequency oscillations to the limiting strain rate. Because of the coupling, the effective constitutive equation is given in a rather implicit way. This, however, is also typical for non-linear problems. For linear equations, the corrector can be eliminated, which produces explicit constitutive equations. The main ingredient that makes elimination possible is an explicit approximation for a solution operator of the corrector equation. In the non-linear case, these approximations are much harder to find than in the linear case.

Prior to this study, similar acoustic homogenization problems were investigated only for linear constitutive equations. The interest in such problems is motivated by their importance for applications in geophysics and biological sciences. These problems are also interesting from the mathematical point of view because the structure of the effective equations is different from the structure of the ε-problems. The main feature that emerges in the limit, is the long-term, memory dependence appearing in the constitutive equations. Namely, a mixture of an elastic solid and a Newtonian fluid has effectively viscoelastic behavior. In [363, 362, 18, 267, 268] such viscoelastic equations were obtained formally by means of two-scale asymptotics, and convergence was not proved. Rigorous works [8, 20, 63, 135] employed two-scale convergence to deal with periodic geometries. A class of more general two-scale non-periodic geometries was considered in [138]. Recently, a quite general framework of G-convergence was used in [340] to deal with the even more general disordered (non-periodic) geometries, including geometries that cannot be modeled as stationary random. Furthermore, [340] takes into account the motion of the interface, albeit in a somewhat limited way. Since all previous works only dealt with the stationary interface case, incorporating interfacial motion in [340] hints at the possibility of developing a G-convergence homogenization theory for moving interface problems. Another remarkable fact proved in [340] is that the viscoelastic effective behavior is a generic feature of a broad class of solid-fluid incompressible composites under most general geometric assumptions.

The initial basis for the development of numerical homogenization approaches is provided in works studying linear composites [62, 190, 191, 246, 305, 402]; whereas, the approaches dealing with the modeling of nonlinear composites are of more recent date. The first results dealing with nonlinear heterogeneous materials were obtained by Talbot and Willis [378], who proposed a procedure extending the Hashin-Shtrikman method [182, 183, 184]. The approach of Talbot and Willis suggests the energy bounds of a nonlinear composite obtained by using the concept of a comparison linear homogeneous body. The main disadvantage arising here is the duality gap between the bounds obtained by using principles of minimum potential energy and minimum complementary energy. This problem is overcome in the work by Castañeda presenting the so-called new variational principle [70, 71] where the idea of the linear heterogeneous comparison body is introduced. More recently, Ponte Castañeda [72] and Lopez-Pamies and Ponte Castañeda [282] focused on the investigation of the class of transversely isotropic composites, and generated estimates for the effective behavior and loss of ellipticity in hyperelastic porous materials with random microstructure. In particular, deBotton et al. [40] have considered the response of a transversely isotropic fiber-reinforced composite made of two incompressible Neo-Hookean phases. Nowadays, numerical strategies take an important position in the modeling of heterogeneous materials and their diversity increases continuously. Some of the most used approaches used are the secant method [310, 377], the partitioning method [207, 208, 209], the adaptive hierarchical modeling method [324, 416], the micro-macro domain decomposition method [417, 418] and the multiscale concept [301, 123, 124, 384]. An important aspect of numerical homogenization is the decrease of computational effort. In order to avoid this drawback, Michel and Suquet [298, 299] suggest an approximate model for elastoplastic and elastoviscoplastic composites by using non-uniform, transformation fields. Alternatively, Yvonnet and He [409] developed a reduced multiscale method (R3M) where a reduced basis is obtained using proper orthogonal decompositions. This method provides an efficient way of capturing the dominant components of an infinite-dimensional process with only a finite number of modes [80]. Another efficient approach is presented in the work of Lebensohn [261]. Here the Fast Fourier Transformation (FFT) is used for the determination of the micromechanical fields in plastically deformed 3D polycrystals. Within the framework of the present study, the application of the numerical homogenization will be demonstrated on examples simulating cancellous bone. There are two single-scale approaches typically used for this tissue. The finite differences time domain (FDTD) method has been used in many contributions for the simulation of 2D and 3D problems [37, 38, 206, 176]. This method belongs to the group of direct methods for the solution of differential equations. Due to its straightforward implementation, it is very suitable for extension of basic models by introducing different additional effects, such as scattering [286, 260, 336, 178, 177]. Another common numerical approach is the finite element method (FEM) and especially its variant, high-resolution, digital-image based FEM. In this method, high-resolution, digital images are transformed into the voxel-based FE models which are used for a further analysis [2, 357].

The multiscale FEM is a new but promising method for simulating cancellous bone. This numerical approach is adapted to the modeling of heterogeneous material with a highly oscillatory microstructure [300, 419, 223, 220, 332, 126] and has already been used for modeling alternative composites [237, 238, 239]. The multiscale FEM is characterized by several important advantages: it is suitable for modeling the nonlinear composites and is highly flexible in the choice of the problems which can be simulated at separate levels of the multiscale approach. When the linear material behavior of the bone constituents is assumed, the simulations do not require much calculation time. Even if the representative volume element (RVE) varies in a macroscopic sample, the calculation can be accomplished within a realistic time.

Many problems in geophysics and engineering involve propagation of acoustic waves in granular materials such as soils or marine sediments. Existing theories of granular media fall mainly into two categories: statistical and phenomenological. The statistical approach is based on the analogy between a granular flow and a dense gas. The asymptotic expansions of the kinetic theory are used to derive effective equations (see, for instance, Jenkins & Savage (1983)[228]). This approach is relatively well developed, but produces good results only for fast rarefied flows with instantaneous inter-particle collisions. In the phenomenological theories of Goodman & Cowin [164], Harris [181], Kirchner [236], the equations of mass, momentum and energy balance are supplemented with the balance equations for additional state variables. These variables are difficult to access experimentally, so there is no consensus as to which variables adequately represent the state of a granular material. Without the benefit of direct measurements, it is also difficult to specify the required constitutive laws for the extra variables.

A notable representative of phenomenological theories which do not use extra variables is hypoplasticity, proposed by Kolymbas (1987) [245]. It seeks to model slow deformations of highly packed granular materials, in which grain-to-grain interactions play a dominant role. In hypoplastic models, the constitutive equation for the stress $\mathbf{T}$ takes the form of a differential equation

$$\dot{\mathbf{T}} = \mathbf{A}(\mathbf{T},c)e(\mathbf{v}) + \mathbf{b}(\mathbf{T},c) \parallel e(\mathbf{v}) \parallel, \tag{0.0.1}$$

where $\mathbf{A},\mathbf{b}$ are nonlinear functions of the stress and the solids volume fraction c, $e(\mathbf{v})$ is the strain rate (symmetric part of the velocity gradient), and $\parallel e(\mathbf{v}) \parallel = \sqrt{e(\mathbf{v})_{ij}e(\mathbf{v})_{ij}}$. Furthermore, $\dot{\mathbf{T}}$ denotes the Jaumann stress rate defined by

$$\dot{\mathbf{T}} = \frac{d\mathbf{T}}{dt} + \mathbf{TW} - \mathbf{WT}.$$

Here d/dt denotes the material time derivative, and **W** is the skew part of the velocity gradient. Nonlinear equations such as (9.9.9) present serious mathematical difficulties. It is also thought [189] that hypoplastic models are not applicable to small acoustic deformations.

In Schaeffer [369] the constitutive equations are of the form $\frac{d\mathbf{T}}{dt} = \mathbf{A}(\mathbf{T})\nabla\mathbf{v}$. The function **A**, which is different for loading and unloading, plays the role of a yield criterion, so that for large stresses the governing system of equations becomes ill-posed. The ill-posedness is thought to be responsible for the formation of shear bands observed experimentally in steady shearing of granular materials.

It seems that acoustics of highly packed granular materials cannot be modeled within the statistical framework because of high concentrations and prolonged contacts, while the phenomenological theories referenced above are complicated and may be inapplicable to wave propagation.

Recently, a linear phenomenological theory of wave propagation in unconsolidated solid-fluid mixtures was proposed by Buckingham (1997, 1998, 2000) [59, 60, 61]. Buckingham's theory is based on the assumption that the grain-to-grain shearing is the primary mechanism for energy dissipation. Postulating a viscoelastic stress-strain relation for a pair of grains in frictional contact, Buckingham obtained linear viscoelastic acoustic equations. The main feature of his theory is the slow time decay of the memory kernels (power-type instead of the exponential). Buckingham's theory correctly predicts dependence of attenuation on frequency. However, results of recent experiments with water-saturated sand in Isakson *et al.* (2003) suggest that part of the material behaves as a solid frame, while Buckingham's theory assumes that a sediment is completely unconsolidated. For a partially consolidated medium, elasticity of the frame must play an important role. This leads to a possibility that some terms might be missing in Buckingham's equations.

Several authors (Sunchez-Palencia [363] (1980), Burridge & Keller [63] (1981), Levy [269] (1987), Bakhvalov & Panasenko [20] (1989), Gilbert & Mikelic [135] (2000), Gilbert & Panchenko [137] (1999)) obtained effective acoustic equations of porous media using homogenization. The homogenization approach allows one to derive effective equations for composite materials from the so-called microscopic balance equations, satisfied locally in each phase, and conditions on the interface between the phases. In Sanchez-Palencia [363], Burridge & Keller [63], Clopeau, Ferrín, Gilbert & Mikelic [] (2003 it was assumed, often implicitly, that the interface between the solid and fluid phases is stationary and the solid phase is connected. The stationary interface assumption is consistent with the typical assumption in acoustics linearization of the microscopic equations.[1] It is also convenient for technical reasons, since homogenization of the moving interface problems is much more difficult. When the solid phase is connected and the interface is static, friction between different parts of the solid phase is impossible, and the energy dissipation is due exclusively to fluid viscosity. Assuming further that the fluid viscosity is not negligible,

[1]In the referential description, the interface is static. Linearized equations are derived assuming that spatial and referential descriptions are identical.

one can average the linearized microscopic acoustic equations to obtain viscoelastic equations of motion. These equations are similar to Buckingham's equations, but the significance of this similarity is not clear, because the history-dependence in Buckingham's equations is due to a different physical phenomenon. The memory kernels in the averaged acoustic equations seem to decay exponentially in time, whereas the kernels in Buckingham's theory decay at a much slower power rate. This discrepancy, together with the experimental results from Isakson *at al.* (2003) [224], shows that the central issue, namely the role played by inter-granular friction in modeling the effective, constitutive equations, is not well understood at present.

To separate this issue from the problems of flow analysis, we study an idealized granular composite which consists of a consolidated viscoelastic matrix with a large number of periodically spaced holes containing rigid particles. The matrix is chosen to be viscoelastic, rather than perfectly elastic, because in geophysical materials the pores of the elastic skeleton are typically filled with fluid. The particles, representing the unconsolidated part of the solid phase, are in frictional contact with the matrix. The frictional forces are modeled initially by the point-wise Coulomb-type law with normal compliance (Kikuchi & Oden (1988)) [235]. The contact conditions of Coulomb type are formulated as inequalities involving tangential forces on the contact surface. The corresponding micro-scale problem has the form of a variational inequality. The most likely outcome of averaging such a model would be another variational inequality, or more generally, an inclusion. To guarantee that the averaged model has the form of an equation, rather than inequality, one needs to work with a micro-scale model of the same type. To obtain this simplified model, we approximate Coulomb contact conditions by nonlinear equations. This idea is borrowed from the papers on analysis of variational inequalities, in particular Kuttler & Shillor (1999) [247].

ACKNOWLEDGEMENTS

This book is a result of years of research collaborations, which in addition to the authors, includes Ming Fang, Philippe Guyenne, George Hsiao, Andro Mikelic, Robert Ronkese, Michael Shoushani, Armand Wirgin and Xuming Xie. References to their contributions are recorded in the text. Robert Gilbert and Klaus Hackl are grateful to the Mathematiches Institut Oberwolfach for granting them three weeks residency for a Research in Pairs Stay to plan the manuscript.

1 Introductory Remarks

In this chapter we introduce some of the basic concepts which will be used in subsequent chapters. These are elements of Functional Analysis. To this end we define some basic notions and theorems. We present essential theorems concerning concepts such as strong and weak derivatives, the Trace Theorem, Korn's lemma, the Lax-Milgram Theorem, etc. The reader should see in addition the books [350, 347, 303, 4, 133].

1.1 SOME FUNCTIONAL SPACES

Let Ω be a domain, i.e. an open connected set in $\mathbb{R}^n$, and let $\bar{\Omega}$ stand for the closure of Ω in $\mathbb{R}^n$. We define C_0^∞ to be the space of infinitely differentiable functions having compact support with respect to Ω. We define $C^k(\bar{\Omega})$ to be the class of k-differentiable functions, i.e. the functions possessing all derivatives of order up to and including k. These functions clearly obey the Hlder condition as long as $k - \lfloor k \rfloor > 0$, where $\lfloor k \rfloor$ is the maximum integer not bigger than k.

Also recall that the $L^p(\Omega)$-space, where $1 \leq p \leq \infty$, is the space of measurable functions defined on Ω, such that the associated norms

$$\|f\|_{L^p(\Omega)} := \left(\int_\Omega |f(x)|^p \, d\mathbf{x} \right)^{\frac{1}{p}} \quad \text{for} \quad 1 \leq p < \infty,$$

$$\|f\|_{L^\infty(\Omega)} := \text{ess sup}_\Omega |f|, \quad \text{if} \quad p = \infty$$

are bounded. These spaces are Banach spaces, i.e. they are complete normed spaces. Recall that if $\mathbf{X}$ is a complex vector space, a norm on $\mathbf{X}$ is a function $\|\cdot\| : \mathbf{X} \to [0, \infty)$, such that

1. $\|x\| = 0$, if and only if $x = 0$,
2. $\|\lambda x\| = |\lambda| \|x\|$, for every $x \in \mathbf{X}, \lambda \in \mathbb{C}$.
3. $\|x + y\| \leq \|x\| + \|y\|$, for every $x, y \in \mathbf{X}$.

A normed space $\mathbf{X}$ is complete if every Cauchy sequence in $\mathbf{X}$ converges in $\mathbf{X}$. That is, if $\{x_n\} \subset \mathbf{X}$, and $\lim\limits_{m,n \to \infty} \|x_n - x_m\| = 0$, then there exists a limit $x \in \mathbf{X}$ such that $\lim\limits_{n \to \infty} \|x_n - x\| = 0$.

If $p = 2$, we get the Hilbert space $L^2(\Omega)$ which has a scalar product in $H^0(\Omega)$

$$(u, v)_0 := \int_\Omega u(\mathbf{x}) v(\mathbf{x}) \, d\mathbf{x}, \ \forall u, v \in L^2(\Omega),$$

where $H^0(\Omega)$ is the completion of $C(\bar{\Omega})$ with respect to the norm

$$\|u\|_{H^0(\Omega)} = \int_\Omega |u(\mathbf{x})| \, d\mathbf{x}.$$

Definition 1.1. *$C^{m,*}(\Omega)$ is the subset of functions $C^m(\Omega)$ whose norm (1.1.1), given below, is bounded.*

Definition 1.2. *A locally integrable function u is said to have a weak derivative u^α, if u^α is locally integrable on Ω and*

$$\int_\Omega \phi u^\alpha \, dx = (-1)^{|\alpha|} \int_\Omega u D^\alpha \phi \, dx, \quad \forall \phi \in C_0^\infty(\Omega),$$

where $C_0^\infty(\Omega)$ is the class of infinitely differentiable functions whose support lies in compact subsets of Ω.

Definition 1.3. *A function u is said to have strong derivatives of up to order p if there exists a sequence $\{u_k\} \subset C^{p,*}$ such that $D^\alpha u_k$ is a Cauchy sequence for all $|\alpha| \leq p$ which converges to $D^\alpha u$ in $L^2(\Omega)$.*

It is easy to see that all strong derivatives are weak derivatives and moreover, all weak derivatives are unique. Hence, strong derivatives are also unique.

Subsequently we introduce the space $H^p(\Omega)$, for an integer $p \geq 0$, as the completion of $C^{p,*}(\Omega)$ with respect to the norm

$$\|u\|_{H^p(\Omega)} = \left(\int_\Omega \sum_{|\alpha| \leq p} \|D^\alpha u(\mathbf{x})\|^2_{L^2(\Omega)} d\mathbf{x} \right)^{\frac{1}{2}}. \tag{1.1.1}$$

In the above, we have used the notation

$$D^\alpha := \frac{\partial^{|\alpha|} u(\mathbf{x})}{\partial_{x_1}^{\alpha_1} \cdots \partial_{x_n}^{\alpha_n}},$$

where $\alpha := (\alpha_1, \alpha_2, \ldots, \alpha_n)$ is a multi-index, $|\alpha| = \alpha_1 + \alpha_2 + \cdots + \alpha_n$, and $\alpha_1, \alpha_2, \ldots, \alpha_n$ are non-negative integers.

In particular $H^1(\Omega)$ is the closure of C^1 functions with the norm

$$\|u\|^2_{H^1(\Omega)} := \int_\Omega \left(\frac{\partial u}{\partial x_i} \frac{\partial u}{\partial x_i} + u^2 \right) dx.$$

The repeated index means the summation convention is applied. It should also be noted that for physics applications, all the quantities need to be in dimensionless form for the equations to make sense.

We denote the boundary of the domain Ω by $\partial\Omega$. We will assume that it is sufficiently smooth, that is, there is a local representation where the functions describing the boundary have the desired regularity. Every point on $\mathbf{x}_0 \in \partial\Omega$ has a neighborhood $\mathcal{N}(x_0)$ that, after an affine mapping, the set $\Omega \cap \mathcal{N}(x_0)$ can be described by the equation $x_n = \phi(x_1, x_2, \cdots, x_{n-1})$ where ϕ is a C^m function. Moreover, the space $H_0^m(\Omega)$ is the completion of the space $C^m(\bar{\Omega})$ such that these functions vanish in a neighborhood of $\partial\Omega$. Likewise, we will say that a domain is Lipschitz if the functions ϕ are Lipschitz.

The closure of $C_0^1(\Omega)$, the space of $C^1(\Omega)$ functions which vanish in a neighborhood of the boundary, is denoted by $H_0^1(\Omega)$. Hence, intuitively we may write

$$H_0^1(\Omega) = \{v : \ v \in H^1(\Omega), \ v = 0 \text{ on } \partial\Omega\}.$$

Finally $C_b(\mathbb{R}^n)$ is the space of continuous bounded functions in $\mathbb{R}^n$.

Next, we recall the Sobolev Imbedding theorem. For further details on Sobolev spaces and Fourier transforms, see Renardy and Rogers [350].

Definition 1.4. *If* $\mathbf{X}$ *and* $\mathbf{Y}$ *are Banach spaces then* $\mathbf{X}$ *is said to be continuously imbedded in* $\mathbf{Y}$, *denoted by* $\mathbf{X} \hookrightarrow \mathbf{Y}$ *if* $\mathbf{X} \subset \mathbf{Y}$ *and there is a constant* C, *such that*

$$\|u\|_Y \leq C\|u\|_X, \quad \forall u \in \mathbf{X}.$$

Lemma 1.1

If the Fourier transform of u denoted by $\mathscr{F}[u]$ is in the space $L^1(\mathbb{R}^n)$, then u is a continuous bounded function. Moreover, $\|u\|_\infty \leq (2\pi)^{-\frac{n}{2}} \|\mathscr{F}[u]\|_1$. ■

Theorem 1.1: Sobolev Imbedding

Let $s > n/2$. Then

$$H^s(\mathbb{R}^n) \hookrightarrow C_b(\mathbb{R}^n).$$

This means that $H^s(\mathbb{R}^n) \subset C_b(\mathbb{R}^n)$ and there exists a constant C such that $\|u\|_\infty \leq C\|u\|_{s,2}$, for every $u \in H^s(\mathbb{R}^n)$. ■

A variant of the following result was given by Agmon [4].

Definition 1.5. *If* $\Omega \subset \mathbb{R}^m$, Ω *is said to have the restricted cone property if* $\partial\Omega$ *has a locally finite open covering* $\mathscr{O}_i$ *and corresponding cones* C_i *with vertices at the origin, such that* $x + C_i \subset \Omega$ *for* $x \in \Omega \cap \mathscr{O}_i$.

Theorem 1.2

Let Ω have the restricted cone property and let $u \in H^m(\Omega)$ with $m > \frac{n}{2}$. Let $\ell = \lfloor m/2 \rfloor - 1$. Then u can be redefined on a set of measure zero such that $u \in C^\ell(\Omega)$. Moreover, for any $r \geq 1$, $|\alpha| \leq \ell, x \in \Omega$,

$$\|D^\alpha u(x)\| \leq \gamma r^{-(m-\frac{n}{2}-|\alpha|)} \left(\|u\|_{m,\Omega} + r^m \|u\|_{0,\Omega}\right), \tag{1.1.2}$$

where γ is a constant which depends only on Ω and m. ■

Theorem 1.3

Let $x_0 \in \Omega$ and $U \subset\subset \Omega$ be a neighborhood of x_0. Let $\Omega' \subset\subset \Omega$ be a domain such that the cones (of the restricted cone property) with vertices in U are contained in Ω'. Now if we choose a sequence of functions $\{u_k\} \subset C^m(\mathbb{R}^n)$ such that $u_k \to u$ in $H^m(\Omega)$. Then by using the inequality (1.1.2), with $\alpha = 0$ for the difference sequence $\{u - u_k\}$ it is seen that $u_k \to u$ uniformly on U to a continuous function. Likewise, $D^\alpha u_k$ converges uniformly to $D^\alpha u_0$ on U for $|\alpha| \leq \ell$. ■

The above result may be extended to manifolds of dimensions less than n, in particular to the domain's boundary $\partial\Omega$.

Definition 1.6. *If $M \subset \Omega \subset \mathbb{R}^n$ is a smooth $n-1$ dimensional manifold and u is a function defined in Ω, then the restriction $u|_M$ is called the trace of u on M.*

If $u \in H^1(\Omega)$ then the imbedding theorem gives meaning to the trace. Let $\{u_k\}$ be a sequence of $C^1(\Omega)$ functions, such that $\|u_k\|_{1,\Omega} < \infty$ and $u_k \to u \in H^1(\Omega)$. If $u_k|_M$ converges in $L^2(M)$ we define the trace of u on M as this limit function; moreover, from [4] we have further that if $u \in H^\alpha(\Omega)$, where $|\alpha| \leq \ell$, then $D^\alpha u_k|_M$ is converges to a continuous function on M. In particular, if $M = \partial\Omega$, then we may speak of the trace of $D^\alpha u|_{\partial\Omega}$. We refer to the trace of u onto $\partial\Omega$ as $\gamma_0 u$.

The space $H_0^1(\Omega)$ is a closed subspace of $H^1(\Omega)$ by the Trace Theorem, and therefore a Hilbert space [350].

1.1.1 PERIODIC FUNCTIONS

Another space of functions of interest to the problems presented in this book are those which are defined by periodically repeating unit structure on $\mathbb{R}^n$, such as the spaces $H_\#^0$ and $V_\#^1$ defined below. These spaces may be seen to be Hilbert spaces:

$$H_\#^0 := \{u : \ u \in L^2_{loc}(\mathbb{R}^n), \ u \text{ is } Q\text{–periodic } \}, \tag{1.1.3}$$

$$V_\#^1 := \{u : \ u \in H^1_{loc}(\mathbb{R}^n), \ u \text{ is } Q\text{–periodic } \}, \tag{1.1.4}$$

where $Q :=]-1,1[^n$ is the unit cube in $\mathbb{R}^n$.

$$(u,v)_{H_\#^0} := \int_Q u\, v\, dy; \qquad (u,v)_{V_\#^1} := \int_Q \left(\frac{\partial u}{\partial y_i} \frac{\partial v}{\partial y_i} + uv \right) dy.$$

For the sake of simplicity we assumed the period is the unit cube Q, whose size may be changed using an affine transformation. Note that $H_\#^0$ and $V_\#^1$ are *not* subsets of $L^2(\mathbb{R}^n)$ as the norms do not necessarily converge on $\mathbb{R}^n$. However, restrictions of functions in $H_\#^0$ and $V_\#^1$ to a bounded domain Ω belong to $L^2(\Omega)$ and $H^1(\Omega)$, respectively. However, we may define $H_\#^0$ with $L^2(Y)$ and $V_\#^1$ with $H^1(Y)$ by setting the traces on the opposite sides of ∂Y equal and hence these spaces are subsets of $L^2(\Omega)$.

We may identify the functions defined on $H_{\#}^m$ with functions defined on Q, i.e. $H_{\#}^m \subset H^m(Q)$. If $u \in C_{\#}^{\infty}$, then u has a Fourier expansion

$$u(x) = \sum_{\xi} a_{\xi} e^{2\pi i \xi \cdot x};$$

moreover, this series and the derived series

$$D^{\alpha}u(x) = (2\pi i)^{|\alpha|} \sum_{\xi} a_{\xi} \xi^{\alpha} e^{2\pi i \xi \cdot x}, \quad |\alpha| \leq m$$

are uniformly convergent. The norm on $H^m(Q)$ is given by

$$\|u\|_m^{\#} := \sum_{|\alpha|<m} \|D^{\alpha}u\|_{0,Q}^2. \tag{1.1.5}$$

It is then easy to show that the following theorem holds [4].

Theorem 1.4

There exists a constant $C > 0$ that depends on m, such that

$$\frac{1}{C}\|u\|_m^{\#} \leq \|u\|_{m,Q} \leq C\|u\|_m^{\#}. \tag{1.1.6}$$

Hence the completions of $C_{\#}^{\infty}$ in these two norms are equivalent. ■

1.1.2 LAX-MILGRAM THEOREM

Theorem 1.5: Lax-Milgram

Let $B[u,v]$ be a bilinear form on a Hilbert space H with norm $\|\cdot\|$. If there exist constants $c_1 > 0$ and $c_2 > 0$ such that for all $u, v \in V$

$$|B[u,v]| \leq c_1 \|u\| \, \|v\|, \quad \text{(continuity condition)}$$

and

$$|B[u,u]| \geq c_2 \|u\|^2, \quad \text{(coercivity condition)}$$

then for every bounded linear functional F on H, there exists a unique $w \in H$ such that

$$F(x) = B[x,w], \quad \forall x \in H.$$

■

The class of functions

$$H(\Omega,\nabla\cdot) := \left\{ u:\ u \in \left(L^2(\Omega)\right)^3,\ \nabla\cdot u \in L^2(\Omega) \right\}$$

forms a Hilbert space with the inner product

$$(u,v)_{H(\Omega,\nabla\cdot)} := (u,v)_{\left(L^2(\Omega)\right)^3} + (\nabla\cdot u, \nabla\cdot v)_{L^2(\Omega)}.$$

Theorem 1.6: Trace Theorem

Let n be the exterior normal to Ω. Then the trace operator $u \mapsto u\cdot n$ is defined on $H(\Omega,\nabla\cdot)$ and is a continuous operator from $H(\Omega,\nabla\cdot)$ to $\left(H^{-1/2}(\partial\Omega)\right)^3$. In particular, if $u \in \left(L^2(\Omega)\right)^3$ and $\nabla\cdot u = 0$, then $u\cdot n|_{\partial\Omega}$, the trace of u, makes sense and is an element of $\left(H^{-1/2}(\partial\Omega)\right)^3$. ■

$H(\Omega,\nabla\cdot)$ contains $H^1(\Omega)$, but the classical Trace Theorem does not hold; since the product $u\cdot n$ makes no sense on $\partial\Omega$; however, $\gamma_0 u \in \left(H^{-1/2}(\partial\Omega)\right)^3$.

Proof. Let $\varphi \in H^{1/2}(\partial\Omega)$. Then there exists a (generally not unique) continuous lift of φ from $H^{1/2}(\partial\Omega)$ to $H^1(\Omega)$; that is, there exists a function $\phi \in H^1(\Omega)$ such that $\phi|_{\partial\Omega} = \varphi$ and

$$\|\phi\|_{H^1(\Omega)} \leq c\,\|\varphi\|_{H^{1/2}(\partial\Omega)}.$$

Integrating by parts, we get

$$\int_\Omega u_i \frac{\partial\phi}{\partial x_i}\,dx + \int_\Omega \frac{\partial u_i}{\partial x_i}\,\phi\,dx = \int_{\partial\Omega} n_i\,u_i\,\varphi\,ds.$$

Note that the right-hand side is the duality inner product between $H^{-1/2}(\partial\Omega)$ and $H^{1/2}(\partial\Omega)$. Now

$$\int_{\partial\Omega} n_i\,u_i\,\varphi\,ds = < n_i\,u_i\,,\,\varphi >_{H^{-1/2}(\partial\Omega)\,,\,H^{1/2}(\partial\Omega)} \quad (1.1.7)$$

$$= (u,\nabla\phi)_{\left(L^2(\Omega)\right)^3} + (\nabla\cdot u\,,\,\phi)_{L^2(\Omega)}. \quad (1.1.8)$$

so

$$\left| < n_i\,u_i\,,\,\varphi >_{H^{-1/2}(\partial\Omega)\,,\,H^{1/2}(\partial\Omega)} \right| \leq 2\,\|u\|_{H(\Omega,\nabla\cdot)}\,\|\phi\|_{H^1(\Omega)} \quad (1.1.9)$$

$$\leq C\,\|u\|_{H(\Omega,\nabla\cdot)}\,\|\varphi\|_{H^{1/2}(\partial\Omega)}, \quad (1.1.10)$$

therefore

$$\|n_i\,u_i\|_{H^{-1/2}(\partial\Omega)} \leq C\,\|u\|_{H(\Omega,\nabla\cdot)}.$$

Remark. Smooth functions form a dense subset of $H(\Omega, \nabla\cdot)$. We define $n_i\, u_i$ by continuity for any function $u \in H(\Omega, \nabla\cdot)$.

Theorem 1.7

$$H = \left\{ u :\ u \in \left(L^2(\Omega)\right)^3,\ \nabla \cdot u = 0,\ u \cdot n|_{\partial\Omega} = 0 \right\},$$

and H is a closed subspace of both $H(\Omega, \nabla\cdot)$ and $\left(L^2(\Omega)\right)^3$. Moreover, $\left(L^2(\Omega)\right)^3$ admits the orthogonal decomposition

$$\left(L^2(\Omega)\right)^3 = H \oplus H^{\perp},$$

and

$$H^{\perp} = \left\{ u :\ u \in \left(L^2(\Omega)\right)^3,\ u = \nabla\phi \ \text{ for some } \ \phi \in H^1(\Omega) \right\}.$$

■

Proof. Let $u \in H$. Then since $\dfrac{\partial u_i}{\partial x_i} = 0$, integration by parts leads to

$$\int_\Omega \nabla\phi\ u\, dx = \int_\Omega \frac{\partial}{\partial x_i}(\phi\ u_i)\, dx = \int_{\partial\Omega} \phi\ u_i\, n_i\, ds = 0.$$

Theorem 1.8

$$V = \left\{ u :\ u \in \left(H_0^1(\Omega)\right)^3,\ \nabla \cdot u = 0 \right\},$$

and with $\langle \cdot\, ,\, \cdot \rangle$ denoting the inner product between $H^{-1}(\Omega)$ and $H_0^1(\Omega)$, a necessary and sufficient condition for

$$\langle g, u \rangle = 0, \qquad \forall u \in V$$

is that g is a gradient; that is, there exists $\phi \in L^2(\Omega)$ such that $g = \nabla\phi$. ■

Remarks.

1. Recall that with $L^2(\Omega)$ identified with its dual, the dual of $H_0^1(\Omega)$ is $H^{-1}(\Omega)$,
2. Function ϕ in the above theorem is unique up to an additive constant. It can be made unique by imposing an additional condition on it, which is typically the zero-average condition

$$< \phi > = \frac{1}{|\Omega|} \int_\Omega \phi\ dx = 0.$$

Lemma. *For every $u \in H_0^1(\Omega)$, the distributions Δu belong to $H^{-1}(\Omega)$.*

Proof. Let $w \in H_0^1(\Omega)$. By definition of distributional derivative

$$-<\Delta v, w>= \int_\Omega \frac{\partial v_i}{\partial x_k} \frac{\partial w}{\partial x_k}\, dx;$$

let us define $\mathbf{p} := \nabla u$, then

$$-\left\langle \frac{\partial p}{\partial x_i}, w_i \right\rangle = \int_\Omega p\, \nabla \cdot w\, dx,$$

by the Cauchy inequality the integrals above represent bounded, linear functionals on $H_0^1(\Omega)$. That is, they represent elements of $\left(H_0^1(\Omega)\right)' = H^{-1}(\Omega)$.
Let us consider the homogeneous Dirichlet problem for the Poisson equation

$$\begin{cases} -\Delta u = f & \text{in} \quad \Omega, \\ u = 0 & \text{on} \quad \partial\Omega \end{cases}$$

Theorem 1.9

If $f \in H^{-1}(\Omega)$, and in particular, if $f \in L^2(\Omega)$, then the equation above has a solution $v \in V$. ■

Proof. By virtue of Friedrich's inequality [229], the continuous bilinear form

$$a[u,v] := \int_\Omega \frac{\partial u_i}{\partial x_k} \frac{\partial w_i}{\partial x_k}\, dx$$

on $H_0^1(\Omega)$ is coercive. Choosing $w \in V$ as a test function, we have

$$-\int_\Omega \nabla p \cdot w\, dx + \int_\Omega \Delta v \cdot w\, dx + \int_\Omega f \cdot w\, dx = 0;$$

therefore by the Lax-Milgram Theorem, there exists a unique $v \in V$ such that

$$a[v,w] = \langle f, w \rangle, \qquad \forall w \in V.$$

Once v is found, we can determine p from

$$\langle -\nabla p + \Delta v + f\ ,\ w \rangle = 0, \qquad \forall w \in H_0^1(\Omega),$$

and we can show by the regularity theory that $p \in L^2(\Omega)$.

Lemma 1.2: Korn

For every Ω, a bounded domain with smooth boundary, there exists $\gamma > 0$ such that for all $v \in \left(H^1(\Omega)\right)^3$

$$\int_\Omega e_{ij}(v)e_{ij}(v)\,dx + \int_\Omega v_i\,v_i\,dx \geq \gamma\,\|u\|_{H^1(\Omega)}$$

■

1.2 VARIATIONAL FORMULATION

Let V be a linear space such that

$$H_0^m(\Omega) \subset V \subset H^m(\Omega),$$

and define a bilinear form by

$$B[u,v] := \sum_{\substack{|\alpha| \leq m \\ |\beta| \leq m}} \left(D^\alpha u\,,\, a_{\alpha\beta}\,D^\beta v \right)_{m,\Omega}$$

where $a_{\alpha\beta}$ are sufficiently smooth functions on $\bar{\Omega}$.

The bilinear form $B[u,v]$ is said to be *coercive* over V if there exist constants $c_0 > 0$ and $\lambda_0 \geq 0$ such that

$$\Re(B[u,u]) \geq c_0\,\|u\|_{m,\Omega}^2 - \lambda_0\,\|u\|_{0,\Omega}^2; \qquad \forall u \in V,$$

and is called *strongly coercive* over V if the inequality holds when $\lambda_0 = 0$.

By the Lax-Milgram Theorem, if B is a coercive bilinear form over V, then for every functional f on V, there *exists* a *unique* $u \in V$ such that

$$B[u,v] = (v,f)_{0,\Omega}; \qquad \forall v \in V.$$

Remarks.

1. The Dirichlet bilinear form

$$a[u,v] = \int_\Omega \frac{\partial u}{\partial x_i}\frac{\partial v}{\partial x_i}\,dx$$

satisfies the continuity condition but not the coercivity condition, in the Lax-Milgram Theorem. However, the bilinear form

$$b[u,v] := \int_Y a_{ij}\,\frac{\partial u}{\partial y_i}\,\frac{\partial v}{\partial y_j}\,dy + \int_Y u\,v\,dy$$

is coercive on V_Y.

2. On $H_0^1(\Omega)$, the Dirichlet bilinear form (stated above) is coercive. To show this, we need to use Friedrich's inequality, which asserts the existence of a constant $\gamma > 0$ such that

$$a[u,u] := \int_\Omega \nabla u \, \nabla u \, dx \geq \gamma \, \|u\|^2_{L^2(\Omega)}, \qquad \forall u \in H_0^1(\Omega).$$

Using this inequality we can write

$$\begin{aligned} a[u,u] &= \frac{1}{2}\int_\Omega \nabla u \, \nabla u \, dx + \frac{1}{2}\int_\Omega \nabla u \, \nabla u \, dx && (1.2.1)\\ &\geq \frac{1}{2}\|\nabla u\|^2_{(L^2(\Omega))^3} + \frac{\gamma}{2}\|u\|^2_{L^2(\Omega)} && (1.2.2)\\ &\geq \min\left(\frac{1}{2}, \frac{\gamma}{2}\right)\|u\|^2_{H_0^1(\Omega)}. && (1.2.3) \end{aligned}$$

Let B be the operator associated with the bilinear form

$$b[u,v] := \int_Y a_{ij} \frac{\partial u}{\partial y_i} \frac{\partial v}{\partial y_j} \, dy + \int_Y u \, v \, dy$$

and consider the Neumann problem of finding $u \in H^1(\Omega)$ such that for all test functions $v \in H^1(\Omega)$ we have

$$b[u,v] = \int_\Omega f \, v dx.$$

This corresponds to the classical Neumann problem of finding u satisfying

$$\begin{cases} -\Delta u + u = f & \text{in} \quad \Omega \\ \dfrac{\partial u}{\partial n} = 0 & \text{on} \quad \partial\Omega \end{cases}$$

with n the outer normal to $\partial\Omega$. The above variational formulation is obtained by multiplying both sides of the equation by an arbitrary test function $v \in H^1(\Omega)$ (which does not need to satisfy any boundary condition) and integrating by parts.

Remark. There is a very important difference between the Dirichlet boundary condition "$u = 0$ on $\partial\Omega$" and the Neumann boundary condition "$\partial u/\partial n = 0$ on $\partial\Omega$." The Dirichlet condition is automatically imposed on all functions in $H_0^1(\Omega)$, which is a closed subspace of $H^1(\Omega)$ by the Trace Theorem, and therefore a Hilbert space. The Neumann condition, however, is not imposed on the functions in H^1. The space

$$V^* := \left\{ v : \ v \in H^1(\Omega), \ \left.\frac{\partial u}{\partial n}\right|_{\partial\Omega} = 0 \right\}$$

is a subspace of $H^1(\Omega)$ but *not* a closed subspace of $H^1(\Omega)$; therefore, V^* is not a Hilbert space and hence the Lax-Milgram Theorem cannot be applied for it.

Now consider the Neumann problem

$$\begin{cases} -\Delta u = f & \text{in} \quad \Omega \\ \dfrac{\partial u}{\partial n} = 0 & \text{on} \quad \partial\Omega \end{cases}$$

For $f \equiv 0$, one solution is $u \equiv 0$, so the solution is not unique. In general, if u is a solution, then integration by parts yields

$$\int_\Omega f\, dx = -\int_\Omega \frac{\partial}{\partial x_i}\left(\frac{\partial u}{\partial x_i}\right) dx = -\int_{\partial\Omega} n_i \frac{\partial u}{\partial x_i}\, ds = -\int_{\partial\Omega} \frac{\partial u}{\partial n}\, ds = 0.$$

In other words, for this Neumann problem to have a solution, function f must satisfy the *compatibility condition*:

$$\int_\Omega f\, dx = 0;$$

that is, function f must have zero average value over Ω.

Suppose B is the operator associated with the bilinear form

$$b[u,v] := (u,v)_{H^1(\Omega)} := \int_\Omega \left(\frac{\partial u}{\partial x_i}\frac{\partial v}{\partial x_i} + uv\right) dx.$$

Then in $L^2(\Omega)$, the boundary value problem

$$\begin{cases} -\Delta u = f & \text{in} \quad \Omega \\ \dfrac{\partial u}{\partial n} = 0 & \text{on} \quad \partial\Omega \end{cases}$$

may be written in the abstract form $(B - I)u = f$. The eigenfunctions of B associated with the eigenvalue 1 satisfy $(B - I)e = 0$ or $Be = e$. Multiplying both sides by $e \in H^1(\Omega)$ with $\partial e/\partial n = 0$ on $\partial\Omega$, and integrating by parts yields

$$\int_\Omega (Be - e)e\, dx \quad = \quad \int_\Omega \frac{\partial}{\partial x_i}\left(\frac{\partial e}{\partial x_i}\right) e\, dx \tag{1.2.4}$$

$$\tag{1.2.5}$$

$$= \quad \int_{\partial\Omega} \frac{\partial u}{\partial n} e\, ds - \int_\Omega \left(\frac{\partial e}{\partial x_i}\right)^2 dx, \tag{1.2.6}$$

so

$$\int_\Omega \left(\frac{\partial e}{\partial x_i}\right)^2 dx = 0;$$

therefore, eigenfunctions e of B associated with eigenvalue 1 are constant on Ω. It turns out that eigenvalue 1 is simple, and the compatibility condition on f for existence of a solution for the boundary value problem becomes

$$f \perp \text{constant functions.}$$

To determine the value of the constant function, which should be orthogonal to f, we add the condition that the solution should have zero average value on Ω:

$$\int_{\Omega} u\, dx = 0.$$

1.3 GEOMETRY OF THE TWO-PHASE COMPOSITE

The porous medium we propose to study is obtained by a periodic arrangement of the pores. The formal description goes along the following lines:

First we define the geometrical structure inside the unit cell $\mathscr{Y} =]0,1[^n$, where $n = 2,3$. We take $\mathscr{Y}_s$ (the solid part) to be a closed subset of $\bar{\mathscr{Y}}$ and define $\mathscr{Y}_f = \mathscr{Y} \backslash \mathscr{Y}_s$ (the fluid part). Next we take the periodic repetition of $\mathscr{Y}_s$ all over $\mathbb{R}^n$ and define $\mathscr{Y}_s^k = \mathscr{Y}_s + k$, $k \in \mathbb{Z}^n$. Obviously the set $E_s = \bigcup_{k \in \mathbb{Z}^n} \mathscr{Y}_s^k$ obtained by taking the union of all solid parts is a closed subset of $\mathbb{R}^n$ and $E_f = \mathbb{R}^n \backslash E_s$ in an open set in $\mathbb{R}^n$. Following Allaire [6] we make the same assumptions on $\mathscr{Y}_f$ and E_f:

1. $\mathscr{Y}_f$ is an open connected set of strictly positive measure, with a Lipschitz boundary, and $\mathscr{Y}_s$ has strictly positive measure in $\bar{\mathscr{Y}}$ as well .
2. E_f and the interior of E_s are open sets with the boundary of class C^1, and are locally located on one side of their boundary. Moreover, the set E_f is connected.

Figure 1.3.1 represents a unit cell. The shaded area represents one phase and the clear area the other phase. For more examples of domains that satisfy the requirements on the boundary between two phases, see Allaire [6].

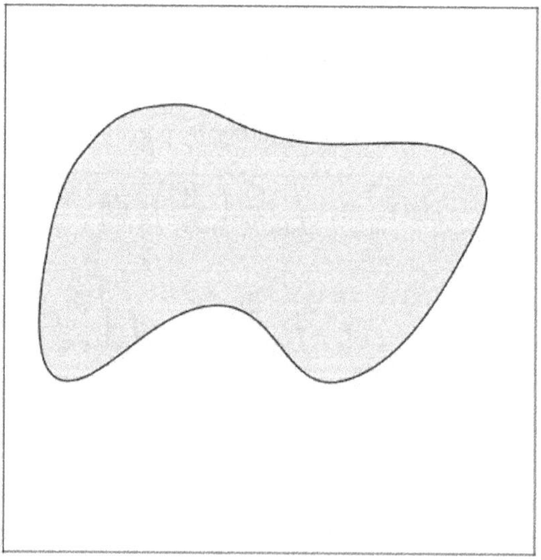

FIGURE 1.3.1 The unit cell for a two-phase system.

Next we define $\Omega :=]0, L[^n$. It can be seen that such domain Ω will be covered with a regular mesh of size ε, each cell being a cube $\mathscr{Y}_i^\varepsilon$, $1 \leq i \leq N(\varepsilon)$, where $N(\varepsilon) = |\Omega|\varepsilon^{-n}[1 + o(1)]$. Each cube $\mathscr{Y}_i^\varepsilon$ in Ω is homeomorphic to $\mathscr{Y}$, by linear homeomorphism Π_i^ε, being composed of translation and a homothety of ratio $1/\varepsilon$. We now define

$$\mathscr{Y}_{s_i}^\varepsilon = (\Pi_i^\varepsilon)^{-1}(\mathscr{Y}_s) \qquad \text{and} \qquad \mathscr{Y}_{f_i}^\varepsilon = (\Pi_i^\varepsilon)^{-1}(\mathscr{Y}_f).$$

For sufficiently small $\varepsilon > 0$ (for simplicity we suppose $L/\varepsilon \in \mathbb{N}^n$), we consider the sets

$$T_\varepsilon = \{k \in \mathbb{Z}^n | \ \mathscr{Y}_{S_k}^\varepsilon \subset \Omega\}$$
$$K_\varepsilon = \{k \in \mathbb{Z}^n | \ \mathscr{Y}_{S_k}^\varepsilon \cap \partial\Omega \neq \emptyset\}$$

and define

$$\Omega_s^\varepsilon = \bigcup_{k \in T_\varepsilon} \mathscr{Y}_{S_k}^\varepsilon, \quad \Omega_f^\varepsilon = \Omega \setminus \Omega_s^\varepsilon, \quad S^\varepsilon = \partial\Omega_s^\varepsilon.$$

The domains Ω_s^ε and Ω_f^ε represent, respectively, the solid and fluid parts of a porous medium Ω. Obviously, $\partial\Omega_f^\varepsilon = \partial\Omega \cup S^\varepsilon$.

Figure 1.3.2 represents a periodically repeated unit cell over the whole $\mathbb{R}^n$.

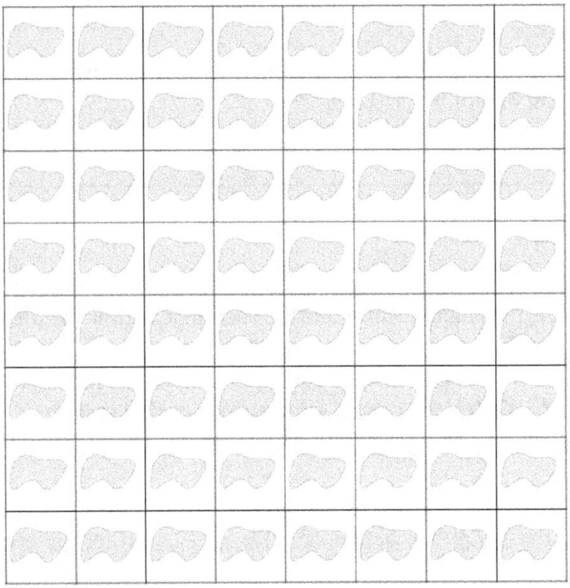

FIGURE 1.3.2 Repeated periodic unit cells.

1.4 TWO-SCALE CONVERGENCE METHOD

A very useful concept in homogenization theory, which we use in several chapters of this book, is that of **two-scale convergence**. Two-scale convergence was introduced by Nguetseng [315] and developed further by Allaire [8] and Zhikov [412]. The two-scale convergence is easier to use than some of the more general homogenization procedures such as $\Gamma-$, $G-$, or $H-$ convergence. However, it is restricted to the case where the medium has a periodic pore structure. A class of more general two-scale non-periodic geometries was considered in [138]. In addition, a quite general framework of G-convergence was used in [340] to deal with the even more general disordered (non-periodic) geometries, including geometries that cannot be modeled as stationary random. A remarkable fact proved in [340] is that the viscoelastic effective behavior is a generic feature of a broad class of solid-fluid incompressible composites under most general geometric assumptions.

We recall classical results from [315],[8], and [412].

Definition 1.7. *The sequence* $\{w^{\varepsilon}\} \subset L^2(\Omega)$ *is said to* **two-scale converge** *to a limit* $w \in L^2(\Omega \times \mathcal{Y})$ *iff for any* $\sigma \in C^{\infty}(\Omega; C^{\infty}_{\#}(\mathcal{Y}))$ *("#" denotes Q-periodicity) one has*

$$\lim_{\varepsilon \to 0} \int_{\Omega} w^{\varepsilon}(x)\sigma(x, \frac{x}{\varepsilon})\, dx = \int_{\Omega}\int_{\mathcal{Y}} w(x,y)\sigma(x,y)\, dy\, dx.$$

Lemma 1.3

From each bounded sequence in $L^2(\Omega)$ one can extract a subsequence which two-scale converges to a limit $w \in L^2(\Omega \times \mathcal{Y})$. ■

Lemma 1.4

(i) Let w^{ε} and $\varepsilon \nabla_x w^{\varepsilon}$ be bounded sequences in $L^2(\Omega)$. Then there exists a function $w \in L^2(\Omega; H^1_{\#}(\mathcal{Y}))$ and a subsequence such that both w^{ε} and $\varepsilon \nabla_x w^{\varepsilon}$ two-scale converge to w and $\nabla_y w$, respectively.

(ii) Let w^{ε} and $\nabla_x w^{\varepsilon}$ be bounded sequences in $L^2(\Omega)$. Then there exist functions $w \in L^2(\Omega)$, $v \in L^2(\Omega; H^1_{\#}(\mathcal{Y}))$ and a subsequence such that both w^{ε} and $\nabla_x w^{\varepsilon}$ two-scale converge to w and $\nabla_x w(x) + \nabla_y v(x,y)$, respectively. ■

For the proof of this result see [7, 8] and [315].

Remark. Let $\sigma \in L^2_{\#}(\mathcal{Y})$, define $\sigma^{\varepsilon}(x) = \sigma(\frac{x}{\varepsilon})$, and let the sequence $\{w^{\varepsilon}\} \subset L^2(\Omega)$ two-scale converge to a limit $w \in L^2(\Omega \times \mathcal{Y})$. Then $\{\sigma^{\varepsilon} w^{\varepsilon}\}$ two-scale converges to a limit σw.

Proposition 1.1. *(See Allaire [8] and Fasano, Mikelić, Primicerio [120])* *Let* $\{w^{\varepsilon}\} \subset H^1_{\#}(\Omega)^n$ *be a sequence such that* $||w^{\varepsilon}||_{L^2(\Omega)^n} \le C$, $||\nabla w^{\varepsilon}||_{L^2(\Omega^{\varepsilon}_s)^{n^2}}$

$\leq C$ and $||\nabla w^{\varepsilon}||_{L^2(\Omega_f^{\varepsilon})^{n^2}} \leq \frac{C}{\varepsilon}$. Then there exist functions $w \in H^1_{\#}(\Omega)^n$, $v \in L^2(\Omega; H^1_{\#}(\mathscr{Y})^n)$, $v = 0$ on $\overline{\mathscr{Y}}_s$ and $u^1 \in L^2(\Omega; H^1_{\#}(\mathscr{Y}_s)^n/\mathbb{R})$ such that, up to a subsequence,

$$w^{\varepsilon} \to w(x) + \chi_{\mathscr{Y}_f}(y)v(x,y) \text{ in the 2-scale sense,}$$

$$\chi_{\Omega_s^{\varepsilon}} \nabla w^{\varepsilon} \to \chi_{\mathscr{Y}_s}(y)[\nabla_x w(x) + \nabla_y u^1(x,y)] \text{ in the 2-scale sense,}$$

$$\varepsilon \chi_{\Omega_f^{\varepsilon}} \nabla w^{\varepsilon} \to \chi_{\mathscr{Y}_f}(y) \nabla_y v(x,y) \text{ in the 2-scale sense.}$$

1.5 THE CONCEPT OF A HOMOGENIZED EQUATION

The equivalent to solving the partial differential boundary value problem

$$-\frac{\partial}{\partial x_i}\left(a_{ij}(\frac{x}{\varepsilon})\,\frac{\partial \mathbf{u}}{\partial x_j}\right) = f_i(\mathbf{x}) \tag{1.5.1}$$

on the cell X with periodic boundary conditions is equivalent to solving variational problem: find $\mathbf{u} \in V_X$ such that

$$B[\mathbf{u},\mathbf{v}] := \int_X a_{ij}(\frac{x}{\varepsilon})\,\frac{\partial \mathbf{u}}{\partial x_j}\,\frac{\partial v}{\partial x_j}\,dx = \int_X f\,v\,dy; \qquad \forall v \in V_X,$$

which is obtained by multiplying both sides of the above equation by a test function $v \in V_Y$, integrating by parts, and using the fact that

$$\int_{\partial X} n_i\,a_{ij}\,\frac{\partial \mathbf{u}}{\partial x_j}\,v\,ds = 0$$

because of X-periodicity of v, as discussed before.

Conversely, if the variational formulation holds, then we can use the Y-periodicity property and interior regularity theory for elliptic equations in H^2_{loc} to obtain the differential formulation. Note that the bilinear form in the variational formulation is *not* coercive.

We now consider the weak form of equation (1.5.1), where the coefficients $a^{\varepsilon}_{ij}(x) := a_{ij}(\frac{x}{\varepsilon})$ depend on a small parameter ε

$$\int_X a^{\varepsilon}_{ij}(x)\,\frac{\partial \mathbf{u}}{\partial x_j}\,\frac{\partial v}{\partial x_j}\,dy = \int_X f\,v\,dx; \qquad \forall v \in V_X. \tag{1.5.2}$$

The standard technique[1] for seeking a solution to (1.5.2) is to seek an asymptotic solution in powers of $\varepsilon \to 0$, namely

$$\mathbf{u}(\mathbf{x}) = \mathbf{u}^0(\mathbf{x}) + \varepsilon^1\mathbf{u}^1(\mathbf{x},\mathbf{y}) + \varepsilon^2\mathbf{u}^2(\mathbf{x},\mathbf{y}) + \cdots \quad \text{where} \quad \mathbf{y} = \frac{\mathbf{x}}{\varepsilon}.$$

[1]Our discussion here is by now fairly standard and may be found in many books, for example Sanchez-Palencia [363, 362], Burridge and Keller [63], Bakhvalov and Panasenko [20], Jikjov, Kozlov and Oleinik [229], Oleinik, Shamaev and Yosifian [330].

If we substitute this equation into (1.5.2), recalling that

$$\frac{d\mathbf{u}}{dx_i} = \frac{\partial \mathbf{u}}{\partial x_i} + \frac{1}{\varepsilon}\frac{\partial \mathbf{u}}{\partial y_i}$$

and hence, we obtain

$$\frac{d\mathbf{u}^\varepsilon}{dx_i} = \left(\frac{\partial \mathbf{u}^0}{\partial x_i} + \frac{\partial \mathbf{u}^1}{\partial y_i}\right) + \varepsilon\left(\frac{\partial \mathbf{u}^1}{\partial x_i} + \frac{\partial \mathbf{u}^2}{\partial y_i}\right) + O(\varepsilon^2).$$

Moreover, if we define $\mathbf{p}$ component-wise as

$$p_i^\varepsilon(\mathbf{x}) := a_{ij}^\varepsilon \frac{\partial \mathbf{u}}{\partial x_j}(\mathbf{x}), \tag{1.5.3}$$

then expanding in powers of ε

$$p_i^\varepsilon(\mathbf{x}) = p_i^0(\mathbf{x},\mathbf{y}) + \varepsilon p_i^1(\mathbf{x},\mathbf{y}) + \varepsilon^2 p_i^2(\mathbf{x},\mathbf{y}) + O(\varepsilon^3)$$

it follows from (1.5.3 that

$$\frac{\partial p_i^0}{\partial y_i} = 0 \quad \text{and} \quad \frac{\partial}{\partial y_i}\left[a_{ij}(\mathbf{y})\left(\frac{\partial \mathbf{u}^0}{\partial x_j}(\mathbf{x}) + \frac{\partial \mathbf{u}^1}{\partial y_j}(\mathbf{x},\mathbf{y})\right)\right] = 0. \tag{1.5.4}$$

The second equation of (1.5.4) may be rewritten as

$$-\frac{\partial}{\partial y_i}\left(a_{ij}(\mathbf{y})\frac{\partial \mathbf{u}^1}{\partial y_j}\right) = \frac{\partial \mathbf{u}^0}{\partial x_j}(\mathbf{x})\frac{\partial(a_{ij}(\mathbf{y}))}{\partial y_i} \tag{1.5.5}$$

where we consider $\mathbf{u}^0$ as a known. This suggests we consider the solution of an alternate weak formulation, namely find $w^k \in V_Y$ with zero average $< w^k > = 0$ on Y, such that

$$\int_Y a_{ij}^\varepsilon \frac{\partial w^k}{\partial y_j}\frac{\partial v}{\partial y_i}\, dy = \int_Y \frac{\partial a_{ik}}{\partial y_i}\, v\, dy.$$

By linearity of equation (1.5.4) the solution can then be written in the form

$$\mathbf{u}(x,y) = \frac{\partial u^0(x)}{\partial x_k}\, w^k(y) + c(x),$$

where $c(\mathbf{x})$ is a function of x only. From (1.5.3) then

$$p_i^0(x,y) = a_{ij}(y)\left(\frac{\partial u^0(x)}{\partial x_j} + \frac{\partial u^0}{\partial x_k}\frac{\partial w^k}{\partial y_j}\right) \tag{1.5.6}$$

$$= \left(a_{ik} + a_{ij}\frac{\partial w^k}{\partial y_j}\right)\frac{\partial u^0(x)}{\partial x_k}. \tag{1.5.7}$$

Averaging over Y (with respect to the variable y) gives

$$< p_i^0 > (x) = a_{ik}^h \frac{\partial u^0(x)}{\partial x_k},$$

with the *homogenized coefficients* a^h_{ij} defined by

$$a^h_{ik}(x) \quad := \quad \left\langle a_{ik}(y) + a_{ij}(y)\, \frac{\partial w^k}{\partial y_j}(y) \right\rangle \tag{1.5.8}$$

(1.5.9)

$$= \quad \left\langle a_{ij}(y) \left(\delta_{jk} + \frac{\partial w^k}{\partial y_j}(y) \right) \right\rangle; \tag{1.5.10}$$

and from

$$-\frac{\partial}{\partial x_i} < p^0_i > (x) = f(x)$$

we arrive at the homogenized equation

$$-\frac{\partial}{\partial x_i} \left(a^h_{ik}(x)\, \frac{\partial u^0}{\partial x_k}(x) \right) = f(x).$$

Returning to the equation for w^k we have

$$\int_Y a_{ij}\, \frac{\partial w^k}{\partial y_j}\, \frac{\partial v}{\partial y_i}\, dy = \int_Y \frac{\partial a_{ik}}{\partial y_i}\, v\, dy.$$

Integration by parts yields

$$\int_Y a_{ij}\, \frac{\partial w^k}{\partial y_j}\, \frac{\partial v}{\partial y_i}\, dy = \int_{\partial Y} n_i\, a_{ik}\, v\, ds - \int_Y a_{ik}\, \frac{\partial v}{\partial y_i}\, dy.$$

The integral over the boundary of Y is zero by the Y-periodicity of the a_{ik}. Hence we have

$$0 \quad = \quad \int_Y a_{ij} \left(\frac{\partial w^k}{\partial y_j} + \delta_{jk} \right) \frac{\partial v}{\partial y_i}\, dy \tag{1.5.11}$$

(1.5.12)

$$= \quad \int_Y a_{mj}\, \frac{\partial}{\partial y_j} \left(w^k + y_k \right) \frac{\partial v}{\partial y_m}\, dy. \tag{1.5.13}$$

Taking $v = w^i$ and noting that

$$\frac{\partial \mathbf{u}}{\partial y_j} = \frac{\partial u^0}{\partial x_k}\, \frac{\partial w^k}{\partial y_j}$$

we get

$$\frac{1}{|Y|} \int_Y a_{mj}\, \frac{\partial u^0}{\partial x_i}\, \frac{\partial u^0}{\partial x_k}\, \frac{\partial}{\partial y_j} \left(w^k + y_k \right) \frac{\partial}{\partial y_m} \left(w^i + y_i \right) dy$$

$$= \frac{1}{|Y|} \int_Y a_{mj} \left(\frac{\partial \mathbf{u}}{\partial y_j} + \delta_{kj}\, \frac{\partial u^0}{\partial y_j} \right) \left(\frac{\partial \mathbf{u}}{\partial y_m} + \delta_{mi}\, \frac{\partial u^0}{\partial x_i} \right) dy.$$

To show that the sequence $\mathbf{u}^\varepsilon$ remains bounded as $\varepsilon \to 0$ starting with the variational formulation

$$\int_\Omega a_{ij}^\varepsilon \frac{\partial u^\varepsilon}{\partial x_i} \frac{\partial v}{\partial x_j}\, dx = \int_\Omega f\, v\, dx, \qquad \forall v \in H_0^1(\Omega)$$

we argue as follows: by setting $v = u^\varepsilon$ in the above we get

$$\gamma \int_\Omega \frac{\partial u^\varepsilon}{\partial x_i} \frac{\partial u^\varepsilon}{\partial x_i}\, dx \;\leq\; \int_\Omega a_{ij}^\varepsilon \frac{\partial u^\varepsilon}{\partial x_i} \frac{\partial u^\varepsilon}{\partial x_j}\, dx \tag{1.5.14}$$

$$\leq\; C' \, \|u^\varepsilon\|_{L^2(\Omega)} \tag{1.5.15}$$

$$\leq\; C'' \, \|u^\varepsilon\|_{H_0^1(\Omega)} \tag{1.5.16}$$

so

$$\gamma' \, \|u^\varepsilon\|^2_{H_0^1(\Omega)} \leq C'' \, \|u^\varepsilon\|_{H_0^1(\Omega)} \qquad \Longrightarrow \qquad \|u^\varepsilon\|_{H_0^1(\Omega)} \leq C.$$

This shows that the functions u^ε remain bounded in $H_0^1(\Omega)$. Consequently, by pre-compactness we can extract a subsequence of $\{u^\varepsilon\}_{\varepsilon>0}$ that converges weakly to some $u^* \in H_0^1(\Omega)$. The theorem is proven if we establish that $u^* = u^0$ for every sequence. Since u^ε is bounded in $H_0^1(\Omega)$, the $\partial u^\varepsilon / \partial x_i$ are bounded in $L^2(\Omega)$ so

$$\|p_i^\varepsilon(x)\|_{L^2(\Omega)} = \left\| a_{ij}^\varepsilon(x) \frac{\partial u^\varepsilon}{\partial x_j}(x) \right\|_{L^2(\Omega)} \leq C.$$

By pre-compactness, we can extract a subsequence of $\{p_i^\varepsilon\}_{\varepsilon>0}$ that converges weakly to some $p_i^* \in L^2(\Omega)$. Now from

$$\int_\Omega a_{ij}^\varepsilon \frac{\partial u^\varepsilon}{\partial x_i} \frac{\partial v}{\partial x_j}\, dx = \int_\Omega p_j^\varepsilon \frac{\partial v}{\partial x_j}\, dx = \int_\Omega f\, v\, dx, \qquad \forall v \in H_0^1(\Omega)$$

by passing to the limit as $\varepsilon \to 0$, for every fixed $v \in L^2(\Omega)$ we obtain

$$\int_\Omega p_i^* \frac{\partial v}{\partial x_i}\, dx = \int_\Omega f\, v\, dx, \qquad \forall v \in H_0^1(\Omega).$$

Let us suppose that the following holds

$$p_i^* = a_{ij}^h \frac{\partial u^*}{\partial x_j} \qquad \text{in } \Omega;$$

then this would show that $u^* \in H_0^1(\Omega)$ satisfies the variational formulation of the problem, and therefore by uniqueness of solution we would have $u^* = u^0$. So it only remains to prove the above equality. This is proved in Sanchez-Palencia following the approach of Tartar [363] Part II. We direct the reader to this reference for further details.

1.6 TWO-SCALE CONVERGENCE WITH TIME DEPENDENCE

Definition 1.8. $\{\mathbf{u}^{\varepsilon}(x,t)\} \subset [L^2([0,T]\times\Omega)]^n$ *two-scale converges to* $\mathbf{u}(t,x,y) \in [L^2([0,T]\times\Omega\times\mathbf{Y})]^n$ *iff for any* $\phi(t,x,y) \in [C^{\infty}([0,T]\times\Omega, C^{\infty}_{\#}(\mathbf{Y}))]^n$*, one has*

$$\lim_{\varepsilon\to 0}\int_0^T\int_{\Omega}\mathbf{u}^{\varepsilon}(t,x)\cdot\phi(t,x,\frac{x}{\varepsilon})\,dx\,dt = \int_0^T\int_{\Omega}\int_{\mathbf{Y}}\mathbf{u}(t,x,y)\cdot\phi(t,x,y)\,dy\,dx\,dt. \quad (1.6.1)$$

One can establish the following theorem with the obvious modifications of the argument in [315] and [8]:

Theorem 1.10

The following implications hold:

(i) If $\{\mathbf{u}^{\varepsilon}(x,t)\}$ is a bounded sequence in $[L^2([0,T],L^2(\Omega))]^n$, then there exists $\mathbf{u}_0(t,x,y) \in [L^2([0,T]\times\Omega, L^2_{\#}(\mathbf{Y}))]^n$ such that a subsequence of $\{\mathbf{u}^{\varepsilon}(t,x)\}$ two-scale converges to $\mathbf{u}_0(t,x,y)$ in the sense of Definition 1.8.

(ii) If $\{\mathbf{u}^{\varepsilon}(x,t)\}$ is a bounded sequence in $[L^2([0,T],H^1(\Omega))]^n$, then there exist $\mathbf{u}_0(t,x) \in [L^2([0,T],H^1(\Omega))]^n$ and $\mathbf{u}_1(t,x,y) \in [L^2([0,T]\times\Omega, H^1_{\#}(\mathbf{Y}))]^n$ such that a subsequence of $\{\mathbf{u}^{\varepsilon}(t,x)\}$ two-scale converges to $\mathbf{u}_0(t,x)$ and a subsequence of $\nabla_x\mathbf{u}^{\varepsilon}$ two-scale converges to $\nabla_x\mathbf{u}_0 + \nabla_y\mathbf{u}_1$ in the sense of Definition 1.8.

(iii) If $\{\mathbf{u}^{\varepsilon}(x,t)\}$ and $\{\varepsilon\nabla_x\mathbf{u}^{\varepsilon}(x,t)\}$ are bounded sequences in $[L^2([0,T],L^2(\Omega))]^n$, then there exists $\mathbf{u}_0(t,x,y) \in [L^2([0,T]\times\Omega, H^1_{\#}(\mathbf{Y}))]^n$ such that a subsequence of $\{\mathbf{u}^{\varepsilon}(t,x)\}$ and $\{\varepsilon\nabla_x\mathbf{u}^{\varepsilon}(x,t)\}$ two-scale converges to $\mathbf{u}_0(t,x,y)$ and $\nabla_y\mathbf{u}_0(t,x,y)$ in the sense of Definition 1.8.

■

1.7 POTENTIAL AND SOLENOIDAL FIELDS

A vector field $\mathbf{f} = (f_1, f_2, \dots, f_n)$, $f_i \in L^2_{\text{loc}}(\mathbb{R}^n)$ is called *vortex-free* in $\mathbb{R}^n$ if

$$\int_{\mathbb{R}^n}\left(f_i\frac{\partial\phi}{\partial x_j} - f_j\frac{\partial\phi}{\partial x_i}\right)dx = 0, \quad \forall\phi \in C_0^{\infty}(\mathbb{R}^n).$$

Recall that any vortex-free field $\mathbf{f}$ has a potential function $u \in H^1_{\text{loc}}(\mathbb{R}^n)$, so that $\mathbf{f} = \nabla u$.

A vector field $\mathbf{f}$ is said to be solenoidal in $\mathbb{R}^n$ if

$$\int_{\mathbb{R}^n} f_i\frac{\partial\phi}{\partial x_i}\,dx = 0, \quad \forall\phi \in C_0^{\infty}(\mathbb{R}^n).$$

The spaces of potential and solenoidal vector fields denoted by $\mathbf{L}^2_{\text{pot}}(\Omega)$ and $\mathbf{L}^2_{\text{sol}}(\Omega)$ respectively, form closed sets in $\mathbf{L}^2(\Omega) := (L^2(\Omega))^n$. In a random setting we shall refer to a vector field $\mathbf{f} \in \mathbf{L}^2(\Omega)$ as *potential* (*solenoidal*, respectively) if almost all of its realizations $\mathbf{f}(T(x)\omega)$ are potential (solenoidal, respectively) in $\mathbb{R}^n$.

It is known that the convergence in $\mathbf{L}^2(\Omega)$ implies the convergence of almost all realizations in $\mathbf{L}^2_{\text{loc}}(\mathbb{R}^n)$ [229]. In the random setting we use the following spaces:

$$\mathscr{V}^2_{\text{pot}}(\Omega) = \left\{\mathbf{f} \in \mathbf{L}^2_{\text{pot}}(\Omega), \mathbb{E}\{\mathbf{f}\} = 0\right\}, \tag{1.7.1}$$

$$\mathscr{V}^2_{\text{sol}}(\Omega) = \left\{\mathbf{f} \in \mathbf{L}^2_{\text{sol}}(\Omega), \mathbb{E}\{\mathbf{f}\} = 0\right\}, \tag{1.7.2}$$

where $\mathbb{E}\{\mathbf{f}\}$ is the expectation of $\mathbf{f}$ with respect to the probability measure μ, $\mathbf{L}^2_{\text{pot}}(\Omega)$, respectively $\mathbf{L}^2_{\text{sol}}(\Omega)$, is the set of all $\mathbf{f} \in (L^2(\Omega))^n$ such that almost all (a.a.) realizations of $\mathbf{f}(\mathscr{T}(x)\omega)$ are potential, respectively solenoidal, in $\mathbb{R}^n$. Moreover, the following result holds (see [229] for more details and the proof).

Lemma 1.5: Weyl's Decomposition

The following orthogonal decompositions are valid:

$$\mathbf{L}^2(\Omega) = \mathscr{V}^2_{\text{pot}}(\Omega) \oplus \mathscr{V}^2_{\text{sol}}(\Omega) \oplus \mathbb{R}^n$$

$$\mathbf{L}^2(\Omega) = \mathscr{V}^2_{\text{pot}}(\Omega) \oplus \mathbf{L}^2_{\text{sol}}(\Omega). \tag{1.7.3}$$

■

2 The Homogenization Technique Applied to Soft Tissue

2.1 HOMOGENIZATION OF SOFT TISSUE

For tissue in vivo there is a hysteresis relationship between stress and strain [128, 129], which has been adopted for modeling the viscoelastic behavior of soft tissue. Collagen is a major constituent of soft tissue [295]. The remaining material consists of elastin, reticulum and a hydrophilic gel. It is known that there is a hysteresis loop in cyclic loading and unloading of connective tissue. However, from curves in Fig. 2.1.1, it is clear that the mechanical properties of soft tissue are more dependent on their structure rather than the relative amounts of their constituents. For more details concerning elastin, collagen, and intestinal smooth muscle (see [129] pp. 265–266).

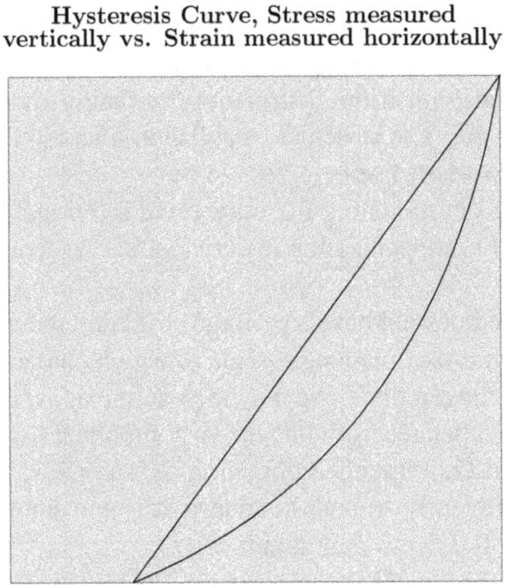
Hysteresis Curve, Stress measured
vertically vs. Strain measured horizontally

FIGURE 2.1.1 Hysteresis loops all follow this general form.

Elastin is almost perfectly elastic and the hysteresis loop is very small. Collagen is a viscoelastic tissue and the hysteresis loop is moderate. These loops occur for strains that are much larger than usually encountered by the material under normal conditions.

The hydrophilic gel varies with the type of tissue but it contains mucopolysaccharides and tissue fluid, i.e. ground material. The collagen fibrils are oriented roughly parallel to one another in the case of tendon and form a meshwork in the case of skin. Similar to tendons, muscle is composed of fascicles containing bundles of fibers as shown in Fig. 2.1.2.

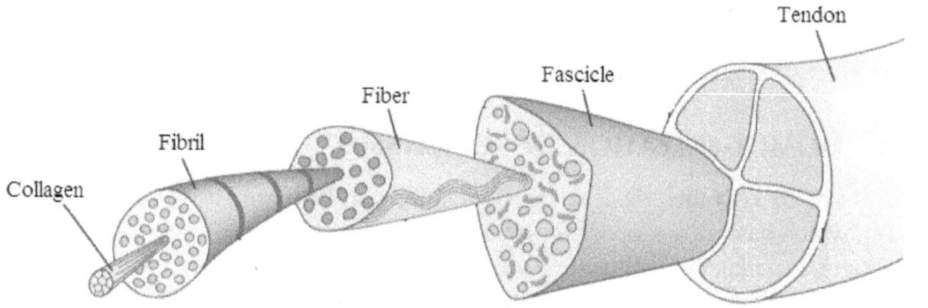

FIGURE 2.1.2 The fibrous structure of a tendon. Public domain image. https://commons.wikimedia.org/wiki

The collagen fibrils are assembled into bundles to form fibers which have diameters ranging from 0.2 to 12 μm. In tendon's collagen fibers are packaged into parallel-fibered bundles or fascicles. This suggests the reasonability of modeling tendons and ligaments as a material with a microstructure consisting of interspersed collagen fibers in a hydrophilic gel. The structure of collagen fibers in the skin is more like a three-dimensional network of fibrils [129].

As we are interested in modeling the ultrasound interrogation of tissue we shall employ the methods of homogenization to derive effective acoustic equations for the composite tissue.

Although soft tissue does not have a periodic structure, there is a scale separation. Suppose that there is microscopic length-scale ε, which characterizes some physical measurement, say, a collagen fibril. We may then define a fast variable $\mathbf{y} = \frac{\mathbf{x}}{\varepsilon}$, which allows us to record fast changes from one micro component to another. Let the tissue sample be contained in Ω, a bounded domain in $\mathbb{R}^n, n = 2,3$. For simplicity we assume that Ω has a sufficiently smooth boundary. It is convenient that we let the fast variable $\mathbf{y}$ live in $\mathbf{Y} = [0,1]^n, n = 2,3$, a unit cell.

As the purpose of the model is the acoustic interrogation of soft tissue, we introduce as the tissue displacement, $\mathbf{u}^{\varepsilon}(t,x)$. The assumption of a small parameter, ε, implies that $\mathbf{u}^{\varepsilon}$ must depend on two spatial displacement vectors on widely separated scales: namely $\mathbf{x}$, the "macroscopic variable", and "microscopic" variable $\mathbf{y} = \frac{\mathbf{x}}{\varepsilon}$. As we did earlier, we expand the displacement field $\mathbf{u}$ as an asymptotic expansion:

$$\mathbf{u}(t,\mathbf{x}) = \mathbf{u}^{\varepsilon}(t,\mathbf{x}) = \mathbf{u}_0(t,\mathbf{x},y) + \varepsilon \mathbf{u}_1(t,\mathbf{x},\mathbf{y}) + \varepsilon^2 \mathbf{u}_2(\mathbf{x},\mathbf{y}) + \cdots$$

In this expansion we can consider $\mathbf{u}_0$ as the macroscopic displacement, while $\mathbf{u}_1$, $\mathbf{u}_2$, ... are the microscopic displacements.

The soft tissue composite is an elastic collagen with interstitial viscoelastic gel. Following Fung [129], we introduce viscoelastic, history-dependent, constitutive tensor-equations, namely

$$\sigma = \mathbb{A} : \left(\mathbf{x}, \frac{\mathbf{x}}{\varepsilon}\right) \mathbf{e} + \int_0^t \mathbb{B}\left(\mathbf{x}, \frac{\mathbf{x}}{\varepsilon}, t-s\right) : \mathbf{e}(\mathbf{u}(s))\, ds$$

or in component form

$$\sigma_{ij} = a_{ijkl}\left(\mathbf{x}, \frac{\mathbf{x}}{\varepsilon}\right) e_{kl}(\mathbf{u}(t)) + \int_0^t b_{ijkl}\left(x, \frac{\mathbf{x}}{\varepsilon}, t-s\right) e_{kl}(\mathbf{u}(s))ds, \tag{2.1.1}$$

where $\mathbf{e}(\mathbf{u})$ is the strain tensor.
The elastic modulus $\mathbb{A}(\mathbf{x}, \frac{\mathbf{x}}{\varepsilon})$ represents an instantaneous elastic effect, which is assumed to be dominant in the collagen. It is assumed that the elastic coefficients satisfy the usual symmetry and the ellipticity conditions:

$$a_{ijkl} = a_{ijlk} = a_{jikl} = a_{klij}, \tag{2.1.2}$$
$$a_{ijkl}h_{ij}h_{kl} \geq \alpha h_{ij}h_{ij}, \qquad \alpha > 0. \tag{2.1.3}$$

As the acoustic displacements are small, the elastic response is dominant, with just small viscoelastic components in the collagen region. In the hydrophilic gel the material is primarily viscoelastic with memory. Homogenization, as shall be seen, leads to physically reasonable effective acoustic equations. The coefficients $b_{ijkl}(\mathbf{x}, \mathbf{y}, t)$, which are dominant in the hydrophilic gel, are assumed to be bounded, smooth, periodic in y and satisfy:

$$b_{ijkl} = b_{jikl}, \tag{2.1.4}$$

and the regularity assumption, namely, that the $\partial_t b_{ijkl}$ are bounded.

The equation of vibratory motion describing the acoustic displacement is given by

$$\ddot{\mathbf{u}} = \text{div}\, \sigma^{\varepsilon} + \mathbf{f} \text{ in } \Omega \tag{2.1.5}$$

On the boundary $\partial\Omega$, we assume a homogeneous Dirichlet condition

$$\mathbf{u}^{\varepsilon}|_{\partial\Omega} = 0, \tag{2.1.6}$$

and the initial conditions are assumed to be

$$\partial_t \mathbf{u}^{\varepsilon}(x,0) = 0, \qquad \mathbf{u}^{\varepsilon}(x,0) = 0. \tag{2.1.7}$$

We use the spaces:

$$H = [H_0^1(\Omega)]^n, \qquad V = [L^2(\Omega)]^n. \tag{2.1.8}$$

Hence, the problem may now be formulated as:

2.1. *Find* $\mathbf{u}^\varepsilon \in L^\infty([0,T],H), \partial_t \mathbf{u} \in L^\infty([0,T],V) \cap L^2([0,T],H), \partial_{tt}\mathbf{u} \in L^\infty([0,T],H')$ *so that* $\mathbf{u}^\varepsilon|_{t=0} = 0, \partial_t \mathbf{u}^\varepsilon|_{t=0} = 0$ *and* $\mathbf{u}$ *satisfies*

$$\int_\Omega \partial_{tt}\mathbf{u}^\varepsilon \cdot \phi \, dx + \int_\Omega a_{ijkl} e_{kl}(\mathbf{u}^\varepsilon) e_{ij}(\phi) dx + \int_0^t \int_\Omega b_{ijkl}(t-s) e_{kl}(\mathbf{u}^\varepsilon(s)) e_{ij}(\phi(t)) dx ds$$
$$= \int_\Omega \mathbf{f} \cdot \phi \qquad \text{for any } \phi \in H \tag{2.1.9}$$

where we assume $\mathbf{f} \in L^2([0,T],[L^2(U)]^n)$.

For simplicity of notation, we suppress ε dependence in this section. We first establish the uniqueness theorem.

Theorem 2.1

A weak solution to Problem (2.1) is unique solution. ■

Proof. We can model the proof on that in Duvaut, pp. 186-187, [108] and refer the reader to that paper for further details. □

2.2 GALERKIN APPROXIMATIONS

By analogy with the approach taken in [108, 112], we construct the weak solution of the hyperbolic integro-differential problem using Galerkin's method. More precisely, select smooth functions $\mathbf{w}_k = \mathbf{w}_k(x)$ $(k = 1, 2, \ldots)$ that form an orthogonal basis for H and an orthonormal basis for V. Such bases always exist for a Hilbert space. Now choose a positive integer m. We then look for a function $\mathbf{u}_m : [0,T] \to H$ in the form of a series

$$\mathbf{u}_m(t) := \sum_{k=1}^{m} d_m^k(t) \mathbf{w}_k,$$

where we intend to select the coefficient $d_m^k(t)$ $(0 \le t \le T, k = 1, \ldots, m)$ to satisfy

$$d_m^k(0) = d_m^{k\,\prime}(0) = 0 \qquad (k = 1, \ldots, m)$$

and such that

$$(\partial_{tt}\mathbf{u}_m, \mathbf{w}_j) + \mathscr{A}(\mathbf{u}_m, \mathbf{w}_j) + \left(\int_0^t \mathscr{B}(t-s)\mathbf{w}_m(s) ds, \mathbf{w}_j \right)$$
$$= d_m^{j\,\prime\prime} + \sum_{k=1}^{m} d_m^k \int_\Omega a e(\mathbf{w}_k) e(\mathbf{w}_j) dx + \sum_{k=1}^{m} \int_0^t \int_\Omega b(t-s) e(\mathbf{w}_k) e(\mathbf{w}_j) d_m^k(s) dx ds$$
$$= \int_\Omega \mathbf{f} \cdot \mathbf{w}_j \qquad j = 1, 2, \ldots, m, \tag{2.2.1}$$

where the meaning of the functions $\mathscr{A}$ and $\mathscr{B}$ is clear from (2.1.1). Since this is a linear system of ordinary integro-differential equations subject to homogeneous initial conditions, there exists a unique $\mathscr{C}^1$ function $\mathbf{d}_m(t) = (d_m^1(t), ..., d_m^m(t))$, satisfying homogeneous initial conditions and solving (2.2.1) for $0 \le t \le T$. We also know $\mathbf{d}_m^{''}(t) \in L^2([0,T])$. We now create a sequence of solutions as $m \to \infty$ and show that a subsequence of $\mathbf{u}_m$ converges to the weak solution. To this end, we will need some uniform estimates, as are given below [118].

Theorem 2.2

There exists a constant C, depending only on Ω, T, the coefficients a_{ij} and b_{ij}, such that

$$\max_{0 \le t \le T} (\|\mathbf{u}_m\|_H + \|\partial_t \mathbf{u}_m\|_V) + \|\partial_{tt} \mathbf{u}_m\|_{L^2(0,T;H^{-1})} \le C\|\mathbf{f}\|_{L^2(0,T;V)}$$

for $m = 1, 2, ...$ ∎

Proof. Multiply equation (2.2.1) by $d_m^{j\,'}$, sum $j = 1, ..., m$, and recall the construction of $\mathbf{u}_m$ to discover

$$(\partial_{tt}\mathbf{u}_m, \partial_t \mathbf{u}_m) + a(\mathbf{u}_m, \partial_t \mathbf{u}_m) = (\mathbf{f}, \partial_t \mathbf{u}_m) - \left(\int_0^t b(t-s)\mathbf{u}_m(s)ds, \partial_t \mathbf{u}_m\right) \quad (2.2.2)$$

for a.e. $0 \le t \le T$. It follows that

$$\frac{d}{dt}\left(\|\partial_t \mathbf{u}_m\|_{L^2(U)} + a(\mathbf{u}_m, \mathbf{u}_m)\right) \le (\|\mathbf{f}\|^2_{L^2(U)} + \|\partial_t \mathbf{u}_m\|^2_{L^2(U)}) - 2\left(\int_0^t b(t-s)\mathbf{u}_m(s)ds, \partial_t \mathbf{u}_m(t)\right) \quad (2.2.3)$$

and therefore

$$\begin{aligned} &\|\partial_t \mathbf{u}_m\|_{L^2(U)} + a(\mathbf{u}_m, \mathbf{u}_m) \\ \le\ & \int_0^t (\|\mathbf{f}\|^2_{L^2(U)} + \|\partial_t \mathbf{u}_m\|^2_{L^2(U)})ds - 2\int_0^t \left(\int_0^{s_1} b(s_1 - s)\mathbf{u}_m(s)ds, \partial_{s_1}\mathbf{u}_m(s_1)\right) ds_1 \end{aligned}$$

We note that, $\forall \lambda > 0$, there exists $\alpha > 0$ such that

$$a(\mathbf{u}_m, \mathbf{u}_m) + \lambda(\mathbf{u}_m, \mathbf{u}_m) \ge \alpha \|\mathbf{u}_m\|_H$$

and

$$\int_0^t ds_1 \left(\int_0^{s_1} B(s_1 - s)\mathbf{u}_m(s)ds, \partial_t \mathbf{u}_m(s_1)\right) \le C\left(\int_0^t \|\mathbf{u}_m\|^2 ds + \|\mathbf{u}_m\| \int_0^t \|\mathbf{u}_m\| ds\right),$$

which implies

$$\|\partial_t \mathbf{u}_m\|^2_{L^2(U)} + \|\mathbf{u}_m\|^2_H$$

$$\leq C\int_0^t \left(||\partial_t \mathbf{u}_m||_{L^2(U)} + ||\mathbf{u}_m||_H^2\right) ds + \int_0^T ||\mathbf{f}||^2_{L^2(U)} ds$$

Gronwall's inequality then yields the estimate

$$||\partial_t \mathbf{u}_m||^2_{L^2(U)} + ||\mathbf{u}_m||^2_H \leq C\int_0^T (||\mathbf{f}||^2_{L^2(U)}.ds$$

Finally, fix any $\mathbf{v} \in H$ with $||v||_H \leq 1$, and write $\mathbf{v} = \mathbf{v}^1 + \mathbf{v}^2$, where $\mathbf{v}^1 \in \text{span}\ \{\mathbf{w}_k\}_{k=1}^m$ and $\mathbf{v}^2, \mathbf{w}_k) = 0$ $(k = 1, ..., m)$. Then

$$(\partial_{tt}\mathbf{u}_m, v) = (\partial_{tt}\mathbf{u}_m, v^1) = (\mathbf{f}, v^1) - a(\mathbf{u}_m, v^1) - (\int_0^t b(t-s)\mathbf{u}_m(s)ds, v^1) \qquad (2.2.4)$$

Thus

$$||\partial_{tt}\mathbf{u}_m, \mathbf{v}|| \leq C(||\mathbf{f}||_{L^2(U)} + ||\mathbf{u}_m||_H)$$

since $||\mathbf{v}^1||_H \leq 1$. Consequently

$$\int_0^T ||\partial_{tt}\mathbf{u}_m||^2_{H^{-1}} dt \leq C\int_0^T ||\mathbf{f}||_{L^2(U)} + ||\mathbf{u}_m||_H dt \leq C||\mathbf{f}||^2_{L^2(0,T;L^2(U))}$$

□

Now we pass to limits in our Galerkin approximations.

Theorem 2.3

There exists a weak solution to Problem (2.1). ■

Proof. According to the energy estimates, there exists a subsequence $\{\mathbf{u}_{m_h}\}_{h=1}^\infty \in \{\mathbf{u}_m\}$ and $\mathbf{u} \in L^2(0,T;H_0^1(U))$ with $\partial_t \mathbf{u} \in L^2(0,T;L^2(U))$, $\partial_{tt}\mathbf{u} \in L^2(0,T;H^{-1}(U))$, such that

$$\begin{cases} \mathbf{u}_{m_h} \rightharpoonup \mathbf{u} & \text{weakly in } L^2(0,T;H_0^1(U)), \\ \mathbf{u}'_{m_h} \rightharpoonup \mathbf{u}' & \text{weakly in } L^2(0,T;L^2(U)), \\ \mathbf{u}_{m_h}\text{''} \rightharpoonup \mathbf{u}\text{''} & \text{weakly in } L^2(0,T;H^{-1}(U)). \end{cases}$$

Next fix an integer N and choose a function $\mathbf{v} \in C^([0,T];H_0^1(U))$ of the form

$$\mathbf{v} = \sum_{k=1}^N d^k(t)\mathbf{w}_k, \qquad (2.2.5)$$

where $-d^{k\text{''}}$ are smooth functions. We select $m \geq N$, multiply equation (2.2.1) by d^k, sum $k = 1, ..., m$, and then integrate with respect to t, to discover

$$\int_0^T (\partial_{tt}\mathbf{u}_m, \mathbf{v}) + a(\mathbf{u}_m, \mathbf{v})dt = \int_0^T (\mathbf{f}, \mathbf{v})dt - \int_0^T dt\left(\int_0^t B(t-s)\mathbf{u}_m(s)ds, \mathbf{v}(t)\right) \qquad (2.2.6)$$

We set $m = m_h$ and recall the weak convergence, to find in the limit that

$$\int_0^T (\partial_{tt}\mathbf{u}, \mathbf{v}) + a(\mathbf{u}, \mathbf{v})dt = \int_0^T (\mathbf{f}, \mathbf{v})dt - \int_0^T dt \left(\int_0^t B(t-s)\mathbf{u}(s)ds, \mathbf{v}(t) \right). \quad (2.2.7)$$

This equality then holds for all functions $\mathbf{v} \in L^2(0,T;H_0^1(U))$, since functions of the form (2.2.5) are dense in this space. The existence theorem is thereby proved. □

Passing to limits in Theorem (2.2), we deduce the following

Theorem 2.4: Regularity

$$\sup_{0 \le t \le T} (\|\mathbf{u}\|_H + \|\partial_t \mathbf{u}\|_V) \le C\|\mathbf{f}\|_{L^2(0,T;V)}$$

■

The regularity theorem (2.4) suggests that two-scale convergence makes sense for Problem (2.1). Let's consider a scalar function $\psi(t) \in \mathscr{C}^2[0,T]$, satisfying $\psi(T) = \psi'(T) = 0$. Multiplying (2.1.9) by $\psi(t)$ satisfying and integrating in time from 0 to T, we have

$$\begin{aligned} &\int_0^T \psi''(t)dt \int_\Omega \mathbf{u}^\varepsilon \cdot \phi + \int_0^T \psi(t)dt \int_\Omega [A^\varepsilon e(\mathbf{u}^\varepsilon)] : e(\phi) \\ &+ \int_0^T \psi(t)dt \int_0^t ds \int_\Omega [B^\varepsilon(t-s)e(\mathbf{u}^\varepsilon(s))] : e(\phi) \\ &= \int_0^T \psi(t)dt \int_\Omega \mathbf{f} \cdot \phi. \end{aligned} \quad (2.2.8)$$

If we replace the test function by $\phi(x) + \varepsilon\phi_1(x, \frac{x}{\varepsilon})$, then applying Theorem (1.10), we obtain:

$$\int_0^T \psi''(t)dt \int_\Omega \mathbf{u}^\varepsilon \cdot [\phi(x) + \varepsilon\phi_1\left(x, \frac{x}{\varepsilon}\right)] \to \int_0^T \psi''(t)dt \int_\Omega \mathbf{u}_0 \cdot \phi \quad (2.2.9)$$

$$\begin{aligned} &\int_0^T \psi(t)dt \int_\Omega [A^\varepsilon e_x(\mathbf{u}^\varepsilon)] : e_x([\phi(x) + \varepsilon\phi_1\left(x, \frac{x}{\varepsilon}\right)]) \\ &\quad \to \int_0^T \psi(t)dt \int_\Omega \int_{\mathbf{Y}} A[e_x(\mathbf{u}_0) + e_y(\mathbf{u}_1)] : (e_x(\phi) + e_y(\phi_1)) \end{aligned} \quad (2.2.10)$$

$$\int_0^T \psi(t)dt \int_0^t ds \int_\Omega [B^\varepsilon(t-s)e_x(\mathbf{u}^\varepsilon(s))] : e_x([\phi(x) + \varepsilon\phi_1\left(x, \frac{x}{\varepsilon}\right)]) \to$$

$$\int_0^T \psi(t)dt \int_0^t ds \int_\Omega \int_{\mathbf{Y}} B(t-s)[e_x(\mathbf{u}_0)+e_y(\mathbf{u}_1)] : (e_x(\phi)+e_y(\phi_1)) \quad (2.2.11)$$

$$\int_0^T \psi(t)dt \int_\Omega \mathbf{f}\cdot[\phi(x)+\varepsilon\phi_1\left(x,\frac{x}{\varepsilon}\right)] \to \int_0^T \psi(t)dt \int_\Omega \mathbf{f}\cdot\phi, \quad (2.2.12)$$

Lemma 2.1

Let $\{\mathbf{u}_0(t,x), \mathbf{u}_1(t,x,y)\}$ be as in Theorem (1.10), then $\{\mathbf{u}_0(t,x), \mathbf{u}_1(t,x,y)\}$satisfy the following equations in a distributional sense.

$$\frac{d^2}{dt^2}\int_\Omega \mathbf{u}_0\phi + \int_\Omega \int_{\mathbf{Y}_s} A[e_x(\mathbf{u}_0)+e_y(\mathbf{u}_1)] : e_x(\phi)$$
$$+\int_0^t ds \int_\Omega \int_{\mathbf{Y}} B(t-s)[e_x(\mathbf{u}_0(s))+e_y(\mathbf{u}_1(s))] : e_x(\phi) = \int_\Omega \mathbf{f}\cdot\phi \quad (2.2.13)$$

$$\int_\Omega \int_{\mathbf{Y}} A[e_x(\mathbf{u}_0)+e_y(\mathbf{u}_1)] : e_y(\phi_1) + \int_0^t ds \int_\Omega \int_{\mathbf{Y}} B(t-s)[e_x(\mathbf{u}_0(s))$$
$$+e_y(\mathbf{u}_1(s))] : e_y(\phi_1) = 0 \quad (2.2.14)$$

$$\mathbf{u}_0(0) = \partial_t \mathbf{u}_0(0) = 0, \mathbf{u}_1(0) = 0 \quad (2.2.15)$$

■

Proof. Let $\varepsilon \to 0$ in (10.4.4) and use (2.2.9)–(2.2.12) and integration by parts in t:

$$\int_0^T \psi''(t)\int_\Omega \mathbf{u}_0\phi + \int_0^T \psi(t)\int_\Omega \int_{\mathbf{Y}} A[e_x(\mathbf{u}_0)+e_y(\mathbf{u}_1)] : e_x(\phi)$$
$$+\int_0^T \psi(t)\int_0^t ds \int_\Omega \int_{\mathbf{Y}} B(t-s)[e_x(\mathbf{u}_0(s))+e_y(\mathbf{u}_1(s))] : e_x(\phi)$$
$$+\int_0^T \psi(t)\int_\Omega \int_{\mathbf{Y}} A[e_x(\mathbf{u}_0)+e_y(\mathbf{u}_1)] : e_y(\phi_1) \quad (2.2.16)$$
$$+\int_0^T \psi(t)\int_0^t ds \int_\Omega \int_{\mathbf{Y}} B(t-s)[e_x(\mathbf{u}_0(s))+e_y(\mathbf{u}_1(s))] : e_y(\phi_1)$$
$$= \int_\Omega \mathbf{f}\cdot\phi$$

We are now able to establish the lemma by setting $\phi_1 = 0$ and $\phi = 0$ respectively. □

Lemma 2.2

The system (2.2.13)–(2.2.15) has a unique solution. ■

Proof. It is sufficient to prove that for $\mathbf{f} = \mathbf{0}$, $g = 0$ we have only the trivial solution $\mathbf{u}_0 = \mathbf{u}_1 = 0$. Set $\phi = \partial_t \mathbf{u}_0, \phi_1 = \partial_t \mathbf{u}_1$ as a test function in (2.2.13) and (2.2.15), respectively:

$$\int_\Omega \partial_{tt}\mathbf{u}_0 \partial_t \mathbf{u}_0 + \int_\Omega \int_{\mathbf{Y}} A(e_x(\mathbf{u}_0) + e_y(\mathbf{u}_1)) : e_x(\partial_t \mathbf{u}_0) \\ + \int_0^t ds \int_\Omega \int_{\mathbf{Y}} B(t-s)(e_x(\mathbf{u}_0(s)) + e_y(\mathbf{u}_1(s))) : e_x(\partial_t \mathbf{u}_0(t)) = 0, \tag{2.2.17}$$

$$\int_\Omega \int_{\mathbf{Y}} A(e_x(\mathbf{u}_0) + e_y(\mathbf{u}_1)) : e_y(\partial_t \mathbf{u}_1) \\ + \int_0^t ds \int_\Omega \int_{\mathbf{Y}} B(t-s)(e_x(\mathbf{u}_0(s)) + e_y(\mathbf{u}_1(s)) : e_y(\partial_t \mathbf{u}_1(t)) = 0. \tag{2.2.18}$$

Adding (2.2.17) and (2.2.18) and integrating in time:

$$\int_\Omega (\partial_t \mathbf{u}_0)^2 + \int_\Omega \int_{\mathbf{Y}} A(e_x(\mathbf{u}_0) + e_y(\mathbf{u}_1)) : (e_x(\mathbf{u}_0) + e_y(\mathbf{u}_1)) \\ = -2 \int_0^t ds \int_0^s ds_1 \int_\Omega \int_{\mathbf{Y}} B(s-s_1)(e_x(\mathbf{u}_0(s_1)) + e_y(\mathbf{u}_1(s_1))) : (e_x(\partial_s \mathbf{u}_0) + e_y(\partial_s \mathbf{u}_1)). \tag{2.2.19}$$

Note that

$$\left| \int_0^t ds \int_0^s ds_1 \int_\Omega \int_{\mathbf{Y}} B(s-s_1) e(\varphi(s_1)) : e(\partial_t \varphi(s)) \right| \\ \leq C \left[\int_0^t \|\varphi(s)\|^2 ds + \|\varphi(t)\| \int_0^t \|\varphi(s)\| ds \right].$$

Gronwall's inequality, initial conditions $\mathbf{u}_0 \equiv 0$ and $\mathbf{u}_1 \equiv 0$, imply $\mathbf{u}_0 = 0, \mathbf{u}_1 = 0$. □

2.3 DERIVATION OF THE EFFECTIVE EQUATION OF $\mathbf{U}_0$

Since it is known that the system (2.2.13)-(2.2.15) has a unique solution, we seek this solution $\mathbf{u}_1(t,x,y)$ in the form

$$\mathbf{u}_1(t,x,y) = \sum_{ij} \int_0^t \mathbf{K}^{ij}(y,t-s)(e_x(\partial_s \mathbf{u}_0))_{ij}(x,s) ds \tag{2.3.1}$$

Substituting (2.3.1) into (10.4.13)

$$\int_\Omega \int_{\mathbf{Y}} A \left\{ e_x(\mathbf{u}_0) + \sum_{ij} \int_0^t e_y(\mathbf{K}^{ij})(y,t-s)(\partial_s e_x(\mathbf{u}_0))_{ij}(x,s) ds \right\} : e_y(\phi_1) \\ + \int_0^t ds_1 \int_\Omega \int_{\mathbf{Y}} B(t-s_1) \\ \times \left\{ e_x(\mathbf{u}_0(s_1)) + \sum_{ij} \int_0^{s_1} e_y(\mathbf{K}^{ij})(y,s_1-s)(\partial_s e_x(\mathbf{u}_0))_{ij}(x,s) ds \right\} : e_y(\phi_1(t)) \\ = 0 \tag{2.3.2}$$

Interchanging order of integration and integrating by parts with respect to s, we have

$$\begin{aligned}
&\int_0^t ds_1 \int_\Omega \int_{\mathbf{Y}} B(t-s_1) \int_0^{s_1} e_y(\mathbf{K}^{ij})(y, s_1 - s)(\partial_s e_x(\mathbf{u}_0))_{ij}(x,s)ds \\
&= \int_0^t (\partial_s e_x(\mathbf{u}_0))_{ij}(x,s)ds \int_\Omega \int_{\mathbf{Y}} \int_s^t B(t-s_1) e_y(\mathbf{K}^{ij})(y, s_1 - s) ds_1 \\
&= \int_0^t (e_x(\mathbf{u}_0))_{ij}(x,s) \int_\Omega \int_{\mathbf{Y}} \{B(t-s) e_y(\mathbf{K}^{ij})(y,0) \\
&\quad + \int_s^t B(t-s_1) e_y(\partial_2 \mathbf{K}^{ij})(y, s_1 - s) ds_1 \Big\} ds,
\end{aligned}$$

where we use $\partial_2(\cdot)$ to represent the partial derivative with respect to the second argument. Integrating by parts,

$$\begin{aligned}
&\int_0^t e_y(\mathbf{K}^{ij})(y, t-s)(\partial_s e_x(\mathbf{u}_0))_{ij}(x,s)ds = e_y(\mathbf{K}^{ij})(y,0)(e_x(\mathbf{u}_0))_{ij}(x,t) \\
&+ \int_0^t e_y(\partial_2 \mathbf{K}^{ij})(y, t-s)(e_x(\mathbf{u}_0))_{ij}(x,s)ds
\end{aligned}$$

Therefore (2.3.2) becomes

$$\begin{aligned}
&\int_\Omega \int_{\mathbf{Y}} A \Big\{ e_x(\mathbf{u}_0) + \sum_{ij} \Big[e_y(\mathbf{K}^{ij})(y,0)(e_x(\mathbf{u}_0))_{ij}(x,t) \\
&+ \int_0^t e_y(\partial_2 \mathbf{K}^{ij})(y, t-s)(e_x(\mathbf{u}_0))_{ij}(x,s)ds \Big] \Big\} : e_y(\phi_1) \\
&+ \int_0^t ds \int_\Omega \int_{\mathbf{Y}} \Big\{ B(t-s) e_x(\mathbf{u}_0(s)) + \sum_{ij} e_x(\mathbf{u}_0)_{ij}(x,s) \\
&\times \Big\{ B(t-s) e_y(\mathbf{K}^{ij})(y,0) + \int_s^t B(t-s_1) e_y(\partial_2 \mathbf{K}^{ij})(y, s_1 - s) ds_1 \Big\} \Big\} : e_y(\phi_1(t)) \\
&= 0
\end{aligned} \tag{2.3.3}$$

Collecting $e_x(\mathbf{u}_0)$ terms, and choosing $\mathbf{K}^{ij}(0)$ so that

$$\int_\Omega \int_{\mathbf{Y}} A \Big\{ e_x(\mathbf{u}_0) + \sum_{ij} e_y(\mathbf{K}^{ij})(y,0)(e_x(\mathbf{u}_0))_{ij}(x,t) \Big\} : e_y(\phi_1) = 0 \tag{2.3.4}$$

we have

$$\int_{\mathbf{Y}} A \Big\{ \delta_{ij} + e_y(\mathbf{K}^{ij})(y,0) \Big\} : e_y(\phi_1) = 0 \text{ for any } \phi_1 \in [H^1_\#(\mathbf{Y})]^n, \tag{2.3.5}$$

where δ_{ij} are the discrete version of the delta functions. Consequently, $\mathbf{K}^{ij}(0)$ is determined by the following linear elastic system:

$$\begin{aligned}
&\operatorname{div}_y \Big\{ A \left(\delta_{ij} + e_y(\mathbf{K}^{ij})(y,0) \right) \Big\} = 0 \\
&\mathbf{K}^{ij}(y,0) \in [H^1_\#(\mathbf{Y})]^n
\end{aligned} \tag{2.3.6}$$

Collecting time integral terms in (2.3.3) and choosing $\mathbf{K}^{ij}(t,y)$ so that

$$\begin{aligned}\int_0^t ds \int_\Omega \int_{\mathbf{Y}} \Big\{ & B(t-s)e_x(\mathbf{u}_0(s)) + \sum_{ij} e_x(\mathbf{u}_0)_{ij}(x,s) \\ & \times \Big\{ Ae_y(\partial_2 \mathbf{K}^{ij})(y,t-s) + B(t-s)e_y(\mathbf{K}^{ij})(y,0) \\ & + \int_s^t B(t-s_1)e_y(\partial_2 \mathbf{K}^{ij})(y,s_1-s)ds_1 \Big\}\Big\} : e_y(\phi_1) = 0\end{aligned} \tag{2.3.7}$$

implies

$$\begin{aligned}\int_{\mathbf{Y}} \Big\{ & B(t-s)\delta_{ij} + Ae_y(\partial_2 \mathbf{K}^{ij})(y,t-s) + B(t-s)e_y(\mathbf{K}^{ij})(y,0) \\ & + \int_s^t B(t-s_1)e_y(\partial_2 \mathbf{K}^{ij})(y,s_1-s)ds_1 \Big\} : e_y(\phi_1) = 0\end{aligned} \tag{2.3.8}$$

i.e $\mathbf{K}^{ij}(\tau,y)$ is determined by the following system

$$\begin{aligned}& \operatorname{div}_y \Big\{ B(\tau)\delta_{ij} + Ae_y(\partial_2 \mathbf{K}^{ij})(y,\tau) + B(\tau)e_y(\mathbf{K}^{ij})(y,0) \\ & \quad + \int_\tau^0 B(s_2)e_y(\partial_2 \mathbf{K}^{ij})(y,s_2+\tau)ds_2 \Big\} = 0 \\ & \mathbf{K}^{ij}(t,y) \in L^2([0,T],[H^1_\#(\mathbf{Y})]^n) \\ & \mathbf{K}^{ij}(0,y) \text{ is given by (5.1.37)}\end{aligned} \tag{2.3.9}$$

The effective equations are found by substituting (2.3.1) into (2.2.13)

$$\begin{aligned}& \frac{d^2}{dt^2}\int_\Omega \mathbf{u}_0\phi + \int_\Omega \int_{\mathbf{Y}} A \Big\{ e_x(\mathbf{u}_0) + \sum_{ij} \int_0^t e_y(\mathbf{K}^{ij})(y,t-s)(\partial_s e_x(\mathbf{u}_0))_{ij}(x,s)ds \Big\} : e_x(\phi) \\ & + \int_0^t ds \int_\Omega \int_{\mathbf{Y}} B(t-s) \left[e_x(\mathbf{u}_0) + \sum_{ij} \int_0^s e_y(\mathbf{K}^{ij})(y,s-s_1)(e_x(\partial_{s_1}\mathbf{u}_0))_{ij}(x,s_1)ds_1 \right] : e_x(\phi) \\ & = \int_\Omega \mathbf{f}\cdot\phi.\end{aligned} \tag{2.3.10}$$

Performing integration by parts and interchanging order of integration, we have

$$\begin{aligned}
&\frac{d^2}{dt^2}\int_\Omega \mathbf{u}_0\phi + \int_\Omega\int_\mathbf{Y} A\left\{e_x(\mathbf{u}_0) + \sum_{ij} e_y(\mathbf{K}^{ij})(y,0)(e_x(\mathbf{u}_0))_{ij}(x,t)\right\} : e_x(\phi)\\
&+\int_0^t ds \int_\Omega\int_\mathbf{Y}\Big\{B(t-s)e_x(\mathbf{u}_0(s)) + \sum_{ij} e_x(\mathbf{u}_0))_{ij}(x,s)\\
&\times\Big\{B(t-s)e_y(\mathbf{K}^{ij})(y,0) + Ae_y(\partial_2\mathbf{K}^{ij})(y,t-s)\\
&+\int_s^t B(t-s_1)e_y(\partial_2\mathbf{K}^{ij})(y,s_1-s)ds_1\Big\}\Big\} : e_x(\phi)\\
&= \int_\Omega \mathbf{f}\cdot\phi.
\end{aligned} \tag{2.3.11}$$

The above equation may be simplified by introducing the symmetric tensors $\mathscr{A}$ and $\mathscr{B}$ as:

$$\mathscr{A}_{ijkl} = \int_\mathbf{Y} A_{ijkl} + \int_\mathbf{Y}\sum_{mn}\frac{1}{2}A_{klmn}\left(\frac{\partial K_n^{ij}(0)}{\partial y_m} + \frac{\partial K_m^{ij}(0)}{\partial y_n}\right), \tag{2.3.12}$$

$$\begin{aligned}
\mathscr{B}_{ijkl}(\tau) = \int_\mathbf{Y} B_{ijkl} + \int_\mathbf{Y}\sum_{mn}\Big\{&\frac{1}{2}A_{klmn}\left(\frac{\partial(\partial_2 K_n^{ij})}{\partial y_m} + \frac{\partial(\partial_2 K_m^{ij})}{\partial y_n}\right)\\
&+\frac{1}{2}B_{klmn}\left(\frac{\partial K_n^{ij}(0)}{\partial y_m} + \frac{\partial K_m^{ij}(0)}{\partial y_n}\right)\\
&+\int_\tau^0 B(s_2)e_y(\partial_2\mathbf{K}^{ij})(y,s_2+\tau)ds_2\Big\}.
\end{aligned} \tag{2.3.13}$$

The effective equation for $\mathbf{u}_0$ may then be seen to take the form

$$\partial_{tt}\mathbf{u}_0 - \operatorname{div}\mathscr{A}e(\mathbf{u}_0) - \operatorname{div}\int_0^t \mathscr{B}(t-s)e(\mathbf{u}_0)ds = \mathbf{f}. \tag{2.3.14}$$

From the effective equation (2.3.14), we get the effective constitutive relation between stress and strain. We summarize the result as theorem [118]:

Theorem 2.5

Let $\mathbf{u}_0(t,x)$ be as in Theorem 1.10, then $\mathbf{u}_0(t,x)$ satisfies equations (2.3.14) and (2.2.15).

$$\sigma_0 = \mathscr{A}e(\mathbf{u}_0) + \int_0^t \mathscr{B}(t-s)e(\mathbf{u}_0)ds. \tag{2.3.15}$$

■

3 Acoustics in Porous Media

3.1 INTRODUCTION

In this chapter we consider the acoustic response of poroelastic materials. An example of such a medium is cancellous bone. Quantitative ultrasound techniques, (**QUT**), give us the possibility to determine the effects of bone microstructure on its strength. **QUT** could provide an important new diagnostic tool for determining more accurately the actual bone rigidity [257], [121], [127]. Since the loss of bone density and the destruction of the bone microstructure is most evident in osteoporotic cancellous bone, it is natural to investigate the possibility of developing accurate ultrasound models for the illumination of cancellous bone. It would be of enormous clinical advantage if accurate methods could be developed using ultrasound interrogation to diagnose osteoporosis and fractures. Homogenization methods have been successfully used to study bone microstructure [119, 138, 143, 142]. A short discussion of these methods will appear in later chapters. Cancellous (also called trabecular) bone consists of trabeculae and marrow; see Figure 3.1.1.

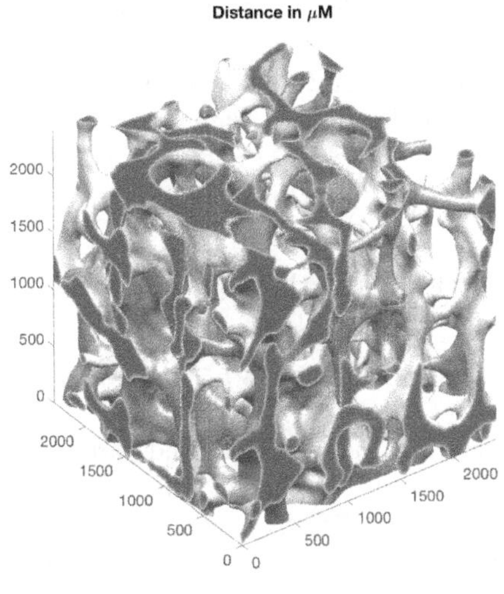

FIGURE 3.1.1 Trabecular bone constructed from CAT scans.

Mathematically, we consider cancellous bone to consist of an elastic frame and an interstitial fluid consisting of blood and marrow. Blood and marrow can be treated as a non-Newtonian fluid; this will be discussed later in Chapter 9. In this chapter we treat only the case of Newtonian interstitial fluids.

In terms of dimensionless physical quantities [135, 91] we may describe the elastic frame by its constitutive equation

$$\sigma^{s,\varepsilon} = A\mathbf{e}(\mathbf{u}^\varepsilon) \text{ in } \Omega_s^\varepsilon \times]0,T[\tag{3.1.1}$$

and its equation of motion

$$\rho_s \frac{\partial^2 \mathbf{u}^\varepsilon}{\partial t^2} - \operatorname{div}(A\mathbf{e}(\mathbf{u}^\varepsilon)) = \rho_s \mathbf{F} \text{ in } \Omega_s^\varepsilon \times]0,T[, \tag{3.1.2}$$

where $\mathbf{e}$ is the strain tensor. The interstitial (pore) fluid has the constitutive equation

$$\sigma^{f,\varepsilon} = -p^\varepsilon \mathbb{I} + 2\mu\varepsilon^{-r}\mathbf{e}(\left(\frac{\partial \mathbf{u}^\varepsilon}{\partial t}\right) \text{ in } \Omega_f^\varepsilon \times]0,T[\tag{3.1.3}$$

and its equation of motion

$$\rho_f \frac{\partial^2 \mathbf{u}^\varepsilon}{\partial t^2} - \operatorname{div}(\sigma^{f,\varepsilon}) = \rho_f \mathbf{F} \text{ in } \Omega_f^\varepsilon \times]0,T[. \tag{3.1.4}$$

For simplicity we first assume that the fluid is incompressible and postpone the case of a slightly compressible fluid until later. Namely we consider

$$\operatorname{div} \frac{\partial \mathbf{u}^\varepsilon}{\partial t} = 0 \text{ in } \Omega_f^\varepsilon \times]0,T[\tag{3.1.5}$$

The initial conditions are chosen to be homogeneous so as to avoid blow up

$$\mathbf{u}^\varepsilon(0) = \frac{\partial \mathbf{u}^\varepsilon}{\partial t}(0) = 0 \text{ in } \Omega, \tag{3.1.6}$$

where Ω_s^ε, Ω_f^ε are respectively the solid and fluid parts of the domain and Γ_ε the interface between the two phases. The choice of the exponent r which appears in the pore fluid constitutive equation is directly related to the contrast of property number $C := \frac{\mu}{a\lambda} = O(\varepsilon^{-r})$, where a is the characteristic size of the elasticity coefficients. Furthermore, the contrast of property number is related to the Strouhal number Sh, by $Sh = O(\varepsilon^{-2}C)$. For the modeling using the Strouhal number we refer to [18]. The porous medium we propose to study in this chapter is obtained by a periodic arrangement of the pores. The formal description goes along the lines introduced in Chapter 1 and the domains Ω_s^ε and Ω_f^ε represent, respectively, the solid and fluid parts of a porous medium Ω, which we assume are connected. For simplicity we suppose $L/\varepsilon \in \mathbb{N}$.

3.2 DIPHASIC MACROSCOPIC BEHAVIOR

The following material is based on the work [135, 91]. We consider the case where the phases move separately, which corresponds to $r = -1$. The solid matrix obeys the equation of motion

$$\rho_s \frac{\partial^2 \mathbf{u}^\varepsilon}{\partial t^2} - \operatorname{div}(\sigma^{s,\varepsilon}) = \rho_s \mathbf{F} \ \text{ in } \Omega_s^\varepsilon \times [0,T], \tag{3.2.1}$$

where the stress is given as

$$\sigma^{s,\varepsilon} = A\mathbf{e}(\mathbf{u}^\varepsilon). \tag{3.2.2}$$

The interstitial fluid part obeys the Stokes' system

$$\rho_f \frac{\partial^2 \mathbf{u}^\varepsilon}{\partial t^2} - \operatorname{div}(\sigma^{\varepsilon}) = \rho_f \mathbf{F} \ \text{ in } \Omega_f^\varepsilon \times [0,T], \tag{3.2.3}$$

where the fluid stress is described by

$$\sigma^{f,\varepsilon} = -p^\varepsilon I + 2\mu\varepsilon^2 e\left(\frac{\partial u^\varepsilon}{\partial t}\right), \tag{3.2.4}$$

$$\operatorname{div} \frac{\partial u^\varepsilon}{\partial t} = 0, \tag{3.2.5}$$

where $\frac{\partial u_t^\varepsilon}{\partial t}$ is the fluid velocity and p^ε is the fluid pressure. At the interface between fluid and solid parts we have

$$[\mathbf{u}^\varepsilon] = 0 \text{ on } \Gamma_\varepsilon \times]0,T[\tag{3.2.6}$$

as the statement of continuity of displacements.[1] The normal stresses across the boundaries are equal

$$\sigma^{s,\varepsilon} \cdot \nu = \sigma^{f,\varepsilon} \cdot \nu \text{ on } \Gamma_\varepsilon \times [0,T]. \tag{3.2.7}$$

At the outer boundary we suppose periodicity, i.e.

$$\{\mathbf{u}^\varepsilon, p^\varepsilon\} \text{ are } L-\text{periodic}. \tag{3.2.8}$$

For simplicity we suppose that there was neither flow nor deformations at $t = 0$, i.e.

$$\begin{aligned} \mathbf{u}^\varepsilon(\mathbf{x},0) &= \mathbf{e}(0, \\ \frac{\partial \mathbf{u}}{\partial t}(\mathbf{x},0) &= 0 \quad \text{in } \Omega. \end{aligned} \tag{3.2.9}$$

The variational formulation which corresponds to (3.2.1)-(3.2.9) is given by [135]:

---

[1] $[\cdot]$ means the jump in the unknown quantity enclosed in the brackets.

Find $\mathbf{u}^\varepsilon \in H^1(0,T;H^1_{\text{per}}(\Omega)^n)$ with $\dfrac{d^2\mathbf{u}^\varepsilon}{dt^2} \in L^2(0,T;L^2(\Omega)^n)$ and $p^\varepsilon \in L^2(0,T;L^2(\Omega_f^\varepsilon))$ such that

$$\frac{d^2}{dt^2}\int_\Omega \rho^\varepsilon \mathbf{u}^\varepsilon(t)\varphi\,d\mathbf{x} + \frac{d}{dt}\int_{\Omega_f^\varepsilon} 2\mu\varepsilon^2 \mathbf{e}(\mathbf{u}^\varepsilon(t)) : \mathbf{e}((\varphi)\,d\mathbf{x}$$

$$+\int_{\Omega_s^\varepsilon} A : \mathbf{e}((\mathbf{u}^\varepsilon(t)) : \mathbf{e}((\varphi)\,d\mathbf{x} - \int_{\Omega_f^\varepsilon} p^\varepsilon \operatorname{div}\varphi\,d\mathbf{x} = \int_\Omega \rho^\varepsilon F\varphi\,d\mathbf{x},$$

$$\forall\varphi \in H^1_{\text{per}}(\Omega)^n, \qquad (a.e.) \text{ in }]0,T[, \tag{3.2.10}$$

where

$$\rho^\varepsilon = \rho_f\chi_{\Omega_f^\varepsilon} + \rho_s\chi_{\Omega_s^\varepsilon} \tag{3.2.11}$$

and

$$\begin{cases} \mathbf{u}^\varepsilon(0) = & 0, \\ \frac{\partial u^\varepsilon}{\partial t}(0) = & 0 \end{cases}, \tag{3.2.12}$$

and

$$\operatorname{div}\frac{\partial u^\varepsilon}{\partial t} = 0 \text{ in } \Omega_f^\varepsilon \times [0,T]. \tag{3.2.13}$$

In [135] it is shown that by taking $\varphi = \frac{\partial u^\varepsilon}{\partial t}$ as the test function in (3.2.10), the following estimates are obtained

$$\begin{aligned}
\left\|\frac{\partial u^\varepsilon}{\partial t}\right\|_{L^\infty(0,T;L^2(\Omega)^n)} &\le C,\\
\|\mathbf{e}((\mathbf{u}^\varepsilon)\|_{L^\infty(0,T;L^2(\Omega_s^\varepsilon)^{n^2})} &\le C,\\
\left\|e\left(\frac{\partial u^\varepsilon}{\partial t}\right)\right\|_{L^2(0,T;L^2(\Omega_f^\varepsilon)^{n^2})} &\le \frac{C}{\varepsilon},\\
\|\mathbf{e}(\mathbf{u}^\varepsilon)\|_{L^\infty(0,T;L^2(\Omega_f^\varepsilon)^{n^2})} &\le \frac{C}{\varepsilon}.
\end{aligned} \tag{3.2.14}$$

It is standard practice to extend the pressure field p^ε to $\Omega^\varepsilon \times [0,T]$ using

$$\tilde{p}^\varepsilon(\mathbf{x},t) = \begin{cases} p^\varepsilon(\mathbf{x},t) - \frac{1}{|\Omega|}\int_{\Omega_f^\varepsilon} p^\varepsilon(\mathbf{x},t)\,d\mathbf{x}, & x \in \Omega_f^\varepsilon,\\ -\frac{1}{|\Omega|}\int_{\Omega_f^\varepsilon} p^\varepsilon(\mathbf{x},t)d\mathbf{x}, & \mathbf{x} \in \Omega_s^\varepsilon. \end{cases} \tag{3.2.15}$$

Then $\int_\Omega \tilde{p}^\varepsilon \operatorname{div}\varphi = \int_{\Omega_f^\varepsilon} p^\varepsilon \operatorname{div}\varphi$, $\forall\varphi \in H^1_{\text{per}}(\Omega)^n$, and we obtain

$$\|\tilde{p}^\varepsilon\|_{H^1(0,T;L^2_0(\Omega))} \le C, \tag{3.2.16}$$

$$\|\nabla\tilde{p}^\varepsilon\|_{H^1(0,T;H^{-1}_{\text{per}}(\Omega)^n)} \le C. \tag{3.2.17}$$

Using compactness arguments, this implies that for $\forall t \in [0,T]$ there are subsequences such that

$$\mathbf{u}^{\varepsilon} \to \mathbf{u}^0(\mathbf{x},t) + \chi_{\mathscr{Y}_f}(y)\mathbf{v}(\mathbf{x},y,t) \text{ in the 2-scale sense,} \tag{3.2.18}$$

$$\chi_{\Omega_s^{\varepsilon}} \nabla \mathbf{u}^{\varepsilon} \to \chi_{\mathscr{Y}_s}(y)[\nabla_x \mathbf{u}^0(\mathbf{x},t) + \nabla_y \mathbf{u}^1(\mathbf{x},y,t)] \text{ in the 2-scale sense,} \tag{3.2.19}$$

$$\varepsilon \chi_{\Omega_f^{\varepsilon}} \nabla \mathbf{u}^{\varepsilon} \to \chi_{\mathscr{Y}_f}(y)\nabla_y v(\mathbf{x},y,t) \text{ in the 2-scale sense,} \tag{3.2.20}$$

$$\tilde{p}^{\varepsilon} \to \tilde{p}^0(\mathbf{x},y,t) \text{ in the 2-scale sense.} \tag{3.2.21}$$

In order to extract particular relations, we choose special test functions and pass to the limit as $\varepsilon \to 0$ in the variational formulation (3.2.10) It is convenient to choose the following special test functions

$$\varphi \in C^{\infty}_{\text{per}}(\overline{\Omega})^n, \psi \in C^{\infty}_{\text{per}}(\Omega \times \mathscr{Y})^n$$

and $\zeta \in C^{\infty}_{\text{per}}(\Omega \times \mathscr{Y}))^n$, where $\operatorname{supp} \zeta \subset \mathscr{Y}_f$ and $\operatorname{div}_y \zeta = 0$ in $\mathscr{Y}_f$.

It can be shown [135] that the variational formulation (3.2.10), in the limit leads to the following expressions depending on which test functions, $\varphi(x)$, $\varepsilon\psi\left(x, \frac{x}{\varepsilon}\right)$ or $\zeta\left(x, \frac{x}{\varepsilon}\right)$ are used:

$$\frac{d^2}{dt^2}\left\{\bar{\rho}\int_{\Omega} \mathbf{u}^0(t)\varphi + \int_{\Omega}\left(\int_{G_f} \rho_f v\right)\varphi\right\} + \int_{\Omega}\int_{G_s} A(\mathbf{e}_x(\mathbf{u}^0) + \mathbf{e}_y(\mathbf{u}^1)) : \mathbf{e}_x(\varphi)$$

$$-\int_{\Omega}\left(\int_{G} \tilde{p}^0 d\mathbf{y}\right) \operatorname{div}_x \varphi = \int_{\Omega} \bar{\rho}\mathbf{F}\varphi\ ,\ \forall \varphi \in H^1_{\text{per}}(\Omega)^n, \tag{3.2.22}$$

$$\int_{\Omega}\int_{G_s} A(\mathbf{e}_x(\mathbf{u}^0) + \mathbf{e}_y(\mathbf{u}^1)) : \mathbf{e}_y(\psi) - \int_{\Omega}\int_{G} \tilde{p}^0 \operatorname{div}_y \psi = 0\ ,$$

$$\forall \psi \in L^2(\Omega, H^1_{\text{per}}(\mathscr{Y}))^n, \tag{3.2.23}$$

$$\frac{d^2}{dt^2}\int_{\Omega}\int_{G_f} \rho_f(\mathbf{u}^0(t) + \mathbf{v}(t))\zeta + 2\mu\int_{\Omega}\int_{G_f} \mathbf{e}_y\left(\frac{\partial \mathbf{v}}{\partial t}\right) : e_y(\zeta)$$

$$-\int_{\Omega}\int_{G_f} \tilde{p}^0 \operatorname{div}_x \zeta = \int_{\Omega}\int_{G_f} \rho_f F\zeta,\ ,\ \forall \zeta \in L^2(\Omega, H^1_{\text{per}}(\mathscr{Y}))^n, \tag{3.2.24}$$

such that $\zeta = 0$ on $\mathscr{Y}_s$ and $\operatorname{div}_y \zeta = 0$ on $\mathscr{Y}_f$,

$$\mathbf{u}^0(0) = 0,\ \frac{\partial \mathbf{u}^0}{\partial t}(0) = 0, \tag{3.2.25}$$

$$\mathbf{u}^1(0) = 0, \tag{3.2.26}$$

$$\mathbf{v}(0) = 0, \quad \frac{\partial \mathbf{v}}{\partial t}(0) = 0, \tag{3.2.27}$$

$$\{\mathbf{u}^0, \mathbf{u}^1, v, \tilde{p}^0\} \text{ are } L\text{-periodic in } x. \tag{3.2.28}$$

Proposition 3.1. *The limit functions* $\mathbf{u}^0$, $\mathbf{u}^1$, $\mathbf{v}$ *and* $\tilde{p}^0$ *satisfy [91]:*

1.

$$\operatorname{div}_y \frac{\partial \mathbf{v}}{\partial t}(\mathbf{x}, \mathbf{y}, t) = 0 \text{ in } \Omega \times \mathscr{Y}_f \times [0, T], \tag{3.2.29}$$

$$\tilde{p}^0 = -\lim_{\varepsilon \to 0} \frac{1}{|\Omega|} \int_{\Omega_f^\varepsilon} p^\varepsilon(x, t) d\mathbf{x} = B(t) \text{ in } \Omega \times \mathscr{Y}_s \times [0, T]; \tag{3.2.30}$$

moreover,

$$\int_\Omega \int_G \tilde{p}^0 \, d\mathbf{y} \, d\mathbf{x} = 0 \text{ on } [0, T]. \tag{3.2.31}$$

2.

$$\tilde{p}^0(x, y, t) = \chi_{\mathscr{Y}_f}(y) p^0(\mathbf{x}, t) + \chi_{\mathscr{Y}_s}(y) B(t) \text{ in } \Omega \times \mathscr{Y} \times [0, T]. \tag{3.2.32}$$

3.

$$\operatorname{div}_x \left\{ |\mathscr{Y}_f| \frac{\partial \mathbf{u}^0}{\partial t}(\mathbf{x}, t) + \int_{G_f} \frac{\partial \mathbf{v}}{\partial t}(\mathbf{x}, \mathbf{y}, t) d\mathbf{y} \right\}$$
$$= \int_{G_s} \operatorname{div}_y \frac{\partial u^1}{\partial t}(\mathbf{x}, \mathbf{y}, t) d\mathbf{y} \text{ in } \Omega \times [0, T]. \tag{3.2.33}$$

and

$$\operatorname{div}_x \int_{G_f} \frac{\partial \mathbf{v}}{\partial t}(\mathbf{x}, \mathbf{y}, t) d\mathbf{y} \in L^2(\Omega \times [0, T]). \tag{3.2.34}$$

See [135, 91] for the proof.

From the above proposition, as $\mathbf{u}^\varepsilon$ is bounded in $H(\Omega, \operatorname{div})$, we have the following weak convergence:

$$\mathbf{u}^\varepsilon \rightharpoonup \mathbf{u}^0 + \int_{G_f} \mathbf{v} d\mathbf{y} \quad \text{in } L^2(\Omega)^n \text{ weakly, } \forall t \in [0, T]$$

and

$$\operatorname{div} \mathbf{u}^\varepsilon \rightharpoonup \operatorname{div}_x \mathbf{u}^0 + \operatorname{div}_x \int_{G_f} v d\mathbf{y} \quad \text{in } L^2(\Omega)^n \text{ weakly, } \forall t \in [0, T].$$

Lemma 3.1

The limit functions $\{\mathbf{u}^0, \mathbf{u}^1, \mathbf{v}, \tilde{p}^0\}$ with

$$\mathbf{u}^0 \in H^3(0,T;L^2(\Omega)^n) \cap H^2(0,T;H^1_{\text{per}}(\Omega)^n),$$
$$\mathbf{u}^1 \in H^3(0,T;L^2(\Omega;H^1_{\text{per}}(\mathscr{Y}_s)/\mathbb{R})^n),$$
$$\mathbf{v} \in H^3(0,T;L^2(\Omega\times\mathscr{Y})^n) \cap H^2(0,T;L^2(\Omega;H^1_{\text{per}}(\mathscr{Y}))^n) \text{ and}$$
$$\tilde{p}^0 \in H^1(0,T;L^2(\Omega;L^2(\mathscr{Y})))$$

satisfy the system (3.2.22)-(3.2.33) with $\mathbf{v}=0$ on $\bar{\mathscr{Y}_s}$. ■

It is shown in [91] that the system (3.2.22)-(3.2.33) has a unique solution.

Thus, we have $\mathbf{u}^0=0$, $\mathbf{v}=0$ and $\mathbf{u}^1=0$. Now, from (3.2.23), (3.2.22) and (3.2.31) we conclude $\tilde{p}^0=0$. ■

3.2.1 DERIVATION OF THE EFFECTIVE EQUATIONS FOR $\mathbf{u}^0$

The system (3.2.22)-(3.2.33) is too complicated to be used directly; hence, we seek a solution in a particular form, namely we decompose $\mathbf{u}^1$ as $\mathbf{u}^1 = \mathbf{z}^1 + \mathbf{w}^1$, where

$$\begin{cases} -\operatorname{div}_y\{A(\mathbf{e}_x(\mathbf{u}^0)+\mathbf{e}_y(\mathbf{z}^1))\}=0 & \text{in } \mathscr{Y}_s \\ A(\mathbf{e}_x(\mathbf{u}^0)+\mathbf{e}_y(z^1))\nu=0 & \text{on } \partial\mathscr{Y}_s/\partial\mathscr{Y}, \end{cases} \tag{3.2.35}$$

and

$$\begin{cases} -\operatorname{div}_y\{A\mathbf{e}_y(\mathbf{w}^1)\}=0 & \text{in } \mathscr{Y}_s, \\ A\mathbf{e}_y(w^1)\nu = -\left(p^0+\frac{|\mathscr{Y}_f|}{|\mathscr{Y}_s||\Omega|}\int_\Omega p^0\,d\mathbf{x}\right)\nu & \text{on } \partial\mathscr{Y}_s/\partial\mathscr{Y}. \end{cases} \tag{3.2.36}$$

This suggests seeking separated solutions in the form

$$\mathbf{z}^1 := \sum_{i,j}\left(\mathbf{e}_x(\mathbf{u}^0)\right)_{ij} W^{ij}(y) \tag{3.2.37}$$

and

$$\mathbf{w}^1 := \left(p^0(x,t)+\frac{|\mathscr{Y}_f|}{|\mathscr{Y}_s|\,|\Omega|}\int_\Omega p^0 d\mathbf{x}\right)\mathbf{w}^0(y). \tag{3.2.38}$$

Substituting (3.2.37)-(3.2.38) into (3.2.35) and (3.2.36) yields [91]

$$\begin{cases} \operatorname{div}_y\left\{A\left(\frac{e_i\otimes e_j+e_j\otimes e_i}{2}+\mathbf{e}_y(W^{ij})\right)\right\}=0 & \text{in } \mathscr{Y}_s, \\ A(\frac{e_i\otimes e_j+e_j\otimes e_i}{2}+\mathbf{e}_y(W^{ij}))\nu=0 & \text{on } \partial\mathscr{Y}_s/\partial\mathscr{Y}, \\ \int_{\mathscr{Y}_s} W^{ij}(y)\,d\mathbf{y}=0 & \end{cases} \tag{3.2.39}$$

and

$$\begin{cases} -\operatorname{div}_y \{A\mathbf{e}_y(\mathbf{w}^0)\} = 0 & \text{in } \mathscr{Y}_s, \\ A\mathbf{e}_y(\mathbf{w}^0)\nu = -\nu & \text{on } \partial\mathscr{Y}_s/\partial\mathscr{Y}, \\ \int\limits_{\mathscr{Y}_s} \mathbf{w}^0(y)\, d\mathbf{y} = 0. \end{cases} \tag{3.2.40}$$

Hence $\mathbf{u}^1$ now has the form

$$\mathbf{u}^1(x,y,t) = p^0(x,t)w^0(y) + \sum_{i,j} \left(\mathbf{e}_x(\mathbf{u}^0(x,t))\right)_{ij} W^{ij}(y) + \frac{|\mathscr{Y}_f|}{|\mathscr{Y}_s|\,|\Omega|}\left(\int\limits_{\Omega} p^0 d\mathbf{x}\right)\mathbf{w}^0(\mathbf{y}). \tag{3.2.41}$$

The differential form of (3.2.24) is

$$\rho_f \frac{\partial^2 \mathbf{u}^0}{\partial t^2} + \rho_f \frac{\partial^2 \mathbf{v}}{\partial t^2} - \mu\Delta_y \frac{\partial \mathbf{v}}{\partial t} + \nabla_x p^0 + \nabla_y \pi = \rho_f \mathbf{F} \text{ in } \mathscr{Y}_f, \tag{3.2.42}$$

which we rewrite as

$$\begin{cases} \rho_f \dfrac{\partial^2 \mathbf{v}}{\partial t^2} - \mu\Delta_y \dfrac{\partial \mathbf{v}}{\partial t} + \nabla_y \pi = \rho_f \phi(\mathbf{x},t), \\ \operatorname{div}_y \frac{\partial \mathbf{v}}{\partial t} = 0, \\ v\,|_{\partial\mathscr{Y}_f/\partial\mathscr{Y}} = 0, \quad \{\mathbf{v},\pi\} \text{ 1-periodic in } y, \end{cases} \tag{3.2.43}$$

where

$$\phi(x,t) := \mathbf{F}(x,t) - \frac{1}{\rho_f}\nabla_x p^0(x,t) - \frac{\partial^2 \mathbf{u}^0}{\partial t^2}.$$

We construct the solution $\mathbf{v}$ by first generating solutions to the homogeneous system as we showed in [91]

$$\begin{cases} \dfrac{\partial \mathbf{w}^i}{\partial t} - \Delta \mathbf{w}^i + \nabla \pi^i = 0, \\ \operatorname{div}_y \mathbf{w}^i = 0, \quad \mathbf{w}^i(y,0) = \mathbf{e}_i, \\ w^i\,|_{\partial\mathscr{Y}_f} = 0, \quad \{\mathbf{w}^i, \pi^i\} \text{ is 1-periodic.} \end{cases} \tag{3.2.44}$$

In terms of the $\{\mathbf{w}^i, \pi^i\}$, $\mathbf{v}$ and π have a representation of its components v_i as in [91],

$$\frac{\partial \mathbf{v}_i}{\partial t} := \sum_j \int_0^t \mathbf{w}_i^j(y, \frac{\rho_f}{\mu}(t-\tau))\phi_j(x,\tau)d\tau, \tag{3.2.45}$$

$$\pi := \mu \sum_j \int_0^t \pi^j(y, \frac{\rho_f}{\mu}(t-\tau))\phi_j(x,\tau)d\tau. \tag{3.2.46}$$

Then

$$\rho_f \frac{\partial^2 \mathbf{v}_i}{\partial t^2} = \frac{d}{dt}\sum_j \int_0^t \mathbf{w}_i^j(y, \frac{\rho_f}{\mu}(t-\tau))\rho_f \phi_j(x,\tau)d\tau \tag{3.2.47}$$

and

$$\begin{aligned} \frac{\partial^2}{\partial t^2}\int_{G_f} \rho_f \mathbf{v}_i d\mathbf{y} &= \frac{d}{dt}\sum_j \int_0^t \int_{G_f} \mathbf{w}_i^j(y, \frac{\rho_f}{\mu}(t-\tau))\rho_f \phi_j(x,\tau)d\tau \\ &= \frac{d}{dt}\sum_j \int_0^t \int_{G_f} \mathbf{w}_i^j(y, \frac{\rho_f}{\mu}(t-\tau))d\mathbf{y} \\ &\quad \left\{\rho_f \mathbf{F}_j(x,\tau) - \frac{\partial p^0}{\partial x_j}(x,\tau) - \rho_f \frac{\partial^2 \mathbf{u}_j^0}{\partial \tau \mathbf{u}^2}(x,\tau)\right\} d\tau. \end{aligned} \tag{3.2.48}$$

Using the strong equivalent of equation (3.2.22) yields, after substituting the expressions for $\mathbf{u}^1$ and $\mathbf{v}$ [91],

$$\begin{aligned} &\bar{\rho}\frac{\partial^2 u^0}{\partial t^2} - \sum_{i,j} e_i \frac{d}{dt}\int_0^t \mathscr{A}_{ij}(t-\tau)\left[\frac{\partial p^0}{\partial x_j}(\mathbf{x},\tau) + \rho_f \frac{\partial^2 \mathbf{u}_j^0}{\partial t^2}(\mathbf{x},\tau)\right] d\tau \\ &\quad - \operatorname{div}_x\{A^H e_x(\mathbf{u}^0)\} - \operatorname{div}_x\{p^0 \mathscr{B}^H\} + |\mathscr{Y}_f|\nabla_x p^0 \\ &\quad = \bar{\rho}\mathbf{F} - \sum_{i,j} e_i \frac{d}{dt}\int_0^t \mathscr{A}_{ij}(t-\tau)\rho_f \mathbf{F}_j(\mathbf{x},\tau)d\tau, \end{aligned} \tag{3.2.49}$$

where

$$A^H_{klij} := \left(\int_{G_s} A\left(\frac{e_i \otimes e_j + e_j \otimes e_i}{2} + \mathbf{e}_y(W^{ij})\right)\right)_{kl}, \tag{3.2.50}$$

$$\mathscr{B}^H := \int_{G_s} \mathbf{A}\mathbf{e}_y(w^0), \tag{3.2.51}$$

$$\mathscr{C}^H_{ij} := \int_{G_s} \operatorname{div}_y W^{ij}(y)\,d\mathbf{y}, \tag{3.2.52}$$

$$\mathscr{A}_{ij}(t) := \int_{G_f} w_i^j(y, \frac{\rho_f}{\mu}t)d\mathbf{y} \tag{3.2.53}$$

and then

$$\mathscr{A}_{ij}(0) = \int_{G_f} w_i^j(y,0)\,d\mathbf{y} = |\mathscr{Y}_f|\delta_{ij}. \tag{3.2.54}$$

On the other hand, (3.2.33) becomes

$$\operatorname{div}_x \left\{ |\mathscr{Y}_f| \rho_f \frac{\partial \mathbf{u}^0}{\partial t} + \sum_{i,j} \mathbf{e}_i \int_0^t \mathscr{A}_{ij}(t-\tau) \left[\rho_f F_j(x,\tau) - \frac{\partial p^0}{\partial x_j}(x,\tau) \right] d\tau - \sum_{i,j} e_i \int_0^t \mathscr{A}_{ij}(t-\tau) \rho_f \frac{\partial^2 \mathbf{u}_j^0}{\partial \tau \mathbf{u}^2}(x,\tau) d\tau \right\}$$

$$= \mathscr{C}^H : \mathbf{e}_x \left(\frac{\partial \mathbf{u}^0}{\partial t} \right) + \left(\frac{\partial p^0}{\partial t} + \frac{|\mathscr{Y}_f|}{|\mathscr{Y}_s|\,|\Omega|} \int_\Omega \frac{\partial p^0}{\partial t} d\mathbf{x} \right) \int_{G_s} \operatorname{div}_y w^0 \, d\mathbf{y}. \quad (3.2.55)$$

Hence in the diphasic case we are left with equations (3.2.54) and (3.2.55) for the unknowns $\mathbf{u}^0$ and p^0.

3.3 WELL-POSEDNESS FOR PROBLEMS (3.2.48) AND (3.2.55)

The following proposition is established in Mikelić [302]

Proposition 3.2. *The cell problem* (3.2.44) *has a unique solution* $\{w^i, \pi^i\} \in L^2(0,T;V_{\text{per}}) \times H^{-1}(0,T;L_0^2(\mathscr{Y}_f))$, *where*

$$V_{\text{per}} = \{\mathbf{z} \in H^1_{\text{per}}; \mathbf{z} = 0 \text{ on } \partial\mathscr{Y}_s \backslash \partial\mathscr{Y} \text{ and } \operatorname{div}\mathbf{z} = 0 \text{ in } \mathscr{Y}_f\}$$

with

$$H^1_{\text{per}} = \{\mathbf{z} \in H^1(\mathscr{Y}_f)^n; \mathbf{z} \text{ is } H^1(\mathscr{Y}) - \text{periodic}\}.$$

Furthermore, $\frac{\partial \mathbf{w}^i}{\partial t} \in L^2(0,T;V'_{\text{per}})$ *and* $\mathbf{w}^i \in C([0,T];L^2(\mathscr{Y}_f)^n)$.

We note that the initial and boundary conditions are not compatible and the solution is not globally regular.

It is also shown in Mikelić [302] that the matrix $\mathscr{A}_{ij}(t)$ given by (3.2.53) is positive definite, symmetric, and tends exponentially to zero, when $t \to +\infty$. The following proposition is established in Sanchez-Palencia [363].

Proposition 3.3. *The tensors* A^H *and* B^H, *defined by* (3.2.50) *and* (3.2.51), *respectively, are positive definite and symmetric.*

Theorem 3.1

Let $\{\mathbf{u}^0, p^0, \mathbf{u}^1, v\}$ be given by (3.2.48), (3.2.55), (3.2.41) and (3.2.45), respectively. Then these define the unique solution for (3.2.48)-(3.2.55). ∎

Proof: It is sufficient to prove that $\{\mathbf{u}^0, p^0\}$ is the unique solution for (3.2.48) and (3.2.55). The existence is obtained by construction, so it only remains to prove that such solution is unique. We use the method of Laplace's transform, applied to (3.2.48)-(3.2.55) with $F = 0$ and $\mathbf{u}^0(0) = 0$. For the definition and properties of the

Laplace transform of a distribution, we refer the reader to the textbook of Dautray and Lions [103].

Then, by Laplace transforming (3.2.48) and (3.2.55), we obtain

$$\gamma^2 \left(\bar{\rho} I - \gamma \rho_f \hat{\mathscr{A}}\right) \hat{\mathbf{u}}^0 + \operatorname{div}_x \left\{ \left(|\mathscr{Y}_f| I - \mathscr{B}^H - \gamma \hat{\mathscr{A}} \right) \hat{p}^0 \right\} - \operatorname{div}_x \left\{ A^H \mathbf{e}_x(\hat{\mathbf{u}}^0) \right\} = 0 \quad (3.3.1)$$

and

$$\gamma \operatorname{div}_x \left\{ \left(|\mathscr{Y}_f| \rho_f I - \gamma \rho_f \hat{\mathscr{A}} \right) \hat{u}^0 \right\} - \operatorname{div}_x \left\{ \hat{\mathscr{A}} \nabla \hat{p}^0 \right\} = \gamma \rho_f \left(\hat{p}^0 + \frac{|\mathscr{Y}_f|}{|\mathscr{Y}_s|\,|\Omega|} \int_\Omega \hat{p}^0 d\mathbf{x} \right) \int_{G_s} \operatorname{div}_y w^0 + \gamma \rho_f \mathscr{C}^H : \mathbf{e}_x(\hat{\mathbf{u}}^0). \quad (3.3.2)$$

At this point we need fact that $\bar{\rho} I - \gamma \rho_f \hat{\mathscr{A}}$ is positive definite; this is verified in [91]. Now, taking $\hat{u}^0$ and $\hat{p}^0$ as a test functions in (3.3.1) and (3.3.2), respectively, we get

$$\gamma^2 \int_\Omega \left(\bar{\rho} I - \gamma \rho_f \hat{\mathscr{A}}\right) \hat{u}^0 \hat{u}^0 + \int_\Omega \left(|\mathscr{Y}_f| I - \gamma \hat{\mathscr{A}} \right) \nabla \hat{p}^0 \hat{u}^0 - \int_\Omega \mathscr{B}^H \nabla \hat{p}^0 \hat{u}^0 + \int_\Omega A^H e_x(\hat{u}^0) : e_x(\hat{u}^0) = 0 \quad (3.3.3)$$

and

$$\gamma \rho_f \int_\Omega \left(|\mathscr{Y}_f| I - \gamma \hat{\mathscr{A}} \right) \nabla \hat{p}^0 \hat{u}^0 = \int_\Omega \hat{\mathscr{A}} \nabla \hat{p}^0 \nabla \hat{p}^0 - \gamma \rho_f \int_\Omega \hat{p}^0 \left(\hat{p}^0 + \frac{|\mathscr{Y}_f|}{|\mathscr{Y}_s|\,|\Omega|} \int_\Omega \hat{p}^0 d\mathbf{x} \right) \int_{G_s} \operatorname{div}_y w^0 - \gamma \rho_f \int_\Omega \mathscr{C}^H : e_x(\hat{u}^0) \hat{p}^0. \quad (3.3.4)$$

Then, by using (3.3.4), (3.3.3) can be written as follows

$$\gamma^2 \int_\Omega \left(\bar{\rho} I - \gamma \rho_f \hat{\mathscr{A}}\right) \hat{u}^0 \hat{u}^0 + \frac{1}{\gamma \rho_f} \int_\Omega \hat{\mathscr{A}} \nabla \hat{p}^0 \nabla \hat{p}^0 - \int_\Omega \mathscr{B}^H \nabla \hat{p}^0 \hat{u}^0 + \int_\Omega A^H e_x(\hat{u}^0) : e_x(\hat{u}^0) - \int_\Omega \hat{p}^0 \left(\hat{p}^0 + \frac{|\mathscr{Y}_f|}{|\mathscr{Y}_s|\,|\Omega|} \int_\Omega \hat{p}^0 d\mathbf{x} \right) \int_{G_s} \operatorname{div}_y w^0 - \int_\Omega \mathscr{C}^H : e_x(\hat{u}^0) \hat{p}^0 = 0. \quad (3.3.5)$$

Now, we take $\hat{u}^1$ as a test function in (3.2.23) and we obtain

$$-\int_\Omega \hat{p}^0 \int_{G_s} \operatorname{div}_y \hat{u}^1 = \int_\Omega \int_{G_s} A(e_x(\hat{u}^0) + e_y(\hat{u}^1)) : e_y(\hat{u}^1). \tag{3.3.6}$$

After adding (3.3.5) and (3.3.6) we get $\hat{u}^0 = 0$. Now it is easy to conclude uniqueness. ■

It was shown in [91] that the following corrector results hold:

Theorem 3.2

Let $\mathbf{u}^0$, $\mathbf{u}^1$ and v be defined as above. Then we have

$$\sqrt{\rho^\varepsilon}\mathbf{u}^\varepsilon(\mathbf{x},t) - \sqrt{\rho^0}\mathbf{u}^0(\mathbf{x},t) - \chi_{\mathscr{Y}_f}\left(\frac{\mathbf{x}}{\varepsilon}\right)\sqrt{\rho_f}\mathbf{v}(\mathbf{x},\frac{x}{\varepsilon},t) \to 0 \quad \text{in } C([0,T];L^2(\Omega)^n), \tag{3.3.7}$$

$$\chi_{\Omega_s^\varepsilon} A^{1/2}\mathbf{e}(\mathbf{u}^\varepsilon) - \chi_{\mathscr{Y}_s} A^{1/2}\mathbf{e}(\mathbf{u}^0) - \chi_{\mathscr{Y}_s} A^{1/2}\mathbf{e}(\mathbf{u}^1) \to 0 \quad \text{in } C([0,T];L^2(\Omega)^{n^2}), \tag{3.3.8}$$

where $\rho^0 = \rho_f \chi_{\mathscr{Y}_f} + \rho_s \chi_{\mathscr{Y}_s}$. ■

3.4 THE SLIGHTLY COMPRESSIBLE DIPHASIC BEHAVIOR

Assuming that the fluid is slightly compressible, the pressure may be removed from equation (3.2.4), by the small variation of pressure from the rest state assumption, i.e. δp is small and $\delta p \sim -c^2\rho_f \operatorname{div}\mathbf{u}^\varepsilon$. This can be considered as a "regularized" version of the incompressible case from Section 3.2 and, therefore, we just give a short outline of the homogenization procedure.

The slightly compressible case has been considered by Nguetseng [316], as an early application of the 2-scale convergence. Here we rephrase some of those results by stating the homogenized system more explicitly. We set $\gamma = c^2\rho_f$, and the variational formulation (3.2.10) becomes:

Find $\mathbf{u}^\varepsilon \in H^1(0,T;H^1_{\text{per}}(\Omega)^n)$ with $\dfrac{d^2\mathbf{u}^\varepsilon}{dt^2} \in L^2(0,T;L^2(\Omega)^n)$ such that

$$\frac{d^2}{dt^2}\int_\Omega \rho^\varepsilon \mathbf{u}^\varepsilon(t)\varphi + 2\mu\frac{d}{dt}\int_{\Omega_f^\varepsilon} \varepsilon^2\mathbf{e}(\mathbf{u}^\varepsilon(t)) : \mathbf{e}(\varphi)$$
$$+ \int_{\Omega_s^\varepsilon} A\mathbf{e}((\mathbf{u}^\varepsilon(t)) : \mathbf{e}(\varphi) + \gamma\int_{\Omega_f^\varepsilon} \operatorname{div}\mathbf{u}^\varepsilon \operatorname{div}\varphi$$
$$= \int_\Omega \rho^\varepsilon F\varphi, \ \forall\varphi \in H^1_{\text{per}}(\Omega)^n, \ (a.e.) \text{ in }]0,T[, \tag{3.4.1}$$

where we again take the initial conditions to be

$$\mathbf{u}^\varepsilon(0) = 0 \tag{3.4.2}$$

and

$$\frac{\partial u^\varepsilon}{\partial t}(0) = 0. \tag{3.4.3}$$

To obtain (3.4.1) we have made the same assumption as Nguetseng [316]. The term with constant γ can be assumed as the pressure with the relation $p^\varepsilon = -\gamma \operatorname{div} \mathbf{u}^\varepsilon$ in the fluid part.

Then, (3.4.1) is well posed and we have the estimates, by taking $\varphi = \frac{\partial u^\varepsilon}{\partial t}$:

$$\left\|\frac{\partial u^\varepsilon}{\partial t}\right\|_{L^\infty(0,T;L^2(\Omega)^n)} \leq C, \tag{3.4.4}$$

$$\left\|\mathbf{e}((\mathbf{u}^\varepsilon)\right\|_{L^\infty(0,T;L^2(\Omega_s^\varepsilon)^{n^2})} \leq C, \tag{3.4.5}$$

$$\left\|\mathbf{e}\left(\frac{\partial u^\varepsilon}{\partial t}\right)\right\|_{L^2(0,T;L^2(\Omega_f^\varepsilon)^{n^2})} \leq \frac{C}{\varepsilon}, \tag{3.4.6}$$

$$\left\|\mathbf{e}((\mathbf{u}^\varepsilon)\right\|_{L^\infty(0,T;L^2(\Omega_f^\varepsilon)^{n^2})} \leq \frac{C}{\varepsilon}. \tag{3.4.7}$$

As in the incompressible case, we extend the pressure field p^ε to $\Omega^\varepsilon \times [0,T]$ by

$$\tilde{p}^\varepsilon(x,t) = \begin{cases} -\gamma \operatorname{div} \mathbf{u}^\varepsilon(x,t) + \frac{\gamma}{|\Omega|} \int_{G_f^\varepsilon} \operatorname{div} \mathbf{u}^\varepsilon(x,t)\, d\mathbf{x}, & x \in \Omega_f^\varepsilon, \\ \\ \frac{\gamma}{|\Omega|} \int_{G_f^\varepsilon} \operatorname{div} \mathbf{u}^\varepsilon(x,t)\, d\mathbf{x}, & x \in \Omega_s^\varepsilon. \end{cases} \tag{3.4.8}$$

Then

$$\left\|\tilde{p}^\varepsilon\right\|_{L^\infty(0,T;L^2(\Omega^\varepsilon))} \leq C. \tag{3.4.9}$$

This implies that for all $t \in [0,T]$ there are subsequences such that

$$\mathbf{u}^\varepsilon \to \mathbf{u}^0(x,t) + \chi_{\mathscr{Y}_f}(y)\mathbf{v}(x,y,t) \text{ in the 2-scale sense,} \tag{3.4.10}$$

$$\chi_{\Omega_s^\varepsilon} \nabla \mathbf{u}^\varepsilon \to \chi_{\mathscr{Y}_s}(y)[\nabla_x \mathbf{u}^0(x,t) + \nabla_y \mathbf{u}^1(x,y,t)] \text{ in the 2-scale sense,} \tag{3.4.11}$$

$$\varepsilon \chi_{\Omega_f^\varepsilon} \nabla \mathbf{u}^\varepsilon \to \chi_{\mathscr{Y}_f}(y) \nabla_y \mathbf{v}(x,y,t) \text{ in the 2-scale sense,} \tag{3.4.12}$$

$$\tilde{p}^\varepsilon \to \tilde{p}^0(x,y,t) \text{ in the 2-scale sense.} \tag{3.4.13}$$

Proposition 3.4. *The limit functions $\mathbf{u}^0$, v, $\mathbf{u}^1$ and $\tilde{p}^0$ satisfy:*

$$\operatorname{div}_y \mathbf{v}(x,y,t) = 0 \text{ in } \Omega \times \mathscr{Y}_f \times [0,T], \tag{3.4.14}$$

and

$$\tilde{p}^0(x,y,t) = \chi_{\mathscr{Y}_s}(y) B_{sc}(t) + \chi_{\mathscr{Y}_f}(y) p^0(x,t) \text{ in } \Omega \times \mathscr{Y} \times [0,T], \tag{3.4.15}$$

with

$$B_{sc}(t) = -\frac{\gamma}{|\Omega|}\int_{\Omega}\int_{G_s} \mathrm{div}_y \mathbf{u}^1 \, d\mathbf{y}\, d\mathbf{x} \tag{3.4.16}$$

and

$$-\frac{|\mathscr{Y}_f|}{\gamma} p^0(x,t) = |\mathscr{Y}_f| \,\mathrm{div}_x \mathbf{u}^0(x,t) + \int_{G_f} \mathrm{div}_x \mathbf{v}(x,y,t)\, d\mathbf{y}$$
$$- \int_{G_s} \mathrm{div}_y \mathbf{u}^1 \, d\mathbf{y}\, d\mathbf{x} \;\; in\; \Omega \times [0,T]. \tag{3.4.17}$$

For the proof the reader is directed to [91]. Then we pass to the limit in (3.4.1) as we did in the previous section. We obtain that the limit functions, $\{\mathbf{u}^0, \mathbf{u}^1, v, \tilde{p}^0\}$, where

$$\begin{aligned}
\mathbf{u}^0 &\in H^3(0,T;\Omega)^n) \cap H^2(0,T;H^1_{\mathrm{per}}(\Omega)^n),\\
\mathbf{u}^1 &\in H^3(0,T;L^2(\Omega;H^1_{\mathrm{per}}(\frac{\mathscr{Y}_s}{\mathbb{R}})^n),\\
\mathbf{v} &\in H^3(0,T;L^2(\Omega\times\mathscr{Y})^n) \cap H^2(0,T;L^2(\Omega;H^1_{\mathrm{per}}(\mathscr{Y}))^n),\\
\mathrm{div}_x \int_{G_f} \mathbf{v}\, d\mathbf{y} &\in H^2(0,T;L^2(\Omega)),\\
\tilde{p}^0 &\in H^1(0,T;L^2(\Omega;L^2(\mathscr{Y})))
\end{aligned}$$

satisfies the system 3.2.22–3.2.24, with (3.4.14)-(3.4.17) and the initial conditions

$$\mathbf{u}^0(0) = 0, \;\; \frac{\partial \mathbf{u}^0}{\partial t}(0) = 0, \tag{3.4.18}$$

$$\mathbf{u}^1(0) = 0, \tag{3.4.19}$$

$$\mathbf{v}(0) = 0, \;\; \frac{\partial \mathbf{v}}{\partial t}(0) = 0, \tag{3.4.20}$$

$$\{\mathbf{u}^0, \mathbf{u}^1, v, \tilde{p}^0\} \text{ are } L\text{-periodic in } x. \tag{3.4.21}$$

As before, in order to show uniqueness, it is sufficient to prove that if the driving force $F = 0$ we only have the trivial solution $\mathbf{u}^0 = 0$, $\mathbf{v} = 0$ and $\mathbf{u}^1 = 0$. First, we take $\varphi = \frac{\partial \mathbf{u}^0}{\partial t}$, $\psi = \frac{\partial u^1}{\partial t}$ and $\zeta = \frac{\partial \mathbf{v}}{\partial t}$ as the test functions in 3.2.22–3.2.24, and after adding the three equations together we obtain

$$\frac{1}{2}\frac{d}{dt}\int_{\Omega}\int_{G_f} \rho_f |\frac{\partial \mathbf{u}^0}{\partial t} + \frac{\partial \mathbf{v}}{\partial t}|^2 \, d\mathbf{y}\, d\mathbf{x} + \frac{1}{2}\frac{d}{dt}\int_{\Omega} |\mathscr{Y}_s| \rho_s |\frac{\partial^2 u^0}{\partial t^2}|^2 \, d\mathbf{y}\, d\mathbf{x}$$
$$+ \frac{1}{2}\frac{d}{dt}\int_{\Omega}\int_{G_s} A(\mathbf{e}(_x(\mathbf{u}^0) + \mathbf{e}_y(\mathbf{u}^1)) : (e_x(\mathbf{u}^0) + e_y(\mathbf{u}^1))\, d\mathbf{y}\, d\mathbf{x}$$

$$-\int_\Omega\int_G \tilde{p}^0 \operatorname{div}_x \frac{\partial \mathbf{u}^0}{\partial t}\, d\mathbf{y}\, d\mathbf{x} - \int_\Omega\int_{G_f} p^0 \operatorname{div}_x \frac{\partial \mathbf{v}}{\partial t}\, d\mathbf{y}\, d\mathbf{x}$$

$$-\int_\Omega\int_{G_s} \tilde{p}^0 \operatorname{div}_y \frac{\partial u^1}{\partial t}\, d\mathbf{y}\, d\mathbf{x} + \int_\Omega p^0(x,t)\left(\int_{G_s} \operatorname{div}_y \frac{\partial u^1}{\partial t} d\mathbf{y}\right) d\mathbf{x}$$

$$+2\mu \int_\Omega\int_{G_f} |e_y\left(\frac{\partial \mathbf{v}}{\partial t}\right)|^2\, d\mathbf{y}\, d\mathbf{x} = 0,\ \forall t \in [0,T]. \quad (3.4.22)$$

Finally, by using (3.4.17), we have

$$-\int_\Omega\int_G \tilde{p}^0 \operatorname{div}_x \frac{\partial \mathbf{u}^0}{\partial t}\, d\mathbf{y}\, d\mathbf{x} - \int_\Omega\int_{G_f} p^0 \operatorname{div}_x \frac{\partial \mathbf{v}}{\partial t}\, d\mathbf{y}\, d\mathbf{x}$$

$$-\int_\Omega\int_{G_s} \tilde{p}^0 \operatorname{div}_y \frac{\partial u^1}{\partial t}\, d\mathbf{y}\, d\mathbf{x} + \int_\Omega p^0(x,t)\left(\int_{G_s} \operatorname{div}_y \frac{\partial u^1}{\partial t} d\mathbf{y}\right) d\mathbf{x}$$

$$= \frac{|\mathscr{Y}_f|}{\gamma}\frac{1}{2}\frac{d}{dt}\int_\Omega\int_G (p^0)^2\, d\mathbf{y}\, d\mathbf{x} + \frac{\gamma}{|\Omega|}\frac{1}{2}\frac{d}{dt}\left(\int_\Omega\int_{G_s} \operatorname{div}_y \mathbf{u}^1\, d\mathbf{y}\, d\mathbf{x}\right)^2, \quad (3.4.23)$$

from which uniqueness follows.

The last step is to identify the effective equation of $\mathbf{u}^0$ and is similar to the incompressible case, and we keep the equation (3.2.49). The equation (3.2.55) does not hold anymore.

Equation (3.4.17) implies

$$-\frac{|\mathscr{Y}_f|}{\gamma}\rho_f \frac{\partial p^0}{\partial t} = \operatorname{div}_x \Big\{ |\mathscr{Y}_f|\rho_f \frac{\partial \mathbf{u}^0}{\partial t} + \int_0^t \mathscr{A}(t-\tau)$$
$$[\rho_f F(x,\tau) - \nabla p^0(x,\tau)]\, d\tau - \int_0^t \mathscr{A}(t-\tau)\rho_f \frac{\partial^2 \mathbf{u}^0}{\partial \tau \mathbf{u}^2} d\tau \Big\}$$
$$-\rho_f \mathscr{C}^H : e\left(\frac{\partial \mathbf{u}^0}{\partial t}\right) - \rho_f \left(\frac{\partial p^0}{\partial t} + \frac{|\mathscr{Y}_f|}{|\mathscr{Y}_s|\,|\Omega|}\int_\Omega \frac{\partial p^0}{\partial t} d\mathbf{x}\right)\int_{G_s} \operatorname{div}_y w^0\, d\mathbf{y}. \quad (3.4.24)$$

System (3.2.49), (3.4.24) has the same structure as (3.2.49), (3.2.55) so we skip the detailed discussion.

4 Wet Ionic, Piezoelectric Bone

4.1 INTRODUCTION

In this chapter we focus on cortical bone, a composite material consisting of a piezoelectric-elastic matrix and a conducting interstitial fluid (Güzelsu and Saha (1981),(1984), Güzelsu (1985)). [169, 172, 173, 174] . As the interstitial fluid is ionic, whenever the bone is stimulated ultrasonically a charge appears on the bone matrix and a streaming potential is created in the fluid. The current which occurs forms an electro-magnetic field. Cortical and cancellous bone respond to fractures in different ways; cancellous bone unites very rapidly because there are many points of contact which are rich in blood and cells. In fractured, cortical bone, depending on whether contact of the severed portions is close and immobilized, healing takes place with very little external callus [79]. On the other hand, if immobilization is not rigid, then there is a build up of external callus. In what follows we assume that the solid phase occupies the region $\Omega^{(s)}$ and is piezoelectric; however, there is also ionic advective transfer in the interstitial fluid part $\Omega^{(f)}$, i.e. the material will consist of a piezoelectric frame with an interstitial, conductive ionic pore fluid. As these equations are highly non-linear [160], using asymptotic expansions we were only able to generalize the work of [309, 264] for the dynamical case to obtain a hierarchical expansion in terms of a small parameter, the pore size. In the present work we concentrate on using a mixture theory method to obtain effective equations for a poroelastic, piezoelectric bone matrix. Using this model we consider excitation of a long bone, such as the tibia. By sending a shear wave along the tibia we excite the electromagnetic field which is measurable [169, 170, 171, 172]. Such a procedure has two purposes, first to detect fracture and its state of healing, second as a method of promoting healing through the stimulation of of osteoblast activity. We investigate this problem after developing the effective equations for poropiezio, elastic equations.

These equations are highly non-linear [160]; however, using asymptotic expansions we obtain effective equations, which generalize the work of Murad [309] and Lemaire [264]. We consider the dynamical case occurring in acoustics and obtain a hierarchical expansion in terms of a small parameter based on the pore size. We considered several special cases concerning large and small Peclet numbers, high-frequencies, etc. In the next section, by eliminating the effect of the ionic movement, we obtain a pair of effective equations for the electromagnetic field plus the poroelastic equations of motion.

4.2 WET BONE WITH IONIC INTERACTION

We assume that a poroelastic material, consisting of a piezoelectric, elastic frame lies in the region denoted by $\Omega^{(s)}$; the interstitial, conductive ionic fluid, lies in the complement $\Omega^{(f)}$. The equations of motion for the solid phase are

$$\partial_t^2 \mathbf{u}^{(s)} = \nabla_x \cdot \sigma^{\mathbf{s},\varepsilon}, \quad \mathbf{x} \in \Omega_s \tag{4.2.1}$$

where the solid stress tensor is given by

$$\sigma^{\mathbf{s},\varepsilon} := \mathbb{C} : \mathbf{e}(\mathbf{u}) + \underline{\Pi}^T \cdot \mathbf{E} \tag{4.2.2}$$

and $\underline{\underline{e}}^{(s)}$ indicates the solid strain tensor; whereas, the $\underline{\Pi}^T$ are the piezioelectric stress constants [17]. If this material is isotropic the constitutive equation reduces to

$$\sigma^{(s)} = \lambda_s \nabla \cdot \mathbf{u}^{(s)} \mathbb{I} + 2\mu_s \mathbf{e}^{(s)}(\mathbf{u}^{(s)}) - \underline{\Pi}^T \cdot \mathbf{E}. \tag{4.2.3}$$

It should be noted that constitutive equation also may be written in terms of the electric displacement field $\mathbf{D}^{(s)}$ as

$$\mathbf{D}^{(s)} = \tilde{\varepsilon}^{(s)} : \mathbf{E} + \underline{\Pi} : \mathbf{e}^{(\mathbf{s})}(\mathbf{u}) \tag{4.2.4}$$

where $\mathbf{E}$ is the electric field. In terms of the summation convention we may write piezoelectric term

$$(\underline{\Pi} \cdot \mathbf{E})_{ij} := \Pi_{ijk} \mathbf{E}_k.$$

In the region $\Omega_f := \Omega / \Omega_s$ the fluid equation of motion is

$$\partial_t^2 \mathbf{u}^{(f)} = \nabla_x \cdot \sigma^{(f)}, \quad \mathbf{x} \in \Omega_f, \tag{4.2.5}$$

where

$$\sigma^{(f)} = -p\mathbb{I} + 2\nu_f \mathbf{e}^{(f)}(\mathbf{u}^{(f)}) + \underline{\mathscr{T}}_M. \tag{4.2.6}$$

$$\underline{\mathscr{T}}_M = \frac{\tilde{\varepsilon}_f}{2}(2\mathbf{E} \otimes \mathbf{E} - \mathbf{E}^2 \mathbb{I}) + \frac{1}{2\mu_f}(2\mathbf{B} \otimes \mathbf{B} - B^2 \mathbb{I}) \tag{4.2.7}$$

and where $\underline{\mathscr{T}}_M$ is known as the electromagnetic stress tensor. Here $\otimes$ indicates a tensor product; $\mathbf{E}$ denotes the electric field and E its magnitude. Likewise $\mathbf{B}^{\mathbf{f}}$ is the magnetic induction field and B^f its magnitude. $\tilde{\varepsilon}_f$ is the fluid permitivity, and $\tilde{\mu}_f$ the fluid permeability. If the fluid is considered incompressible, then

$$\nabla \cdot \mathbf{v} = 0. \tag{4.2.8}$$

Our attention will be restricted to the acoustics case; hence, the fluid must be slightly compressible. [1]

---

[1] See [91, 135] for further details concerning the homogenization of the slightly compressible case.

Here p is the pore pressure, v_f the fluid viscosity and $\underline{\underline{e}}^{(f)}(\mathbf{u}^{(f)})$ is the fluid strain. Using

$$\nabla \cdot \underline{\mathscr{T}}_M = -q_f \nabla \mathbf{E}^f \tag{4.2.9}$$

the Stokes' system then takes on the form

$$\rho_f \partial_t \mathbf{v}^f = -\nabla p^f + \mu_f \Delta \mathbf{v} + q_f \mathbf{E}^f, \quad \nabla \cdot \mathbf{v}_t = 0, \tag{4.2.10}$$

where we have used one of the Maxwell equations, namely $\nabla \cdot B = 0$.

On the other hand, in the piezoelectric, elastic matrix magnetism is neglected; hence, the system obeys the Maxwell-Gauss equation, namely

$$\nabla \cdot \mathbf{D}^{(s)} = q_s, \tag{4.2.11}$$

where q_s is the charge in the solid, and hence from (4.4.3) we have

$$\nabla \cdot \left(\tilde{\varepsilon}^{(s)} : \mathbf{E} + \underline{\Pi} : (\mathbf{u}) \right) = q_s \tag{4.2.12}$$

In addition to the mechanical equations of motion for the solid and fluid phases, there is an ionic transfer to be considered. As the mass of the ions is insignificant with respect to the density of the fluid, we treat this as a diffusion dominated transport.
As we are clearly in the fluid region here, we drop the subscript "f" as it is unnecessary. If $D_{\underset{-}{+}}$ is the binary water-ion diffusion coefficient, T the absolute temperature and R the universal ideal gas constant, the convection-diffusion equations governing ion, $\overset{+}{c}{}^{-}$ transport are given by [309]

$$\partial_t \overset{+}{c}{}^{-} + \nabla \cdot \left(\frac{D_{\underset{-}{+}} c^{D}_{\underset{-}{+}}}{RT} \nabla \overset{+}{\mu}{}^{-} \right) = 0, \tag{4.2.13}$$

where $\overset{+}{\mu}{}^{-}$ are molar electrochemical potentials of cations and anions, according to the dilute solution approximation, are given by [65],[287]

$$\overset{+}{\mu}{}^{-} := \pm F \phi_f + RT \log \overset{+}{c}{}^{-} \tag{4.2.14}$$

Following [309] we introduce a dimensionless electric potential $\overline{\phi_f} = \frac{F\phi_f}{RT}$ and assume mono-valent ions (i.e. $Z = 1$); we have

$$\frac{1}{RT} \nabla \overset{+}{\mu}{}^{-} = \frac{\nabla \overset{+}{c}{}^{-}}{\overset{+}{c}{}^{-}} \pm \nabla \phi_f, \tag{4.2.15}$$

and from which the concentration-transport equation follows

$$\partial_t c^{\pm} + \nabla \cdot \left(c^{\pm} \mathbf{v} \right) = \nabla \cdot \left[D_{\pm} \left(\nabla c^{\pm} \pm c^{\pm} \nabla \overline{\phi_f} \right) \right] =$$

$$\nabla \cdot \left[D_{\pm} \exp(\mp \overline{\phi_f}) \nabla \left(c^{\pm} \exp(\pm \overline{\phi_f}) \right) \right]. \tag{4.2.16}$$

Following the approach of Sasidhar and Ruckenstein [365, 366] and Bike and Prieve [27] we split the effects of the electric potential caused by double layer potentials, induced by the charges on the solid material, from those due to a streaming potential caused by the fluid flow. Hence, we write ϕ_f in the form

$$\phi_f = \phi + \psi_b, \tag{4.2.17}$$

where ϕ varies across the pores and is caused by double layer effects and ψ_b is the streaming potential which strives to maintain electric neutrality. Moyne and Murad then characterize ψ as an electric potential inherent in a species of a bulk solution. Denoting c_b as the bulk concentration associated with the local bulk solution, define $\mu_b^{\pm} := \pm F\psi + RT \log c^{\pm}$, as the corresponding electrochemical potential, which by construction is set as, $\mu_b^{\pm} := \mu^{\pm}$. We also use [309] the dimensionless quantities $\overline{\phi} := \frac{F\phi}{RT}$ and $\overline{\psi} := \frac{F\psi}{RT}$, thereby obtaining

$$\mu_b^{\pm} := F\psi + RT \log c_b = \mu^{\pm} = \pm F\phi_f + RT \log c^{\pm} \tag{4.2.18}$$

and the generalized Boltzman distribution

$$c^{\pm} = c_b e^{\mp \overline{\phi_f} \pm \overline{\Psi})} = c_b e^{(\mp \overline{\phi})}, \quad q = -2F c_b \sinh(\overline{\phi}), \tag{4.2.19}$$

where the barred notation means $\bar{\star} = \frac{\star F}{RT}$, and use this to denote the reduced potentials $\bar{\phi}$ and $\bar{\psi}$, etc. The bulk and real fields are linked as [264]

$$\phi_f = \psi_b + \psi, \quad n^{\pm} = c_b e^{\mp \overline{\phi}},$$

$$p_b = p - \pi \equiv p - 2RT c_b \left(\cosh(\overline{\phi}) - 1 \right). \tag{4.2.20}$$

In the electrodynamic case we may represent $\mathbf{E}$ in terms of a scalar and vector potential, namely

$$\mathbf{E} = -\nabla \phi_f - \frac{\partial \mathbf{A}^{(f)}}{\partial t};$$

$$\rho_f \partial_t \mathbf{v}^f = -\nabla p^f + \mu_f \Delta \mathbf{v} + q_f \mathbf{E}, \quad \nabla \cdot \mathbf{v_t} = \mathbf{0}. \tag{4.2.21}$$

In the fluid there will be a flow of ions and hence a current which induces a magnetic field. Hence, in the dynamic case we use time-dependent Maxwell equations in the fluid

$$\tilde{\varepsilon}_f \nabla \cdot \mathbf{E} = \mathbf{q_f}, \quad \nabla \cdot \mathbf{B^{(f)}} = \mathbf{0}, \quad \nabla \times \mathbf{E} = \frac{\partial \mathbf{B_f^{(f)}}}{\partial \mathbf{t}}. \tag{4.2.22}$$

The induced electromagnetic field must satisfy the Maxwell-Gauss equation, which permits us to obtain

$$\mathbf{E} = -\nabla\phi_f - \partial t A^{(f)}, \tag{4.2.23}$$

where $\mathbf{A}^{(f)}$ is a vector potential related to $\mathbf{B}^{(f)}$ by $\mathbf{B}^{(f)} = \nabla \times \mathbf{A}^{(f)}$.

$$\varepsilon_f \nabla \cdot \mathbf{E} = q_f^{\mathscr{V}}, \tag{4.2.24}$$

where ϕ_f is the electric potential in the fluid and $q_f^{\mathscr{V}}$ the volume charge density.

We split ϕ_f into a streaming potential and the dipole potential to obtain

$$\nabla \cdot \left[\nabla\left(\overline{\psi}_b + \overline{\phi}\right) + \frac{\partial \mathbf{A}^{(f)}}{\partial t}\right] = \frac{1}{L_D^2}\sinh(\overline{\phi}), \tag{4.2.25}$$

where L_D is the Debye length given by $L_D = \sqrt{\tilde{\varepsilon}_f RT/(2F^2 c_b)}$. The Debye length characterizes the thickness of the diffuse ionic layer compensating for negative surface charge.

Following [264], by incorporating (4.2.19) into (4.2.21) yields the Stokes' system, which then may be written as

$$\mu_f \Delta \mathbf{v} = \nabla\left(p + q\phi_f\right) = \nabla p_b, \tag{4.2.26}$$

where the apparent bulk pressure written as (4.2.20) leads to

$$\begin{aligned}\rho_f \partial_t \mathbf{v}^{(f)} &= \mu_f \Delta \partial_t \mathbf{u}^{(f)} - \nabla p_b - 2RT\left(\cosh(\overline{\phi}) - 1\right)\nabla c_b \\ &= 2RTc_b \sinh(\overline{\phi})\nabla\overline{\psi}, \quad \nabla \mathbf{v} = 0.\end{aligned} \tag{4.2.27}$$

By the small displacement assumption and slight compressibility in the acoustic response we eliminate the bulk pressure altogether (4.2.28); i.e. replace the bulk pressure with $p_b = -a_f^2 \rho^{(f)} \nabla \cdot \mathbf{v}$ in (4.2.21). This simplifies the Stokes' system to the single equation

$$\begin{aligned}\rho_f \partial_t^2 \mathbf{u}^{(f)} &= \mu_f \Delta \partial_\mathbf{t} \mathbf{u}^{(f)} + \rho_f a_f^2 \nabla\nabla \cdot \mathbf{u} - 2RT\left(\cosh(\overline{\phi}) - 1\right)\nabla c_b \\ &= 2RTc_b \sinh(\overline{\phi})\nabla\overline{\psi}.\end{aligned} \tag{4.2.28}$$

The vector potential $\mathbf{A}$ is seen to satisfy a non-homogeneous wave equation

$$\Delta \mathbf{A}^{(f)} = \mu^f \tilde{\varepsilon}^f \frac{\partial^2 \mathbf{A}^{(f)}}{\partial t^2} - \mu_f \mathbf{J},$$

where the current $\mathbf{J}$ is given by

$$\mathbf{J} = 2\mathbf{v}^f c_b \sinh\overline{\phi}. \tag{4.2.29}$$

In the time harmonic case, with frequency ω, (4.2.28) reduces to

$$\omega^2 \rho_f \mathbf{u}^f + \nabla\left(\rho_f a_f^2 \nabla \cdot \mathbf{u}\right) + i\omega\mu_f(\mathbf{u}) + q\left(\nabla\phi + i\omega \mathbf{A_f}\right) = 0, \tag{4.2.30}$$

where the vector potential $\mathbf{A}^{(f)}$ now satisfies the reduced wave equation

$$\Delta\mathbf{A}^{(f)} + \frac{1}{a_f^2}\left[\nu_f\tilde{\varepsilon}_f\omega^2 + i\mu_f\sigma\omega\right]\mathbf{A}^{(f)} = 0. \tag{4.2.31}$$

As pointed out by [309] the relation of the volume and surface electric charge densities $q_f^{\mathscr{V}}$ and $q_f^{\mathscr{S}}$ can be related using $q_f^{\mathscr{S}} = q_f^{\mathscr{V}}/\ell$.

Now we postulate the boundary values between the two phases. The solid interface is denoted by $\partial\Omega_{fs}$ and it is oriented by having its normal directed into the solid region Ω_s. The presence of a Stern layer [264],the first layer of the electric double layer, may create a jump in the electric surface charges which we denote as $q_{fs}^{\mathscr{S}}$, which is related to the fluid and solid surface charges by $q_{fs}^{\mathscr{S}} = q_s^{\mathscr{S}} - q_f^{\mathscr{S}}$. As mentioned earlier $q_\alpha^{\mathscr{S}}$, $\alpha = s,f$ are the surface chares in the solid and fluid surfaces respectively. We list below the interface conditions:

$$\phi_f = \phi_s, \tag{4.2.32}$$

$$\mathbf{A}^{(f)} = 0, \tag{4.2.33}$$

$$\mathbf{n}\times\left(\nabla\times\mathbf{A}^{(f)}\right) = \mu_f s^{\mathscr{S}}, \tag{4.2.34}$$

$$-\tilde{\varepsilon}_f\left(\nabla\phi_f - \frac{\partial\mathbf{A}^{(f)}}{\partial t}\right)\cdot\mathbf{n} = q^{\mathscr{S}_f}, \tag{4.2.35}$$

$$-\left(\underline{\Pi} : (\mathbf{u} - \tilde{\varepsilon}^{(s)} : \mathbf{E})\right) = q^{\mathscr{S}_s}, \tag{4.2.36}$$

$$\mathbf{u}^{(f)} = \mathbf{u}^{(s)}, \tag{4.2.37}$$

$$-\mathbf{J}_{\pm}\cdot\mathbf{n} = 0, \tag{4.2.38}$$

$$\sigma = {}^{(s)}. \tag{4.2.39}$$

4.2.1 NONDIMENTIONALIZED EQUATIONS

It is advantageous to introduce reference values for the unknowns and to indicate the magnitude of these values. We assume that the size of the sample is of length L and this is the order of the macro-variables. The pore size in the medium are of length ℓ which is small, i.e. $\ell \approx O(\varepsilon)$; hence, the displacements $\mathbf{u}^{(\alpha)}, \alpha = s,f$ are also $O(\varepsilon)$. We introduce some reference dimensions as follows:

$$\ell \text{ micro-length,} \quad L_{\text{sample}} \text{ macro-length,} \quad t_{\text{ref}} = \frac{L^2}{D_{\pm}} \text{ diffusion time scale,}$$

$$\tau_{\text{ref}} = \frac{1}{\omega} \text{ frequency time scale,} \quad v_{\text{ref}} = \frac{\ell^2 p_{\text{ref}}}{\mu_f L} \text{ reference velocity}$$

$$E_{\text{ref}} = \frac{\sigma}{\tilde{\varepsilon}\tilde{\varepsilon}_0} \text{ reference electric field,} \quad \phi_f = \ell E_{\text{ref}} \text{ reference electric potential}$$

$$c_{\text{ref}} = \frac{\sigma}{F\ell} \text{ reference ion concentration,} \quad \mathbb{C}_{\text{ref}} \text{ reference elastic coefficients.}$$

Π_{ref} reference piezoelectric coefficient.

In compact bone samples the macro size is of the order $L \approx 10^{-3}$ and the canicular size is of the order $\ell \approx 10^{-3}$, leading to a small parameter $\ell_{\text{ref}} = \frac{\ell}{L}$. We now proceed to introduce dimensionless coordinates by adding several other reference values.

$$\rho_0 \text{ reference density,} \quad u_0 = \ell \text{ reference displacement,} \quad p_0 \text{ reference pressure,}$$

$$\blacksquare_{\text{ref}} \text{ reference piezoelastic tensor.}$$

We use a primed notation for the dimensionless unknowns, such as for the solid stress $\sigma^{(s)\prime}$. The constitutive equation in the solid becomes

$$\sigma^{(s)\prime} = \mathbb{C}_{\text{ref}} u_{\text{ref}} \left[\mathbb{C}' : e'(\mathbf{u}') + M_p \blacksquare' : \nabla' \phi^{s,\prime}\right] \tag{4.2.40}$$

and as $\mathbf{u}_{\text{ref}} \approx \ell$ the solid elastic equations may be approximated as

$$-\omega^2 \rho_0 \ell \rho_s' = \nabla' \left(\ell \mathbb{C}_{ref} \mathbb{C}' : \mathbf{e}'(\mathbf{u}') - \phi_{\text{ref}} \Pi_{ref} \underline{\Pi}' \cdot \nabla' \phi_s'\right)$$

or

$$-\frac{\omega^2 \rho_0}{\mathbb{C}_{\text{ref}}} u_s' \rho_s = \nabla' \left(\mathbb{C}' : \mathbf{e}' - M_p \underline{\Pi}' \cdot \nabla' \phi'\right), \tag{4.2.41}$$

where

$$M_p := \frac{\phi_{\text{ref}} \Pi}{\ell \mathbb{C}_{\text{ref}}}$$

is a piezoelectric-elastic number, which compares piezoelectric terms to elastic terms. As $\phi_{\text{ref}} \approx \ell$, this implies that $M_p \approx O(1)$. Since $\frac{\rho_0 \omega^2}{\mathbb{C}_{\text{ref}}} \approx O(\frac{1}{\lambda^2})$ and the wave length λ must be large with respect to cell size equation (4.2.41) for the homogenization method to be applicable, we have

$$\tilde{\omega}^2 \rho_s \mathbf{u}' = \nabla' \cdot \left(\mathbb{C}' : \mathbf{e}' - \underline{\Pi}' \cdot \nabla' \phi'\right), \quad \tilde{\omega} = \sqrt{\frac{\omega^2}{\mathbb{C}_{\text{ref}}}} \tag{4.2.42}$$

In the fluid, the time harmonic version of equation (4.2.25) is

$$\nabla \cdot D^{(f)} = -\tilde{\varepsilon}_f \nabla \cdot \left(\nabla \phi_f - i\omega \mathbf{A}_f^{(f)}\right) = q_f^{\mathscr{V}}$$

We now argue, as Moyne and Murad [309], that if the reference electric field magnitude is represented by $\mathbf{E}_{\text{ref}}$ then local electric neutrality dictates that it is related to the surface charge density $q^{\mathscr{S}\,\text{ref}}$ by $q^{\mathscr{S}\,\text{ref}} \tilde{\varepsilon}_f = \phi_{\text{ref}}/\ell$, which suggests we rescale

$$\nabla \cdot \left[\nabla \left(\overline{\psi}_b + \overline{\phi}\right) + i\omega \mathbf{A}^{(f)}\right] = \frac{1}{L_D^2} \sinh(\overline{\phi}).$$

To this end, by noting that as $c_{ref} = \frac{q\mathscr{S}_{\text{ref}}}{F\ell}$ we have that $L_{D\text{ref}} = O(\ell)$, and the above equation takes the form

$$\tilde{\varepsilon}^2\nabla' \cdot \left(\nabla'\left(\overline{\psi}_b' + \overline{\phi}'\right) + i\omega A_f'\right) = \frac{1}{L_{Dref}^2}\sinh(\overline{\phi}'), \tag{4.2.43}$$

4.2.2 FLUID EQUATIONS WITH SLIGHT COMPRESSIBILITY

In the remainder of this section we assume the small compressibility condition holds in the fluid, namely, we approximate the pressure by $p \approx -\rho_f a_f^2 \operatorname{div} \cdot \mathbf{u}$. Hence, the constitutive equation for the fluid phase now becomes

$$\sigma^{(f)} = a_f^2\rho_f\nabla\cdot\mathbf{u}\mathbb{I} + 2i\omega\nu_f\varepsilon^2 e(\mathbf{u}) + \underline{\mathscr{T}}_M;$$

whereas the fluid equations of motion are given by

$$\rho_f\omega^2\mathbf{u} + \nabla\cdot\sigma^{(f)} = a_f^2\nabla\left(\rho_f\nabla\cdot\mathbf{u}\right) + i\omega\nu_f\varepsilon^2\Delta\mathbf{u} + q\mathbf{E} + \rho^f\omega^2\mathbf{u}^{(f)} = 0.$$

Introducing non-dimensional unknowns into the constitutive equation

$$\sigma^{(f)} = a_f^2\rho_{\text{ref}}u_{\text{ref}}\rho'\nabla'\cdot\mathbf{u}' + 2i\omega\nu_f u_{\text{ref}}\varepsilon^2 e'(\mathbf{u}') + \underline{\mathscr{T}}_{ref}\underline{\mathscr{T}}_M',$$

we are able to define a non-dimensional fluid stress as

$$\sigma^{(f)\prime} = \rho'\nabla'\cdot\mathbf{u}' + iV_p\varepsilon^2 e'(\mathbf{u}') + M_x\underline{\mathscr{T}}_M',$$

where the coefficients V_p and M_x are defined as

$$V_p = \frac{2i\omega\nu_f}{a_f^2\rho_{\text{ref}}},\ \text{and}\ M_x = \frac{\underline{\mathscr{T}}_{\text{ref}}}{a_f^2\rho_{ref}u_{\text{ref}}}.$$

We assume that both of these terms are order $O(1)$. Moreover,

$$\overline{\phi}_{\text{ref}} = \frac{F\phi_{\text{ref}}}{RT} = O(1),\quad \overline{\psi}_{ref} = \frac{F\psi_{\text{ref}}}{RT} = O(1)$$

so

$$\overline{\phi}'_{\text{ref}} \approx \overline{\phi}_{\text{ref}},\quad \overline{\psi}'_{\text{ref}} \approx \overline{\psi}_{\text{ref}}. \tag{4.2.44}$$

Finally by representing the charge density q in terms of the bulk ionic concentration c_b,

$$c_b e^{\overset{+}{-}\overline{\phi}} = \overset{+}{c^-},$$

the equations of motion assume the desired form

$$\nabla\cdot\sigma^{(f)} = a_f^2\nabla\left(\rho_f\nabla\cdot\mathbf{u}\right) + i\omega\nu_f\varepsilon^2\Delta\mathbf{u} + c_b\left(\cosh(\overline{\phi}) - 1\right) + \rho^f\omega^2\mathbf{u}^{(f)} = 0.$$

$$\nabla'\nabla'\cdot\mathbf{u}' + il_e\varepsilon^2\Delta'\mathbf{u}' + \frac{\omega^2}{a_f^2}\mathbf{u}' +$$

$$2D_e R'T' \left[\left(\cosh(\overline{\phi}') - 1 \right) \nabla' c'_b - \sinh(\overline{\phi}') \nabla' \overline{\psi}'_b \right], \tag{4.2.45}$$

where the coefficient D_e is defined as

$$D_e = \frac{R_{\text{ref}} T_{\text{ref}} c_{ref}}{a_f^2 \rho_{\text{ref}}^f}.$$

4.2.3 NERNST-PLANK EQUATIONS

We denote by $q^{\mathscr{V}_{\text{ref}}}$ the volume charge in the fluid, whereas $q^{\mathscr{S}}$ denotes the surface charge on the solid-fluid interface. Choosing the diffusion time-scale $t_{\text{ref}} = \frac{L^2}{D_{\text{ref}}}$, where $D_{\text{ref}} = \max\{D_+, D_-\}$ [264], the non-dimensional Nernst-Plank equation takes the form

$$\partial'_t \left(c'_b \exp(-\overline{\phi}') \right) + Pe \nabla' \left(c'_b \exp(\mp \overline{\phi}') v' \right) + J_d \nabla' \cdot J'_{\pm} = 0. \tag{4.2.46}$$

In the case where diffusion is the main means of transport the Peclet number $Pe = O(\varepsilon^1)$. The ion's mass is insignificant with respect to the mass of the solvent, and hence it appears to be a reasonable assumption for the acoustic case. Lemaire [264] introduces the number $J_d = J_{ref} L_{\text{ref}} / D_{\text{ref}} c_{\text{ref}}$ to compare the electric and diffusive effects due to the reference ionic flux J_{ref}. If diffusion is assumed to be the dominant transport mechanism J_d is scaled as $J_d = O(\varepsilon^0)$ [264]. However, in the acoustic case we expect convection to be dominant and $Pe = O(\varepsilon^{-1})$. Following [264] we assume that the main part of the current is due to ionic movement and J_d is scaled as $J_d = O(\varepsilon^0)$; hence (4.2.46) may be written in non-dimensional form as

$$J_d \mathbf{J}'_{\pm} = -\mathbf{D}'_{\pm} \cdot \left(e^{\mp \overline{\phi}'} \left(\nabla' c'_b \pm c'_b \nabla' \overline{\psi}'_b \right) \right). \tag{4.2.47}$$

4.3 HOMOGENIZATION USING FORMAL POWER SERIES

We now consider formally expanding the solutions to the non-dimensional equations in powers of the parameter ε, i.e.

$$u(x,y,t) = \sum_{k=0}^{\infty} u_{[k]}(x,y,t) \varepsilon^k, \tag{4.3.1}$$

where we take x to be the slow variable and $y = \frac{x}{\varepsilon}$ to be the fast variable. The Taylor series for $\cosh(\overline{\phi})$ and $\sinh(\overline{\phi}))$ [264] suggest that the hyperbolic functions used in expressing the ion concentrations be approximated as

$$\cosh(\overline{\phi}) = \cosh(\overline{\phi}_{[0]}) + \overline{\phi}_{[1]} \sinh(\overline{\phi}_{[0]}) \varepsilon + O(\varepsilon^2) \tag{4.3.2}$$

$$\sinh(\overline{\phi}) = \sinh(\overline{\phi}_{[0]}) + \overline{\phi}_{[1]} \cosh(\overline{\phi}_{[0]}) \varepsilon + O(\varepsilon^2) \tag{4.3.3}$$

The strain then may be expressed in terms of the fast and slow variables as

$$e' := \frac{1}{\varepsilon} e_y + e_x$$

Using the non-dimensional form of the solid equation of variables, we harvest the first few terms in the asymptotic expansion of the solid stress equation (4.2.40)

$$\sigma^{(s)}_{[-1]} = \mathbb{C} e_y(u_{[0]}),$$

$$\sigma^{(s)}_{[0]} = \mathbb{C} :: \left(e_x(u_{[0]}) + e_y(u_{[1]})\right),$$

$$\sigma^{(s)}_{[1]} = \mathbb{C} :: \left(e_x(u_{[1]}) + e_y(u_{[0]}) + \underline{\Pi}'^T \nabla_y \phi^{(s)}_{[0]}\right).$$

The first few terms of the non-dimensional equations (4.2.41) of motion then become

$$\nabla_y \cdot \sigma^{(s)}_{[-1]} = 0,$$

$$\nabla_y \cdot \sigma^{(s)}_{[-0]} + \nabla_x \cdot \sigma^{(s)}_{[-1]} = 0,$$

$$\nabla_y \cdot \sigma^{(s)}_{[1]} + \nabla_x \cdot \sigma^{(s)}_{[0]} + \omega^2 \rho_s u_{[0]} = 0,$$

or in terms of the displacement and potential

$$\nabla_y \cdot \left(\mathbb{C}' : e_y(\mathbf{u}_{[0]} - \underline{\Pi}'^T : \nabla_y \phi_{[0]}\right) = 0, \tag{4.3.4}$$

$$\nabla_y \cdot \left(\mathbb{C}' : e_x(\mathbf{u}_{[0]} - \underline{\Pi}'^T : \nabla_x \phi_{[0]}\right) + \nabla_x \cdot \left(\mathbb{C}' : e_y(\mathbf{u}_{[0]} - \underline{\Pi}'^T : \nabla_y \phi_{[0]}\right) = 0, \tag{4.3.5}$$

$$, \nabla_y \cdot \left(\mathbb{C}' : \left[e_x(\mathbf{u}_{[1]} + e_x(\mathbf{u}_{[2]}\right] - \underline{\Pi}'^T : \left[\nabla_x \phi_{[1]} + \nabla_y \phi_{[2]}\right) +$$
$$\nabla_x \cdot \left(\mathbb{C}' : \left[e_x(\mathbf{u}_{[0]} + e_x(\mathbf{u}_{[1]}\right] - \underline{\Pi}'^T : \left[\nabla_x \phi_{[0]} + \nabla_y \phi_{[1]}\right] = \tilde{\omega}^2 \rho' \tag{4.3.6}$$

It is clear that any magnetic affects are due to ionic movement in the interstitial fluid. However, we assume these effects are minimal in the solid matrix. We treat the solid electrical behavior the same as in Lemaire et al. [264]. Hence, our asymptotic will be the same here, namely, we have

$$\nabla' \cdot \left(\underline{\Pi}' : \nabla' \mathbf{u}' - \mathbf{E}_p \tilde{\varepsilon}'_s \cdot \nabla' \phi'\right) = C_\ell q^{\mathscr{S}}{}', \tag{4.3.7}$$

where

$$\mathbf{E}_p = \frac{\tilde{\varepsilon}_{\text{ref}} \phi_{\text{ref}}}{\Pi_{\text{ref}} u_{\text{ref}}} = O(1)$$

and

$$C_\ell = \frac{q^{\mathscr{S}}_{\text{ref}} L^2}{\Pi_{\text{ref}} u_{\text{ref}}} = O(\varepsilon^{-2}),$$

because of, the electrical neutrality condition balances, the volume charge $q^{\mathscr{V}}_{ref}$ to the surface charge $q^{\mathscr{S}}_{ref}$, namely, $q^{\mathscr{S}}_{ref} = \frac{q^{\mathscr{V}}_{ref}}{\ell}$ [309]. Collecting powers of ε

$$\nabla_y \cdot \left(\underline{\Pi}' : \nabla_y \mathbf{u}_{[0]} - \mathbf{E}_p \tilde{\varepsilon}_s \nabla_y \phi_{s[0]}\right) = q^{\mathscr{S}\prime} \tag{4.3.8}$$

$$\begin{aligned} \nabla_y \cdot \left(\underline{\Pi}' : \left[\nabla_x \mathbf{u}_{[0]} + \nabla_y \mathbf{u}_{[1]}\right] - \mathbf{E}_p \tilde{\varepsilon}_s \left[\nabla_x \phi_{s[0]} + [\nabla_y \phi_{s[1]}]\right)\right) + \\ + \nabla_x \cdot \left(\underline{\Pi}' : \nabla_y \mathbf{u}_{[0]} - \mathbf{E}_p \tilde{\varepsilon}_s \nabla_y \phi_{s[0]}\right) \end{aligned} \tag{4.3.9}$$

The non-dimensional, fluid electricity equation (4.2.43) suggests the asymptotic expansion with terms

$$\nabla_y \cdot \nabla_y \left(\overline{\psi}_{b[0]} + \overline{\phi}_{[0]}\right) = \frac{1}{L^2_{D[0]}} \sinh(\psi_{[0]})$$

$$2\nabla_x \cdot \nabla_y \left(\overline{\psi}_{b[1]} + \overline{\phi}_{[1]}\right) + i\omega \nabla_y \cdot \mathbf{A}^{(f)} = \frac{1}{L^2_{D[0]}} \overline{\phi}_{[1]} \cosh(\overline{\phi}_{[0]})$$

Under the small-displacement assumption, the first few terms in the asymptotic expansion of the fluid constitutive equation becomes

$$\sigma^{(f)}_{[-1]} = \rho' \nabla_y \cdot \mathbf{u}_{[0]},$$

$$\sigma^{(f)}_{[0]} = \rho' \left(\nabla_y \cdot \mathbf{u}_{[1]} + \nabla_x \cdot \mathbf{u}_{[0]}\right) + \underline{\mathscr{T}}_{m[0]},$$

where

$$\underline{\mathscr{T}}_{M[0]} = \frac{\tilde{\varepsilon}_f}{2} \left(2\mathbf{E}_{[0]} \otimes \mathbf{E}_{[0]} - \mathbf{E}^2_{[0]} \mathbb{I}\right) + \frac{1}{2\nu_f} \left(2\mathbf{B}_{[0]} \otimes \mathbf{B}_{[0]} - B^2_{[0]} \mathbb{I}\right).$$

Then the first few terms of the asymptotic expansion of the fluid equations of motion are

$$\nabla_y \cdot \sigma^{(f)}_{[0]} + \nabla_x \cdot \sigma^{(f)}_{[-1]} = 0$$

$$\nabla_y \cdot \sigma^{(f)}_{[1]} + \nabla_x \cdot \sigma^{(f)}_{[0]} + \omega^2 \rho_f u_{[0]} = 0,$$

which become, in the fluid displacement variable expansion

$$\nabla_y \nabla_y \cdot \mathbf{u}^{(f)}_{[0]} = 0,$$

$$(\nabla_x \nabla_y + \nabla_y \nabla_x) \cdot \mathbf{u}^{(f)}_{[1]} + +\omega^2 \rho_f u_{[0]} =,0,$$

In terms of the displacement and potential we hve

$$\nabla_y \cdot \left(\mathbb{C}' : e_y(\mathbf{u}_{[0]} - \underline{\Pi}'^T : \nabla_y \phi_{[0]}\right) = 0, \tag{4.3.10}$$

$$\nabla_y \cdot \left(\mathbb{C}' : e_x(\mathbf{u}_{[0]} - \underline{\Pi}'^T : \nabla_x \phi_{[0]}\right) + \nabla_x \cdot \left(\mathbb{C}' : e_y(\mathbf{u}_{[0]} - \underline{\Pi}'^T : \nabla_y \phi_{[0]}\right) = 0, \tag{4.3.11}$$

$$\begin{aligned} , \nabla_y \cdot \left(\mathbb{C}' : \left[e_x(\mathbf{u}_{[1]} + e_x(\mathbf{u}_{[2]}\right] - \underline{\Pi}'^T : \left[\nabla_x \phi_{[1]} + \nabla_y \phi_{[2]}\right) + \\ \nabla_x \cdot \left(\mathbb{C}' : \left[e_x(\mathbf{u}_{[0]} + e_x(\mathbf{u}_{[1]}\right] - \underline{\Pi}'^T : \left[\nabla_x \phi_{[0]} + \nabla_y \phi_{[1]}\right] = \tilde{\omega}^2 \rho' \end{aligned} \tag{4.3.12}$$

Similarly we can obtain asymptotics of the solid electricity, asymptotics of the fluid electricity, asymptotics of the fluid electricity, and asymptotics in the fluid phase.

4.4 WET BONE WITHOUT IONIC INTERACTION

In this section we ignore the ionic effects, which allows us to formulate a more translucent theory. It might be argued, that in the high-frequency case, this is a reasonable approximation. Our approach here is a mixture theory method which makes use of the Reuss bound for the energy of the system. Below is a unit cell representing a poroelastic material. The darkened material represents the solid frame, whereas the lighter material represents the fluid. The figure was drawn with TikZ in LaTeX.

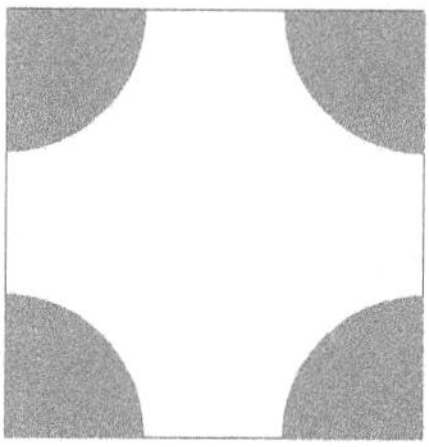

4.4.1 REUSS BOUND ON THE ENERGY

The Reuss bound for the energy may be found by starting with a mixture theory representation for the energy, namely

$$\Psi_{\text{mix}} := \Theta^{(s)} \left[\frac{1}{2} \mathbf{e} : \underline{\underline{\mathbb{C}}} : \mathbf{e} - \mathbf{e} : \underline{\blacksquare} : \cdot \mathbf{E}^{(s)} - \frac{1}{2} \mathbf{E}^{(s)} \cdot \varepsilon^{(s)} \cdot \mathbf{E}^{(s)} \right]$$

$$+ \Theta_F \left[\frac{K^{(f)}}{2} (\zeta - \text{tr}\mathbf{e})^2 - \frac{1}{2} \mathbf{E}^{(f)} \cdot \varepsilon^{(f)} \cdot \mathbf{E}^{(f)} \right], \tag{4.4.1}$$

where $\Theta^{(s)}$ and $\Theta^{(f)}$ are the support of the solid and fluid phases respectively. $\mathbf{E}^{(s)}$ and $\mathbf{E}^{(f)}$ are the electric fields in the solid and fluid respectively. The dielectric tensors in the solid and fluid, respectively, are $\varepsilon^{(s)}$ and $\varepsilon^{(f)}$. Later on we shall consider the isotropic case where these become constants.

Boldface terms are vectors, or second-order, tensors. Underlined, boldface are third-order tensors. Moreover, $\mathbf{e} = \text{sym}\nabla\mathbf{u}$ and $\zeta = -\nabla \cdot \mathbf{w}$, where $\mathbf{w} = \beta\left(\mathbf{u}^{(f)} - \mathbf{u}\right)$. Here $\mathbf{u}^{(f)}$ is the absolute displacement of the fluid and $\mathbf{u}$ is the overall displacement of the frame and fluid [175]. *Blackboard bold* letters are second-order, tensors unless underlined. Two underlines means the tensor is fourth order, etc. The Reuss potential is defined as

$$\Psi_{\text{Reuss}} = \min\left\{\Psi_{\text{mix}} : \Theta^{(s)}\mathbf{E}^{(s)} + \Theta^{(f)}\mathbf{E}^{(f)} = \mathbf{E}\right\} =$$

$$\frac{\Theta^{(s)}}{2}\mathbf{e} : \underline{\underline{\mathbb{C}}} : \mathbf{e} + \Theta_F \frac{K^{(f)}}{2} (\zeta - \operatorname{tr}\mathbf{e})^2 - \Psi_{\text{pe}}. \tag{4.4.2}$$

The constitutive equations are given as

$$\mathbf{D} = \underline{\Pi}^T : \mathbf{e}(\mathbf{u}) + \underline{\underline{\varepsilon}}^{(s)} : \mathbf{E}^{(s)} = \underline{\underline{\varepsilon}}^{(f)} : \mathbf{E}^{(f)}, \tag{4.4.3}$$

where $\mathbf{D}$ is the electric displacement, which we assume is the same in the solid as fluid. $\underline{\underline{\varepsilon}}^{(s)}$ and $\underline{\underline{\varepsilon}}^{(f)}$ are the dielectric, second-order tensors in the solid and fluid respectively. Then the electric force is seen to be

$$\mathbf{E} = \Theta^{(s)} \left(\varepsilon^{(s)}\right)^{-1} \cdot (\mathbf{D} - \underline{\Pi} : \mathbf{e}(\mathbf{u})) + \Theta^{(f)} \left(\varepsilon^{(f)}\right)^{-1} \cdot \mathbf{D}, \tag{4.4.4}$$

which implies

$$\mathbf{D} = \varepsilon_{\text{eff}} \cdot \left[\mathbf{E} + \Theta^{(s)} \left(\varepsilon^{(s)}\right)^{-1} \cdot (\underline{\Pi} : \mathbf{e}(\mathbf{u}))\right], \tag{4.4.5}$$

where

$$\varepsilon_{\text{eff}} = \left(\Theta^{(s)} \left(\varepsilon^{(s)}\right)^{-1} + \Theta^{(f)} \left(\varepsilon^{(f)}\right)^{-1}\right)^{-1}. \tag{4.4.6}$$

Back substituting into the expression for ψ_{pe} yields

$$\Psi_{\text{pe}} = \frac{1}{2}\left(\mathbf{E} + \Theta^{(s)} \left(\varepsilon^{(s)}\right)^{-1} \cdot (\underline{\Pi} : \mathbf{e}(\mathbf{u})\right) : \underline{\underline{\varepsilon}}_{\text{eff}} : \left(\mathbf{E} + \Theta^{(s)} \left(\underline{\underline{\varepsilon}}^{(s)}\right)^{-1} \cdot (\underline{\Pi} : \mathbf{e}(\mathbf{u})\right)$$

$$- \frac{1}{2}\Theta^{(s)} (\underline{\Pi} : \mathbf{e}(\mathbf{u})) \cdot \left(\varepsilon^{(s)}\right)^{-1} \cdot (\underline{\Pi} : \mathbf{e}(\mathbf{u})). \tag{4.4.7}$$

It follows that the Reuss energy has the structure

$$\Psi_{\text{Reuss}} = \frac{1}{2}\mathbf{e}(\mathbf{u}) : \underline{\underline{\mathbb{C}}}_{\text{eff}} : \mathbf{e}(\mathbf{u}) - \mathbf{e}(\mathbf{u}) : \underline{\underline{\varepsilon}}_{\text{eff}} \cdot \mathbf{E} - \frac{1}{2}\mathbf{E} \cdot \underline{\underline{\varepsilon}}_{\text{eff}} \cdot \mathbf{E}$$

$$+ \frac{\Theta^{(f)} K^{(f)}}{2} (\zeta - \operatorname{tr}(\mathbf{e}(\mathbf{u})))^2; \tag{4.4.8}$$

here

$$\underline{\underline{\mathbb{C}}}_{\text{eff}} := \Theta^{(s)} \left(\underline{\underline{\mathbb{C}}} - \underline{\Pi}^T \cdot \left(\varepsilon^{(s)}\right)^{-1} \cdot \underline{\Pi}\right)$$

$$\mathbf{O}\underline{\Pi}_{\text{eff}} := \Theta^{(s)} \varepsilon_{\text{eff}} : \underline{\Pi} \cdot \left(\varepsilon^{(s)}\right)^{-1}.$$

4.4.2 FLUID DISPLACEMENT

We denote spatial averages of a function F over $\Theta^{(f)}$, $\Theta^{(s)}$ and $\Omega := \Theta^{(f)} \cup \Theta^{(s)}$ as $\langle F \rangle^{(s)}$, $\langle F \rangle^{(f)}$ and $\langle F \rangle$ respectively.

The fluid displacement may be written as

$$\mathbf{v}(\mathbf{x})^{(f)} = \mathbf{u} + \mathbf{A}(\mathbf{x}) : \mathbf{e}(\mathbf{x}) + \mathbf{B}(\mathbf{x})\mathbf{w},$$

where

$$\langle \mathbf{B}(\mathbf{x}) \rangle^{(f)} = \mathbb{I}, \quad \langle \mathbf{A}(\mathbf{x}) \rangle^{(f)} = \mathbf{O}, \quad \langle \nabla \mathbf{A} \cdot \nabla \mathbf{B}(\mathbf{x}) \rangle^{(f)} = \mathbf{O}, \quad \left\langle \nabla^{(s)} \mathbf{A}(\mathbf{x}) \right\rangle^{(f)} = \mathbb{I}.$$

Here **u** is the movement of the composite material, and **w** is the relative fluid displacement to the composite displacement. The dissipation is given by

$$\Delta := \frac{\Theta^{(f)}}{2\eta} \left\langle ||\mathrm{dev}\nabla^{(s)} \dot{\mathbf{v}}||^2 \right\rangle^{(f)}$$

$$\mathbf{B} = \frac{\Theta^{(f)}}{2\eta} \left[\dot{e}(\mathbf{u}) : \left\langle (\mathrm{dev}\nabla^{(s)} \mathbf{A})^T \cdot \mathrm{dev}\nabla^{(s)} \mathbf{A}) \right\rangle^{(f)} : \dot{e}(\mathbf{u}) \right.$$
$$\left. + \dot{\mathbf{w}} \cdot \left\langle (\mathrm{dev}\nabla^{(s)} \mathbf{B})^T \cdot (\mathrm{dev}\nabla^{(s)} \mathbf{B}) \right\rangle^2_{(f)} \cdot \dot{\mathbf{w}} \right]$$
$$= \frac{\kappa_1}{2} ||\mathrm{dev}\dot{}||^2 + \frac{\kappa_2}{2} \dot{\mathbf{w}}^2. \tag{4.4.9}$$

4.4.3 KINETIC ENERGY

$$T = \frac{1}{2} \Theta^{(s)} \rho^{(s)} \dot{\mathbf{u}}^2 + \frac{1}{2} \Theta^{(f)} \rho^{(s)} \left\langle |[\dot{\mathbf{u}} + \mathbf{A}(\mathbf{x}) : \dot{\varepsilon} + \mathbf{B}(\mathbf{x}) \dot{\mathbf{w}}|^2 \right\rangle^{(f)}$$
$$= \frac{1}{2} \rho_b \dot{\mathbf{u}}^2 + \Theta^{(f)} \rho^{(f)} \cdot \mathbf{u} \cdot \langle \mathbf{B}(\mathbf{x}) \rangle^{(f)} \cdot \dot{\mathbf{w}} + \frac{1}{2} \Theta^{(f)} \rho^{(f)} \dot{\mathbf{w}} \cdot \left\langle \mathbf{B}^T \mathbf{B} \right\rangle^{(f)} \cdot \dot{\mathbf{w}}, + \text{h.o.t..} \tag{4.4.10}$$

If we define

$$\rho_b := \Theta^{(s)} \rho^{(s)} + \Theta^{(f)} \rho^{(f)},$$

then the kinetic energy can be put in the usual form [28, 29]

$$T + \frac{1}{2} \rho_b \dot{\mathbf{u}}^2 + \Theta^{(f)} \rho^{(f)} \dot{\mathbf{u}} \cdot \dot{\mathbf{w}} = \frac{1}{2} \dot{\mathbf{w}}^2. \tag{4.4.11}$$

4.4.4 CONSTITUTIVE EQUATIONS

The state or constitutive equations may be found by differentiating the energy minus the dissipation, namely

$$\sigma = \frac{\partial \Psi_{\mathrm{Reuss}}}{\partial e} + \frac{\partial \Delta}{\partial \dot{e}}$$
$$= \underline{\underline{\mathbb{C}}}^{\mathrm{eff}} : \mathbf{e} - \underline{\Pi}_{\mathrm{eff}} \cdot \mathbf{E} + \Theta^{(f)} K^{(f)} (\mathrm{tr}\mathbf{e} - \zeta) \tag{4.4.12}$$

$$\mathbf{D} = -\frac{\partial \Psi_{\mathrm{Reuss}}}{\partial \mathbf{E}} = \underline{\Pi}^T_{\mathrm{eff}} : \mathbf{e} + \underline{\underline{\varepsilon}}_{\mathrm{eff}} \cdot \mathbf{E}, \tag{4.4.13}$$

$$p = \frac{\partial \Psi_{\mathrm{Reuss}}}{\partial \zeta} = \Theta^{(f)} K^{(f)} (\zeta - \mathrm{tr}(\underline{\underline{e}})) \tag{4.4.14}$$

The Lagrange equations of motion then become

$$\frac{d}{dt}\frac{\partial T}{\partial \dot{\mathbf{u}}} = \rho_b \ddot{\mathbf{u}} + \Theta^{(f)} \rho^f \ddot{\mathbf{w}} = \nabla \cdot \sigma, \tag{4.4.15}$$

$$\frac{d}{dt}\frac{\partial T}{\partial \dot{\mathbf{w}}} + \frac{\partial \Delta}{\partial \dot{\mathbf{w}}} = \Theta^{(f)} \rho^{(f)} \ddot{\mathbf{u}} + \rho_{22} \ddot{\mathbf{w}} + \kappa_2 \dot{\mathbf{w}} = \nabla p. \tag{4.4.16}$$

4.5 ELECTRODYNAMICS

4.5.1 ELECTRICALLY ISOTROPIC SOLID

A material is electrically isotropic, which means the dielectric and permitttivity tensors are scalars, respectively ε and μ. From the Reuss formulation, $D^{(s)} = D^{(f)} = D$ and $H^{(s)} = H^{(f)} = H$; hence, in the solid phase we construct the magnetic induction $\mathbf{B}^{(s)}$ as follows

$$\nabla \times \mathbf{H} = \dot{\mathbf{D}} = \varepsilon^{(s)} \dot{\mathbf{E}} + \dot{e} : \underline{\Pi}$$

From which we compute, as $\nabla \cdot \mathbf{H} = 0$,

$$\nabla \times \nabla \times \mathbf{H} = -\Delta \mathbf{H} = \varepsilon^{(s)} \nabla \times \dot{\mathbf{E}}^{(s)} + \nabla \times (\dot{e} : \underline{\Pi})$$

and

$$\Delta \mathbf{B}^{(s)} = \varepsilon^{(s)} \mu^{(s)} \ddot{\mathbf{B}}^{(s)} - \mu^{(s)} \nabla \times \left(\underline{\underline{\dot{e}}} : \underline{\Pi} \right). \tag{4.5.1}$$

Next we calculate the equation of the electric force $\mathbf{E}^{(s)}$ beginning with $\nabla \mathbf{E}^{(s)} = -\dot{\mathbf{B}}^{(s)}$ and taking the curl

$$\nabla \times \nabla \times \mathbf{E}^{(s)} = \nabla \nabla \cdot \mathbf{E}^{(s)} - \Delta \mathbf{E}^{(s)} = -\nabla \times \dot{\mathbf{B}}^{(s)} = -\mu^{(s)} \nabla \times \dot{\mathbf{H}}^{(s)} = -\mu^{(s)} \ddot{\mathbf{D}}.$$

If we assume that there is no residual charge in the solid state, then

$$\nabla \cdot \mathbf{D} = 0 = \varepsilon \nabla \cdot \mathbf{E}^{(s)} + \nabla \cdot (\mathbf{e} \underline{\Pi});$$

hence,

$$-\frac{1}{\varepsilon^{(s)}} \nabla \nabla \cdot (\mathbf{e} : \Pi)$$

We arrive at

$$\Delta \mathbf{E}^{(s)} = \varepsilon^{(s)} \mu^{(s)} \ddot{\mathbf{E}}^{(s)} - \frac{1}{\varepsilon^{(s)}} \nabla \nabla \cdot (\mathbf{e} : \underline{\Pi}) + \mu^{(s)} (\ddot{\mathbf{e}} : \underline{\Pi}) \tag{4.5.2}$$

4.5.2 ELECTROMAGNETISM IN THE FLUID

The fluid is assumed to be isotropic; however, this is a simplification. Because of its ionic content, the fluid can maintain a current $\mathbf{J}$; from which it follows that

$$\Delta \mathbf{H}^{(f)} = \varepsilon^{(f)} \nabla \times \dot{\mathbf{E}}^{(f)} - \nabla \times \mathbf{J}^{(f)}, \quad \text{where } \mathbf{J}^{(f)} = \sigma^{(f)} \mathbf{E}^{(f)}$$

This implies,

$$\Delta \mathbf{B}^{(f)} = \varepsilon^{(f)} \mu^{(f)} \ddot{\mathbf{B}}^{(f)} + \mu^{(f)} \sigma^{(f)} \dot{\mathbf{B}}^{(f)}. \tag{4.5.3}$$

Likewise, from

$$\Delta \mathbf{E}^{(f)} = \nabla \times \dot{\mathbf{B}}^{(f)} = \mu^{(f)} \dot{\mathbf{H}}^{(f)} = \mu^{(f)} \ddot{\mathbf{D}} + \mu^{(f)} \sigma^{(f)} \dot{\mathbf{E}}^{(f)},$$

it follows that

$$\Delta \mathbf{E}^{(f)} = \varepsilon^{(f)} \mu^{(f)} \ddot{\mathbf{E}} + \mu^{(f)} \sigma^{(f)} \dot{\mathbf{E}}^{(f)}. \tag{4.5.4}$$

4.5.3 EFFECTIVE ELECTRO-MAGNETIC EQUATIONS

In this section we will assume that the permittivity and permeability are isotropic, i.e. they are scalars and not tensors. Summarizing, in the solid we have the effective constitutive equations for the piezoelectric, poroelastic material

$$\mathbf{B} = \Theta^{(s)} \mathbf{B}^{(s)} + \Theta^{(f)} \mathbf{B}^{(f)} = \left(\Theta^{s)} \mu^{s)} + \Theta^{f)} \mu^{f)} \right) \mathbf{H}$$

$$\begin{aligned} \mu_{\text{eff}} &= \Theta^{(s)} \mu^{(s)} + \Theta^{(f)} \mu^{(f)}, \\ \mathbf{D} &= \varepsilon_{\text{eff}} \cdot \mathbf{E} + \underline{\blacksquare}^{T} : \mathbf{e}, \\ \nabla \cdot \mathbf{D} = 0 &= \nabla \cdot \left(\varepsilon_{\text{eff}} \cdot \mathbf{E} + \underline{\blacksquare}^{T} : \mathbf{e} \right) \end{aligned} \tag{4.5.5}$$

In addition, we have the electromagnetic constitutive equations

$$\mathbf{D} = \varepsilon \mathbf{E}, \quad \nabla \cdot \mathbf{D} = 0, \quad \nabla \cdot \mathbf{B} = 0.$$

$$\nabla \times \mathbf{E} = -\dot{\mathbf{B}}, \quad \nabla \times \mathbf{H} = \dot{\mathbf{D}} + \mathbf{J}^{(f)}, \quad \mathbf{B} = \mu^{(f)} \mathbf{H}, \quad \mathbf{J} = \sigma^{(f)} \mathbf{E},$$

which lead to the effective equations for the electromagnetic field, namely

$$\Delta \mathbf{E} - \varepsilon_{\text{eff}} \mu_{\text{eff}} \partial_t^2 \mathbf{E} = -\frac{1}{\varepsilon_{\text{eff}}} \nabla \nabla \cdot \left(\mathbf{e} : \underline{\blacksquare}^{T} \right) + \mu_{\text{eff}} \left(\partial_t^2 \mathbf{e} : \underline{\blacksquare}^{T} \right), \tag{4.5.6}$$

and

$$\Delta \mathbf{B} - \varepsilon_{\text{eff}} \mu_{\text{eff}} \partial_t^2 \mathbf{B} = -\Theta^{(s)} \mu^{(s)} \nabla \times \left(\partial_t \mathbf{e} : \underline{\blacksquare}^{T} \right) + \Theta^{(f)} \mu^{(f)} \sigma^{(f)} \partial_t \mathbf{B} \tag{4.5.7}$$

5 Viscoelasticity, and Contact Friction between the Phases

5.1 KELVIN-VOIGT MATERIAL

Let $\mathscr{U}$ be a bounded domain in $\mathbb{R}^n$ representing the system of viscoelastic matrix and the voids filled by rigid particles. $\mathscr{U}$ contains a large collection of identical cells periodically arranged .[1] In the unit cell $\mathscr{Y} = [0,1]^n$, let $\mathscr{Y}_p$, the rigid particle part, be a closed subset of $\mathscr{Y}$ and $\mathscr{Y}_s = \mathscr{Y}/\mathscr{Y}_p$ be the viscoelastic solid part. Γ is the interface of $\mathscr{Y}_s$ with $\mathscr{Y}_r$. For simplicity, we assume that Γ is smooth. We construct the periodic microstructure as usual: for any set $\mathscr{D} \subset \mathbb{R}^n$, we define $\varepsilon\mathscr{D} = \{x : \varepsilon^{-1}x \in \mathscr{D}\}$. For any set $\mathscr{T} \subset \mathscr{Y}$, we define $\mathscr{T}^k = \mathscr{T} + \mathbf{k}$ for $\mathbf{k} \in \mathbb{Z}^n$. We define

$$\mathscr{U}_\varepsilon = \cup\{\varepsilon\mathscr{Y}_s^k : \varepsilon\mathscr{Y}_s^k \subset \mathscr{U}, k \in \mathscr{Z}^n\}$$

$$\Gamma_\varepsilon = \cup\{\varepsilon\Gamma^k : \varepsilon\Gamma^k \subset \mathscr{U}, k \in \mathbb{Z}^n\}$$

Let $\mathbf{u}^\varepsilon(t,x)$ be the displacements of the matrix, and $\mathbf{v}^\varepsilon = \dot{\mathbf{u}^\varepsilon} = \partial_t\mathbf{u}$ be the velocity vector. The equation of motion is given by

$$\dot{\mathbf{v}} = \nabla\sigma^\varepsilon + \mathbf{f}, \quad \mathbf{x} \in \mathscr{U}_\varepsilon, \tag{5.1.1}$$

where the constitutive equations are of the Kelvin-Voigt type, namely

$$\sigma^\varepsilon = A^\varepsilon\mathbf{e}(\mathbf{u}^\varepsilon) + B^\varepsilon\mathbf{e}(\mathbf{v}^\varepsilon). \tag{5.1.2}$$

It is assumed that the coefficients in the constitutive equations have both fast and slow variables; hence,

$$A^\varepsilon(x) = \left(A_{ijkl}(x, \frac{x}{\varepsilon})\right), B^\varepsilon(x) = \left(B_{ijkl}(x, \frac{x}{\varepsilon})\right)$$

It is convenient to assume that $A(x,y)$ and $B(x,y)$ are smooth, periodic in y and satisfy the usual symmetry associated with elasticity problems

$$A_{ijkl} = A_{ijlk} = A_{jikl} = A_{klij}, \tag{5.1.3}$$
$$B_{ijkl} = B_{ijlk} = B_{jikl} = B_{klij}; \tag{5.1.4}$$

[1] See Chapter 1 for further details.

moreover the usual ellipticity conditions are assumed to hold

$$A_{ijkl}h_{ij}h_{kl} \geq \lambda_1 h_{ij}h_{ij}, \tag{5.1.5}$$
$$B_{ijkl}h_{ij}h_{kl} \geq \lambda_2 h_{ij}h_{ij}, \tag{5.1.6}$$

where the constants λ_1 and λ_2 are positive and independent of ε.
On the domain boundary $\partial\mathscr{U}$ it is assumed that a homogeneous Dirichlet condition holds

$$\mathbf{u}^\varepsilon|_{\partial\mathscr{U}} = 0. \tag{5.1.7}$$

The contact and friction condition on the interface between the pore and particle Γ^ε, is taken to be a Coulomb-type, law. Suppose $\mathbf{n} = (n_1, n_2, \cdots, n_n)$ is the unit normal on Γ pointing into $\mathscr{Y}_p$, where the normal and tangential components of the displacement are given as

$$u_n = \mathbf{u}\cdot\mathbf{n}, \qquad \mathbf{u}_T = \mathbf{u} - u_n\mathbf{n}.$$

The normal and tangential components of the stress tensor are given by

$$\sigma_n = \sigma_{ij}n_in_j, \text{ and } \sigma_T = \sigma\cdot\mathbf{n} - \sigma_n\mathbf{n},$$

respectively. We linearize the contact condition to make the problem more manageable, namely

$$\sigma_n^\varepsilon = -\varepsilon g_\varepsilon(x) - \varepsilon h_\varepsilon(x)u_n^\varepsilon, \tag{5.1.8}$$

where $g_\varepsilon(x) = g(\frac{x}{\varepsilon}), h_\varepsilon(x) = h(\frac{x}{\varepsilon})$. Here it is assumed that $g(y)$ and $h(y)$ are smooth, positive, and periodic in †.
The friction condition at the interface is also linearized, i.e.

$$\sigma_T^\varepsilon = -\varepsilon\mathbf{v}_T^\varepsilon. \tag{5.1.9}$$

The initial conditions are taken to be homogeneous

$$\mathbf{v}^\varepsilon(x,0) = 0, \quad \mathbf{u}^\varepsilon(x,0) = 0. \tag{5.1.10}$$

In order to formulate the variational form of the problem, we use the Banach spaces:

$$H_\varepsilon = \{\mathbf{u} \in [H^1(\mathscr{U}_\varepsilon)]^n, \text{ and } \mathbf{u}|_{\partial\mathscr{U}} = 0\}, V = [L^2(\mathscr{U}_\varepsilon)]^n. \tag{5.1.11}$$

The variational form, of the problem (5.1.1) (5.1.8) and (5.1.9), then takes the form:

Problem 1: Find $\mathbf{u}^\varepsilon \in L^\infty([0,T],H_\varepsilon), \partial_t\mathbf{u} \in L^\infty([0,T],V) \cap L^2([0,T],H_\varepsilon), \partial_{tt}\mathbf{u} \in L^\infty([0,T],H'_\varepsilon)$ so that $\mathbf{u}^\varepsilon|_{t=0} = 0, \partial_t\mathbf{u}^\varepsilon|_{t=0} = 0$ and $\mathbf{u}$ satisfies

$$\int_{\mathscr{U}_\varepsilon} \partial_{tt}\mathbf{u}^\varepsilon \cdot \phi + \int_{\mathscr{U}_\varepsilon} [Ae(\mathbf{u}^\varepsilon)] : e(\phi) + \int_{\mathscr{U}_\varepsilon} [Be(\partial_t \mathbf{u}^\varepsilon)] : e(\phi)$$
$$+ \int_{\Gamma_\varepsilon} \varepsilon \partial_t \mathbf{u}^\varepsilon \cdot \phi - \int_{\Gamma_\varepsilon} \varepsilon(\partial_t \mathbf{u}^\varepsilon \cdot \mathbf{n})(\mathbf{n}\cdot\phi) + \int_{\Gamma_\varepsilon} \varepsilon g_\varepsilon(\mathbf{n}\cdot\phi) + \int_{\Gamma_\varepsilon} \varepsilon h_\varepsilon(\mathbf{u}^\varepsilon\cdot\mathbf{n})(\mathbf{n}\cdot\phi)$$
$$= \int_{\mathscr{U}_\varepsilon} \mathbf{f}\cdot\phi \text{ for any } \phi \in H_\varepsilon \tag{5.1.12}$$

where we assume $\mathbf{f} \in L^2([0,T],[L^2(U)]^n)$. In [143] Galerkin's method was used to demonstrate that a solution to Problem 1 exists. To this end, we approximate the solution to Problem 1 by $\mathbf{u}_m(x,t)$

$$\mathbf{u}_m(x,t) = \sum_{l=1}^{m} g_{lm}(t)\phi^l(x), \tag{5.1.13}$$

where the $\{\phi^k(x)\}$ in H_ε form a complete system of functions in H_ε. We determine the $g_{lm}(t)$ from the differential equation

$$g''_{jm} + \sum_{l=1}^{m} \alpha_{jl} g'_{lm} + \sum_{l=1}^{m} \beta_{jl} g_{lm} = \sum_{l=1}^{m} \gamma_{jl} \left(\int_{\mathscr{U}_\varepsilon} \mathbf{f}(t)\cdot\phi^j - \varepsilon \int_{\Gamma_\varepsilon} g(\phi^l\cdot\mathbf{n}) \right), 1 \le j \le m, \tag{5.1.14}$$

with the homogeneous initial conditions

$$g_{jm}(0) = g'_{jm}(0) = 0, 1 \le j \le m. \tag{5.1.15}$$

In the above, $(\alpha_{jl}), (\beta_{jl})$ and (γ_{jl}) are constant matrices.
The following lemma [143] concerning the approximants is obtained:

Lemma 5.1

There exists a solution $\mathbf{u}_m(x,t)$ to (10.4.3) such that $\mathbf{u}_m(x,t)$ satisfies (10.4.4) and $\mathbf{u}_m(x,0) = \partial_t \mathbf{u}_m(x,0) = 0$. ■

$$(\mathbf{u},\phi) = \int_{\mathscr{U}_\varepsilon} \mathbf{u}\cdot\phi, \qquad \|\mathbf{u}\|^2 = \|\mathbf{u}\|^2_{L^2} = \int_{\mathscr{U}_\varepsilon} |\mathbf{u}|^2, \tag{5.1.16}$$

the a priori bounds for the approximants are obtained [143]

$$\sup_{t\in[0,T]} \|\nabla \mathbf{u}_m(t)\|^2 \le C\left(\int_0^T \|\mathbf{f}(t)\|^2 dt + \int_0^T \|g(t)\|^2_{L^2(\Gamma)} dt \right) \tag{5.1.17}$$

$$\sup_{t\in[0,T]} ||\partial_t \mathbf{u}_m(t)||^2 \leq C\left(\int_0^T ||\mathbf{f}(t)||^2 dt + \int_0^T ||g(t)||^2_{L^2(\Gamma)} dt\right) \quad (5.1.18)$$

$$\int_0^T ||\nabla\partial_t \mathbf{u}_m(t)||^2 dt \leq C\left(\int_0^T ||\mathbf{f}(t)||^2 dt + \int_0^T ||g(t)||^2_{L^2(\Gamma)} dt\right). \quad (5.1.19)$$

$$\sup_{t\in[0,T]} ||\mathbf{u}_m(t)||^2 \leq C\left(\int_0^T ||\mathbf{f}(t)||^2 dt + \int_0^T ||g(t)||^2_{L^2(\Gamma)} dt\right). \quad (5.1.20)$$

We are next able to show that a solution exists to the original problem which we state as:

Lemma 5.2

There exists a weak solution $\mathbf{u} \in L^\infty([0,T], H_\varepsilon)$ to Problem 1.

■

This is proved by showing there exists an element $\mathbf{u} \in L^\infty([0,T], H_\varepsilon)$ such that for a subsequence $\{\mathbf{u}_m\}$, which we still denote by $\{\mathbf{u}_m\}$,

$$\mathbf{u}_m \text{ converges to } \mathbf{u} \text{ in the weak* topology of } L^\infty([0,T], H_\varepsilon) \quad (5.1.21)$$

that is

$$\int_0^T (\mathbf{u}_m(t) - \mathbf{u}(t), \phi(t))_{H_\varepsilon} dt \to 0 \text{ for any } \phi \in L^1([0,T], H_\varepsilon). \quad (5.1.22)$$

See [143] for further details.

Theorem 5.1

There exists a unique solution $\mathbf{u} \in L^\infty([0,T], H_\varepsilon), \partial_t \mathbf{u} \in L^\infty([0,T], V) \cap L^2([0,T], H_\varepsilon), \partial_{tt} u \in L^\infty([0,T], H'_\varepsilon)$ to Problem 1. Furthermore u satisfies the estimates

$$\sup_{t\in[0,T]} ||\mathbf{u}(t)||^2_{H_\varepsilon} \leq C\left(\int_0^T ||\mathbf{f}(t)||^2 dt + \int_0^T ||g(t)||^2_{L^2(\Gamma)} dt\right) \quad (5.1.23)$$

$$\sup_{t\in[0,T]} ||\partial_t \mathbf{u}(t)||^2 \leq C\left(\int_0^T ||\mathbf{f}(t)||^2 dt + \int_0^T ||g(t)||^2_{L^2(\Gamma)} dt\right) \quad (5.1.24)$$

$$\int_0^T ||\nabla\partial_t \mathbf{u}(t)||^2 dt \leq C\left(\int_0^T ||\mathbf{f}(t)||^2 dt + \int_0^T ||g(t)||^2_{L^2(\Gamma)} dt\right) \quad (5.1.25)$$

where C is positive constant independent of ε. ■

5.1.1 TWO-SCALE CONVERGENCE APPROACH

Having shown the existence of a unique solution to Problem 1, we turn to the two-scale convergence to the effective equations. Allaire et al. [8] and Neuss-Radu [314] have extended two-scale convergence to periodic hypersurfaces, which is exactly what is needed to manage the contact conditions. For expository purposes we state these results below.

Definition 5.1. $\{\mathbf{u}^\varepsilon(t,x)\} \subset [L^2([0,T] \times \Gamma_\varepsilon)]^n$ *two-scale converges to* $\mathbf{u}(t,x,y) \in [L^2([0,T] \times \mathscr{U}, L^2(\Gamma))]^n$ *iff for any* $\phi(t,x,y) \in [C^\infty([0,T] \times \mathscr{U}, C^\infty_\#(\mathscr{Y}))]^n$, *one has*

$$\lim_{\varepsilon \to 0} \varepsilon \int_0^T \int_{\Gamma_\varepsilon} \mathbf{u}^\varepsilon(t,x) \cdot \phi(t,x,\frac{x}{\varepsilon}) d\sigma_\varepsilon dt = \int_0^T \int_{\mathscr{U}} \int_\Gamma \mathbf{u}(t,x,y) \cdot \phi(t,x,y) dt dx d\sigma_y. \tag{5.1.26}$$

It is easy to establish:

Theorem 5.2

Let $\{\mathbf{u}^\varepsilon(x,t)\}$ be a sequence in $[L^2([0,T], L^2(\Gamma_\varepsilon))]^n$ such that

$$\varepsilon \int_{\Gamma_\varepsilon} |\mathbf{u}^\varepsilon(x)|^2 \leq C,$$

where C is a positive constant independent of ε. Then there exists $\mathbf{u}_0(t,x,y) \in [L^2([0,T] \times \mathscr{U}, L^2\#(\Gamma))]^n$ such that a subsequence of $\{\mathbf{u}^\varepsilon(t,x)\}$ two-scale converges to $\mathbf{u}_0(t,x,y)$ in the sense of Definition 5.1. ■

The proof is a simple adaptation of that in [314] and [9]. Two-scale convergence can handle homogenization problem on perforated domains conveniently without requiring any sophisticated extensions such as used in [7]. We only need to use the trivial extension by zero in the hole $\mathscr{Y}_p$ in the following lemma.

Lemma 5.3

Let $\{\mathbf{u}^\varepsilon(x,t)\} \subset L^2([0,T], H_\varepsilon)$ be as in Theorem 5.1, denote by $\tilde{\cdot}$ the extension by zero in $\mathscr{U} - \mathscr{U}_\varepsilon$, then
(i) there exists $\mathbf{u}_0(t,x) \in [L^2([0,T], H^1(\mathscr{U})]^n$ so that up to a subsequence, $\tilde{\mathbf{u}}^\varepsilon(t,x)$ two-scale converges to $\mathbf{u}_0(t,x)\chi(y)$ in the sense of both Definition 1.8 and Definition 5.1. More precisely

$$\lim_{\varepsilon \to 0} \int_0^T \int_{\mathscr{U}_\varepsilon} \mathbf{u}^\varepsilon(t,x) \cdot \phi(t,x,\frac{x}{\varepsilon}) dx dt = \int_0^T \int_{\mathscr{U}} \int_{\mathscr{Y}_s} \mathbf{u}_0(t,x) \cdot \phi(t,x,y) dt dx dy \tag{5.1.27}$$

$$\lim_{\varepsilon\to 0}\varepsilon\int_0^T\int_{\Gamma_\varepsilon}\mathbf{u}^\varepsilon(t,x)\cdot\phi(t,x,\frac{x}{\varepsilon})d\sigma_\varepsilon dt=\int_0^T\int_{\mathscr{U}}\int_{\Gamma}\mathbf{u}_0(t,x)\cdot\phi(t,x,y)dtdxd\sigma_y \tag{5.1.28}$$

for any $\phi(t,x,y)\in[C^\infty([0,T]\times\mathscr{U},C_\#^\infty(\mathscr{Y}))]^n$, where $\chi(y)$ is the characteristic function on $\mathscr{Y}_s$.
(ii). There exists $\mathbf{u}_1(t,x,y)\in[L^2([0,T]\times\mathscr{U},H_\#^1(\mathscr{Y}_s))]^n$ such that a subsequence of the extension of $\nabla_x\mathbf{u}^\varepsilon$ two-scale converges to $\chi(y)[\nabla_x\mathbf{u}_0+\nabla_y\mathbf{u}_1]$.

■

The proof is modeled after the argument given in Theorem 2.9 in [8]; we get (10.4.3) and part (ii). Equation (5.1.28) follows from Proposition 2.6 in [9], which leads to a system of equations for $\{\mathbf{u}_0(t,x),\mathbf{u}_1(t,x,y)\}$.

Lemma 5.4

Let $\{\mathbf{u}_0(t,x),\mathbf{u}_1(t,x,y)\}$ be as in Lemma 5.3, then $\{\mathbf{u}_0(t,x),\mathbf{u}_1(t,x,y)\}$satisfy the following equations in a distributional sense:

$$\begin{aligned}&\frac{d^2}{dt^2}\int_{\mathscr{U}}\mathbf{u}_0\phi+\int_{\mathscr{U}}\int_{\mathscr{Y}_s}A[e_x(\mathbf{u}_0)+e_y(\mathbf{u}_1)]:e_x(\phi)+\frac{d}{dt}\int_{\mathscr{U}}\int_{\mathscr{Y}_s}B[e_x(\mathbf{u}_0)+e_y(\mathbf{u}_1)]:e_x(\phi)\\&+\int_{\mathscr{U}}\int_{\Gamma}\partial_t\mathbf{u}_0\cdot\phi-\int_{\mathscr{U}}\int_{\Gamma}\partial_t\mathbf{u}_0\cdot\mathbf{n}(\phi\cdot\mathbf{n})+\int_{\mathscr{U}}\int_{\Gamma}g\phi\cdot\mathbf{n}+\int_{\mathscr{U}}\int_{\Gamma}h(\mathbf{u}_0\cdot\mathbf{n})(\phi\cdot\mathbf{n})\\&=\int_{\mathscr{U}}\mathbf{f}\cdot\phi.\end{aligned} \tag{5.1.29}$$

$$\int_{\mathscr{U}}\int_{\mathscr{Y}_s}A[e_x(\mathbf{u}_0)+e_y(\mathbf{u}_1)]:e_y(\phi_1)+\int_{\mathscr{U}}\int_{\mathscr{Y}_s}B[e_x(\partial_t\mathbf{u}_0)+e_y(\partial_t\mathbf{u}_1)]:e_y(\phi_1)=0, \tag{5.1.30}$$

$$\mathbf{u}_0(0)=\partial_t\mathbf{u}_0(0)=0,\mathbf{u}_1(0)=0. \tag{5.1.31}$$

■

Moreover, we have [143]

Lemma 5.5

The system (10.4.12)-(10.4.14) has a unique solution. ■

It is sufficient to prove that for $\mathbf{f} = \mathbf{0}$, $g = 0$ we have only the trivial solution $\mathbf{u}_0 = \mathbf{u}_1 = 0$. Set $\phi = \partial_t \mathbf{u}_0, \phi_1 = \partial_t \mathbf{u}_1$ as a test function in (10.4.12) and (10.4.13):

$$\begin{aligned} &\int_{\mathscr{U}} \partial_{tt}\mathbf{u}_0 \partial_t \mathbf{u}_0 + \int_{\mathscr{U}}\int_{\mathscr{Y}_s} A(e_x(\mathbf{u}_0) + e_y(\mathbf{u}_1)) : e_x(\partial_t \mathbf{u}_0) \\ &+ \frac{d}{dt}\int_{\mathscr{U}}\int_{\mathscr{Y}_s} B(e_x(\mathbf{u}_0) + e_y(\mathbf{u}_1)) : e_x(\partial_t \mathbf{u}_0) \\ &+ \int_{\mathscr{U}}\int_{\Gamma} \partial_t \mathbf{u}_0 \cdot \partial_t \mathbf{u}_0 - \int_{\mathscr{U}}\int_{\Gamma} (\partial_t \mathbf{u}_0 \cdot \mathbf{n})^2 + \int_{\mathscr{U}}\int_{\Gamma} \varepsilon h(\mathbf{u}_0 \cdot \mathbf{n})(\partial_t \mathbf{u}_0 \cdot \mathbf{n}) = 0, \end{aligned} \tag{5.1.32}$$

$$\int_{\mathscr{U}}\int_{\mathscr{Y}_s} A(e_x(\mathbf{u}_0) + e_y(\mathbf{u}_1) : e_y(\partial_t \mathbf{u}_1) + \int_{\mathscr{U}}\int_{\mathscr{Y}_s} B(e_x(\partial_t \mathbf{u}_0) + e_y(\partial_t \mathbf{u}_1) : e_y(\partial_t \mathbf{u}_1) = 0. \tag{5.1.33}$$

Adding (5.1.32) and (5.1.33) and integrating in time:

$$\begin{aligned} &\int_{\mathscr{U}} (\partial_t \mathbf{u}_0)^2 + \int_{\mathscr{U}}\int_{\mathscr{Y}_s} A e_x(\mathbf{u}_0) : e_x(\mathbf{u}_0) \\ &+ 2\int_0^t \int_{\mathscr{U}}\int_{\mathscr{Y}_s} B(e_x(\partial_t \mathbf{u}_0) + e_y(\partial_t \mathbf{u}_1)) : (e_x(\partial_t \mathbf{u}_0) + e_y(\partial_t \mathbf{u}_1)) \\ &+ \int_0^t \int_{\mathscr{U}}\int_{\Gamma} \partial_t \mathbf{u}_0 \cdot \partial_t \mathbf{u}_0 - \int_0^t \int_{\mathscr{U}}\int_{\Gamma} (\partial_t \mathbf{u}_0 \cdot \mathbf{n})^2 + \frac{1}{2}\int_{\mathscr{U}}\int_{\Gamma} \varepsilon h(\mathbf{u}_0 \cdot \mathbf{n})^2 = 0 \end{aligned} \tag{5.1.34}$$

Note that

$$\int_0^t \int_{\mathscr{U}}\int_{\Gamma} \partial_t \mathbf{u}_0 \cdot \partial_t \mathbf{u}_0 - \int_0^t \int_{\mathscr{U}}\int_{\Gamma} (\partial_t \mathbf{u}_0 \cdot \mathbf{n})^2 \geq 0,$$

and Equation (5.1.34) implies $\partial_t \mathbf{u}_0 = 0, \partial_t \mathbf{u}_1 = 0$, and using the initial condition,s we have $\mathbf{u}_0 \equiv 0, \mathbf{u}_1 \equiv 0$. We use the scheme suggested by [161] to find a $\mathbf{u}_1(t,x,y)$, namely

$$\mathbf{u}_1(t,x,y) = \sum_{ij} \left\{ \mathbf{M}^{ij}(y) e_x(\mathbf{u}_0)_{ij}(t,x) + \int_0^t \mathbf{K}^{ij}(y,t-s)(e_x(\mathbf{u}_0))_{ij}(x,s)ds \right\}, \tag{5.1.35}$$

where we have to determine the kernels $\mathbf{M}^{ij}(y), \mathbf{K}^{ij}(y,t)$. By substituting (10.4.16) into (10.4.13) it turns out, after integration by parts and collecting terms, that $\mathbf{M}^{ij}$ is determined by the following linear elastic system:

$$\begin{aligned} &\nabla_y \left\{ B\left(\frac{1}{2}(e_i \otimes e_j + e_j \otimes e_i) + e_y(\mathbf{M}^{ij})\right) \right\} = 0 \\ &B\left(\frac{1}{2}(e_i \otimes e_j + e_j \otimes e_i) + e_y(\mathbf{M}^{ij})\right) \cdot \mathbf{n}| = 0 \text{ on } \Gamma, , \\ &\mathbf{M}^{ij}(y) \in [H^1_{\#}(\mathbf{Y}_s)]^n. \end{aligned} \tag{5.1.36}$$

$\mathbf{K}^{ij}(0)$ is determined by the following linear elastic system:

$$
\begin{aligned}
&\nabla_y\left\{A\left(\frac{1}{2}(e_i\otimes e_j+e_j\otimes e_i)\right)+Ae_y(\mathbf{M}^{ij})+Be_y(\mathbf{K}^{ij}(0)\right\}=0\\
&\left\{A\left(\frac{1}{2}(e_i\otimes e_j+e_j\otimes e_i)\right)+Ae_y(\mathbf{M}^{ij})+Be_y(\mathbf{K}^{ij}(0)\right\}\cdot\mathbf{n}=0 \text{ on } \Gamma,\\
&\mathbf{K}^{ij}(0,y)\in[H^1_\#(\mathscr{Y})]^n
\end{aligned}
\tag{5.1.37}
$$

$\mathbf{M}^{ij}(y)$ is given by (5.1.36) and $\mathbf{K}^{ij}(t,y)$ is determined by the following system

$$
\begin{aligned}
&\nabla_y\left\{Be_y(\partial_t\mathbf{K}^{ij})+Ae_y(\mathbf{K}^{ij})\right\}=0\\
&\left\{Be_y(\partial_t\mathbf{K}^{ij})+Ae_y(\mathbf{K}^{ij})\right\}\cdot\mathbf{n}=0, \text{ on } \Gamma,\\
&\mathbf{K}^{ij}(t,y)\in L^2([0,T],[H^1_\#(\mathscr{Y})]^n)\\
&\mathbf{K}^{ij}(0,y) \text{ is given by (5.1.37)}
\end{aligned}
\tag{5.1.38}
$$

We proceed to get the effective equation.
We define the symmetric tensors $\mathscr{A}$, $\mathscr{B}$ and $\mathscr{C}$ by

$$
\mathscr{A}_{ijkl}=\int_{\mathscr{Y}_s}A_{ijkl}+\int_{\mathscr{Y}_s}\sum_{mn}\frac{1}{2}A_{klmn}\left(\frac{\partial M_n^{ij}}{\partial y_m}+\frac{\partial M_m^{ij}}{\partial y_n}\right)+\frac{1}{2}B_{klmn}\left(\frac{\partial K_n^{ij}(0)}{\partial y_m}+\frac{\partial K_m^{ij}(0)}{\partial y_n}\right)
\tag{5.1.39}
$$

$$
\mathscr{B}_{ijkl}=\int_{\mathscr{Y}_s}B_{ijkl}+\int_{\mathscr{Y}_s}\sum_{mn}\frac{1}{2}B_{klmn}\left(\frac{\partial M_n^{ij}}{\partial y_m}+\frac{\partial M_m^{ij}}{\partial y_n}\right)
\tag{5.1.40}
$$

$$
\mathscr{C}_{ijkl}(t)=\int_{\mathscr{Y}_s}\sum_{mn}\left\{\frac{1}{2}A_{klmn}\left(\frac{\partial K_n^{ij}}{\partial y_m}+\frac{\partial K_m^{ij}}{\partial y_n}\right)+\frac{1}{2}B_{klmn}\left(\frac{\partial(\partial_t K_n^{ij})}{\partial y_m}+\frac{\partial(\partial_t K_m^{ij})}{\partial y_n}\right)\right\}
\tag{5.1.41}
$$

$$
N_{ij}=|\Gamma|\delta_{ij}-\int_\Gamma(\mathbf{n}\otimes\mathbf{n})_{ij},\ H_{ij}=\int_\Gamma h(\mathbf{n}\otimes\mathbf{n})_{ij},\mathbf{f}_1=\int_\Gamma g\mathbf{n}
\tag{5.1.42}
$$

By (2.3.10), the effective equation for $\mathbf{u}_0$ is

$$
\partial_{tt}\mathbf{u}_0-\nabla[\mathscr{A}e(\mathbf{u}_0)+\mathscr{B}e(\partial_t\mathbf{u}_0)]-\nabla\int_0^t\mathscr{C}(t-s)e(\mathbf{u}_0)ds+N\cdot\partial_t\mathbf{u}_0+H\cdot\mathbf{u}_0=\mathbf{f}-\mathbf{f}_1
\tag{5.1.43}
$$

Therefore we obtain the following theorem:

Theorem 5.3

Let $\mathbf{u}_0(t,x)$ be as in Lemma 5.3, then $\mathbf{u}_0(t,x)$ satisfies equation (5.1.43) and (10.4.14). ■

From effective equation (5.1.43), we get the effective stress tensor and drag force:

$$\sigma_0 = \mathscr{A} e(\mathbf{u}_0) + \mathscr{B} e(\partial_t \mathbf{u}_0)] - \int_0^t \mathscr{C}(t-s) e(\mathbf{u}_0) ds, \tag{5.1.44}$$

$$\mathbf{D}_0 = N \cdot \partial_t \mathbf{u}_0 + H \cdot \mathbf{u}_0. \tag{5.1.45}$$

The well-posedness for the cell problems and effective problem is established in [143]. See this reference for further details.

5.2 RIGID PARTICLES IN A VISCO-ELASTIC MEDIUM

In this section we discuss the possibility of particle interaction. Expecting small deviations from the effective displacements near the particle-matrix interface, we linearize the contact conditions about the effective displacement and velocity. These modified contact conditions still contain nonlinear functions of the effective variables. The nonlinear functions are further approximated by their Taylor polynomials. This procedure allows one to construct a family of models of increasing complexity, as needed. In particular, the simplest model corresponds to completely linearized contact conditions. The effective equations are obtained using the method of two-scale asymptotic expansions and homogenization. The approach adopted here is similar to the one in Sanchez-Palencia (1980), where the focus is on derivation of the effective equations, rather than on proving convergence of the asymptotic expansions. The averaged equations here differ from the acoustic equations of porous media obtained by homogenization in Sanchez-Palencia (1980), Burridge & Keller (1981), Gilbert & Mikelic (2000) and Gilbert & Panchenko (2003). For the general non-linear contact conditions, the effective equations of motion are history dependent and nonlinear. The constitutive equations depend not only on the strain and strain rate, but also on the velocity and displacement. This is an essential feature reflecting the nature of the contact conditions. Moreover, the effective equation contains a term which may be interpreted as a macroscopic body force, similar in nature to the Stokes' drag force. The effective model provides two constitutive equations: one for the effective stress, and another for the drag force. Simplified periodic geometry allows us to obtain explicit formulas for the constitutive equations. The effective stress contains the same nonlinear functions of the effective velocity as the contact conditions, and nonlinearities in the drag force are similar, but more complicated.

5.3 EQUATIONS OF MOTION AND CONTACT CONDITIONS

The resulting periodic geometry consists of the perforated, connected matrix domain Ω_M^ε and the particulate domain Ω_P^ε, which is a union of many disjoint simply connected components (particles). The boundary of Ω_M^ε is a union of the external boundary $\partial\Omega$ and the particle-matrix interface Γ^ε.

1. Particles

We assume that the particles are rigid bodies, moving with the same prescribed velocity $\dot{\mathbf{U}}(t,\mathbf{x})$. The (rigid) displacement of each particle is denoted by $\mathbf{U}$.

2. Matrix

Since the particulate motion is prescribed, the only unknown in the model is the deformation state of the matrix.

The matrix domain Ω_M^ε is occupied by a viscoelastic material with the constitutive equation given by a Kelvin-Voigt law

$$\sigma = A\mathbf{e}(\mathbf{u}) + B\mathbf{e}(\mathbf{v}), \tag{5.3.1}$$

where σ is the stress tensor, A, B are constant constitutive tensors, $e = \frac{1}{2}(\nabla + \nabla^T)$ is the symmetric part of the gradient matrix, and $\mathbf{u}, \mathbf{v}$ are displacement and velocity, respectively. The momentum balance equation for the matrix is

$$\dot{\mathbf{v}} = \nabla\,\sigma + \mathbf{f}, \quad bx \in \Omega_M^\varepsilon. \tag{5.3.2}$$

5.3.1 BOUNDARY CONDITIONS

Boundary conditions on the contact surface Γ^ε are prescribed for the traction $\boldsymbol{\sigma}\mathbf{n}$ where $\mathbf{n}$ is the exterior, unit normal to Γ^ε, with respect to the domain Ω_M^ε. To keep track of normal and tangential forces, we write

$$\sigma\mathbf{n} = \sigma_n\mathbf{n} + \sigma_T,$$

where $\sigma_n = \sigma_{ij}n_in_j$. The magnitude of the normal force is given by

$$\sigma_n = -p(u_n - \varepsilon g) \quad \text{on } \Gamma^\varepsilon, \tag{5.3.3}$$

where εg is a given function (gap) between the matrix and a particle, measured in the normal direction. The function p is the pressure function. The tangential traction on Γ^ε is prescribed by the following conditions.

$$|\sigma_T| \le F(u_n - \varepsilon g), \tag{5.3.4}$$

$$|\sigma_T| < F(u_n - \varepsilon g) \Rightarrow \mathbf{v}_T - \dot{\mathbf{U}}_T = 0, \tag{5.3.5}$$

$$|\sigma_T| = F(u_n - \varepsilon g) \Rightarrow \mathbf{v}_T - \dot{\mathbf{U}}_T = -\lambda\sigma_T. \tag{5.3.6}$$

In (5.3.5), (5.3.6), $\mathbf{v}_T = \mathbf{v} - v_n\mathbf{n}$. This is a mathematical formulation of the Coulomb friction law. According to this law, no sliding can occur until the tangential traction is greater than $F(U_n - \varepsilon g)$. Once this critical value is reached the velocity is proportional to the tangential component of traction in the opposite direction. The inequality

(5.3.4) means that $|\sigma_T|$ does not exceed the friction bound F, so the maximal magnitude of the tangential force is prescribed in terms of the normal deformation. Condition (5.3.5) means that there is no sliding of particles relative to the matrix when the tangential force is smaller than the friction bound. According to (5.3.6), sliding occurs when the tangential force reaches the friction bound. The relative velocity of sliding is proportional to the tangential force with the friction coefficient $\lambda > 0$. The set of contact conditions (5.3.3)-(5.3.6) is the so-called Coulomb friction law with normal compliance (Kikuchi & Oden (1988)). The functions p and F are non-linear. Based on experimental evidence, it is reasonable to choose (Kikuchi & Oden (1988))

$$p(z) = \begin{cases} z^m, & \text{when } z \geq 0, \\ 0 & \text{otherwise}, \end{cases} \tag{5.3.7}$$

where m is positive. The function F is prescribed similarly, with a larger value of m. On the external boundary $\partial\Omega$ we prescribe periodic boundary conditions.

Initial conditions

The initial conditions are

$$\mathbf{u}(0,x) = \mathbf{u}_0(x), \qquad \mathbf{v}(0,x) = \mathbf{v}_0(x), \quad x \in \Omega_M^\varepsilon. \tag{5.3.8}$$

Because particulate motion is prescribed, the formulation of the problem may be simplified by introducing the relative displacement

$$\xi = \mathbf{u} - \mathbf{U} \tag{5.3.9}$$

and the relative velocity

$$\zeta = \mathbf{v} - \dot{\mathbf{U}} \tag{5.3.10}$$

of the matrix. In terms of ξ, ζ the contact conditions on Γ^ε are written as follows.

$$\sigma_n = -p(\xi_n - \varepsilon\tilde{g}), \tag{5.3.11}$$

$$|\sigma_T| \leq F(\xi_n - \varepsilon\tilde{g}), \tag{5.3.12}$$

$$|\sigma_T| < F(\xi_n - \varepsilon\tilde{g}) \Rightarrow \zeta_T = 0, \tag{5.3.13}$$

$$|\sigma_T| = F(\xi_n - \varepsilon\tilde{g}) \Rightarrow \zeta_T = -\lambda\sigma_T. \tag{5.3.14}$$

where $\tilde{g}$ is a given relative gap function.

5.3.2 APPROXIMATION OF THE CONTACT CONDITIONS

Since dealing with the inequality-type contact conditions (5.3.12)-(5.3.14) is still difficult, one can approximate them by an equation; hence, we use the approximation

$$\sigma_T = -F(\xi_n - \varepsilon\tilde{g})\frac{\zeta_T}{(|\zeta_T|^2 + \delta)^{1/2}}, \quad \text{where } \delta > 0, \tag{5.3.15}$$

from Kuttler & Shillor [247]. Convergence of solutions of the approximate problem to the solution of the exact problem as $\delta \to 0$ was proved in Kuttler & Shillor (1999).

We do not pass to the limit $\delta \to 0$. Instead, we fix a small $\delta > 0$ (which is a "good approximate model" in the above sense), and homogenize the problem with the simplified contact conditions (5.3.15). The dependence on δ is therefore suppressed in what follows. Next, we regularize the functions p, F. These functions, given by formulas similar to (9.8.14), are smoothed out near $z = 0$. The regularized p and F are further approximated by polynomials. This leads to the simplified contact conditions

$$\sigma_T = -F(\xi_n - \varepsilon\tilde{g})G(\zeta_T)\zeta_T, \tag{5.3.16}$$

in which p, F are polynomial functions, and $(|\zeta_T|^2 + \delta)^{-1/2}$ is also approximated by a polynomial function G of ζ_T. Since ξ_n (respectively, ζ_T) are linear functions of components of the vectors ξ (ζ), polynomial functions of $\xi_n - \varepsilon\tilde{g}$ are also polynomial functions of components of ξ. It is natural to linearize (5.3.16) about effective ξ^0, ζ^0 to be determined from the asymptotic procedure, and keep only terms of orders $\varepsilon^0, \varepsilon$. The chief purpose of linearization is to separate the first two terms in the asymptotics of $\sigma_{\mathbf{n}}$.

5.3.3 MICROSCALE EQUATIONS

Writing the equation of motion (5.3.2) and the initial conditions (5.3.8) in terms of ξ, ζ, and linearizing contact conditions (5.3.16), we obtain the micro-scale problem to be averaged. It consists of the following:

1. Momentum balance equation for the matrix.

$$\dot{\zeta} = \nabla\,\sigma + \mathbf{F} \quad \text{in } \Omega_M^\varepsilon. \tag{5.3.17}$$

The $\mathbf{F}$ is not the function in the contact conditions.

2. Contact conditions on Γ^ε.

$$\begin{aligned}
\sigma_n &= -p(\xi_n^0) - \varepsilon p'(\xi_n^0)(\xi_n^1 - \tilde{g}),\\
\sigma_T &= -F(\xi_n^0)G(\zeta_T^0)\zeta_T^0 - \\
&\quad \varepsilon\left[F'(\xi_n^0)G(\zeta_T^0)\zeta_T^0 + F(\xi_n^0)\nabla G(\zeta_T^0)\cdot\zeta_T^1\zeta_T^0 + F(\xi_n^0)G(\zeta_T^0)\zeta_T^1\right],
\end{aligned} \tag{5.3.18}$$

where $\xi = \xi^0 + \varepsilon\xi^1$, $\zeta = \zeta^0 + \varepsilon\zeta^1$.

3. Initial conditions.

$$\xi(0, \mathbf{x}) = \mathbf{u}_0 - \mathbf{U}(0, \cdot), \qquad \zeta(0, \mathbf{x}) = v_0 - \dot{\mathbf{U}}(0, \cdot). \tag{5.3.19}$$

In the equation (5.3.17),

$$\sigma = A\mathbf{e}(\xi) + B\mathbf{e}(\zeta), \tag{5.3.20}$$

and the non-homogeneous term is

$$\mathbf{F} = \mathbf{f} - \ddot{\mathbf{U}}. \tag{5.3.21}$$

To obtain (5.3.17), we used the fact that the gradients of $\mathbf{U}, \dot{\mathbf{U}}$ are constant. On the outer boundary $\partial\Omega$, we prescribe periodic boundary conditions.

5.4 TWO-SCALE EXPANSIONS AND FORMAL HOMOGENIZATION

The starting point of the formal homogenization procedure (Sanchez-Palencia (1980)) is to postulate two-scale asymptotic expansions

$$\begin{aligned}\xi(t,\mathbf{x},\varepsilon) &= \xi^0(t,\mathbf{x}) + \varepsilon\xi^1(t,\mathbf{x},\tfrac{\mathbf{x}}{\varepsilon}) + \varepsilon^2\xi^2(t,\mathbf{x},\tfrac{\mathbf{x}}{\varepsilon}) + \cdots, \\ \zeta(t,\mathbf{x},\varepsilon) &= \zeta^0(t,\mathbf{x}) + \varepsilon\zeta^1(t,\mathbf{x},\tfrac{\mathbf{x}}{\varepsilon}) + \varepsilon^2\zeta^2(t,\mathbf{x},\tfrac{\mathbf{x}}{\varepsilon}) + \cdots,\end{aligned} \tag{5.4.1}$$

where ξ^k, ζ^k are periodic functions of the *fast variable* $\mathbf{y} = \frac{\mathbf{x}}{\varepsilon}$. Since

$$\frac{\partial f}{\partial x_k}(t,\mathbf{x},\frac{\mathbf{x}}{\varepsilon}) = (\partial_{x_k} + \frac{1}{\varepsilon}\partial_{y_k})f(t,\mathbf{x},\mathbf{y})_{|\mathbf{y}=\frac{\mathbf{x}}{\varepsilon}},$$

we can write

$$\mathbf{e}(\zeta) = \mathbf{e}_x(\zeta^0) + \mathbf{e}_y(\zeta^1) + \varepsilon[\mathbf{e}_x(\zeta^1) + \mathbf{e}_y(\zeta_2)] + \cdots \tag{5.4.2}$$

$$\mathbf{e}(\xi) = \mathbf{e}_x(\xi^0) + e_y(\xi^1) + \varepsilon[\mathbf{e}_x(\xi^1) + \mathbf{e}_y(\xi_2)] + \cdots \tag{5.4.3}$$

$$\sigma = \sigma^0 + \varepsilon\sigma^1 + \varepsilon^2\sigma^2 + \cdots, \tag{5.4.4}$$

$$\sigma^0 = A\mathbf{e}_x(\xi^0) + A\mathbf{e}_y(\xi^1) + B\mathbf{e}_x(\zeta^0) + B\mathbf{e}_y(\zeta^1) \tag{5.4.5}$$

$$\sigma^1 = A\mathbf{e}_x(\xi^1) + A\mathbf{e}_y(\xi_2) + B\mathbf{e}_x(\zeta^1) + B\mathbf{e}_y(\zeta_2). \tag{5.4.6}$$

Both sides in (5.4.2)-(5.4.6) are functions of $t, \mathbf{x}, \frac{\mathbf{x}}{\varepsilon}$. Substituting (5.4.2)-(5.4.6) into the momentum equation (5.3.17) and keeping only the terms of orders $\varepsilon^{-1}, \varepsilon^0$ we obtain the equations

$$\nabla_y \sigma^0 = 0, \tag{5.4.7}$$

(term of order ε^{-1}), and

$$\dot{\zeta}^0 = \nabla_x \sigma^0 + \nabla_y \sigma^1 + \mathbf{F}, \tag{5.4.8}$$

(term of order ε^0). The contact conditions for $\underline{\underline{\sigma}}^0, \underline{\underline{\sigma}}^1$ are obtained by separating terms of orders $\varepsilon^0, \varepsilon^1$, respectively in (5.3.18):

$$\underline{\underline{\sigma}}^0\mathbf{n} = -p(\xi^0\cdot\mathbf{n})\mathbf{n} - F(\xi^0\cdot\mathbf{n})G(\zeta_T^0)\zeta_T^0, \tag{5.4.9}$$

$$\begin{aligned}\underline{\underline{\sigma}}^1\mathbf{n} \quad = \quad & -p'(\xi^0\cdot\mathbf{n})(\xi_n^1 - \tilde{g})\mathbf{n} - F'(\xi^0\cdot\mathbf{n})G(\zeta_T^0)\zeta_T^0 - \\ & F(\xi^0\cdot\mathbf{n})\nabla G(\zeta_T^0)\cdot\zeta^1\zeta_T^0 - F(\xi^0\cdot\mathbf{n})G(\zeta_T^0)\zeta_T^1.\end{aligned} \tag{5.4.10}$$

The next step is to "lift" (5.4.7)-(5.4.10) to a space of double dimensions so that $\frac{\mathbf{x}}{\varepsilon}$ is replaced by the independence of **x**, the fast variable **y**. This is useful for two reasons. One is that **x** and **y** separate, and the second is that the geometry of the problem becomes independent of ε. In the equations (5.4.7) and (5.4.8), **x** is a point in Ω_M^ε, a perforated domain with perforations dependent on ε. In the lifted equations, $\mathbf{x} \in \Omega$ (domain with no holes), and $\mathbf{y} \in Y$. Using formulas (5.4.5), (5.4.6), from (5.4.7) and (5.4.9) we obtain

$$\nabla_y\left[A\mathbf{e}_y(\xi^1) + B\mathbf{e}_y(\zeta^1)\right] = 0, \quad \text{in } \Omega\times Y\times[0,T], \tag{5.4.11}$$

$$
\begin{aligned}
(A\mathbf{e}_y(\xi^1)+B\mathbf{e}_y(\zeta^1))\mathbf{n}(\mathbf{y}) \quad = \quad & -\left[A\mathbf{e}_x(\xi^0)+B\mathbf{e}_x(\zeta^0)\right](t,\mathbf{x})\mathbf{n}(\mathbf{y})- \\
& p(\xi^0(t,\mathbf{x})\cdot\mathbf{n}(\mathbf{y}))\mathbf{n}(\mathbf{y})-F(\xi^0_n)G(\zeta^0_T)\zeta^0_T, \\
& \mathbf{y}\in\Gamma_Y,\ \mathbf{x}\in\Omega,\ t\in[0,T].
\end{aligned}
\tag{5.4.12}
$$

For future reference, we also state the weak formulation of (5.4.11), (5.4.12):

$$
\begin{aligned}
&\int_0^T\int_Y(A\mathbf{e}_y(\xi^1)+B\mathbf{e}_y(\zeta^1))\cdot e(\mathbf{w})d\mathbf{y}dt+\int_0^T\int_{\Gamma_Y}(A\mathbf{e}_x(\xi^0)+B\mathbf{e}_x(\zeta^0))\mathbf{n}\cdot\mathbf{w}d\alpha dt+ \\
&\quad\int_0^T\int_{\Gamma_Y}[p(\xi^0\cdot\mathbf{n})\mathbf{n}+F(\xi^0\cdot\mathbf{n})G(\zeta^0_T)\zeta^0_T]\cdot\mathbf{w}\,d\alpha\,dt=0,
\end{aligned}
\tag{5.4.13}
$$

for each smooth function $\mathbf{w}(t,\mathbf{y})$, vanishing at $t=T$ and satisfying periodic boundary conditions on ∂Y. The unknowns in (5.4.11), (5.4.12) and (5.4.13) are ξ^1,ζ^1, viewed as functions of $\mathbf{y}$ which also depend on the parameters $t,\mathbf{x}$. We use (5.4.11), (5.4.12) to express ξ^1,ζ^1 in terms of ξ^0,ζ^0. This is done by first postulating a representation (ansatz)

$$\xi^1(t,\mathbf{x},\mathbf{y})=\mathbf{A}(t,\mathbf{y})\xi^0(t,\mathbf{x}), \tag{5.4.14}$$

where $\mathbf{A}$ is an operator involving several unknown *fast-variable functions* of $t,\mathbf{y}$. If the ansatz is chosen correctly, then after substitution of (5.4.14), $\mathbf{x}$ and $\mathbf{y}$ in (5.4.11), (5.4.12) separate. Next, one requires that (5.4.11),(5.4.12) hold for any choice of ξ^0,ζ^0. Then the *slow-variable functions* containing components of ξ^0,ζ^0 can be "factored out," and (5.4.11), (5.4.12) reduces to several cell problems for fast-variable functions. Despite the fact that the homogenization procedure outlined above is well-known (Sanchez-Palencia (1980)), its implementation for the problem at hand is not trivial. The choice of the ansatz is the heart of the matter. It should reflect the nature of the problem, in particular the contact conditions (5.4.12). There are no general methods for selecting the structure of $\mathbf{A}$ in (5.4.14). Direct computation shows that the standard homogenization substitution

$$\xi^1=A_j(\mathbf{y})\partial_{x_j}\xi^0(\mathbf{x},t) \tag{5.4.15}$$

does not work. The difficulty arises because of the nonlinearity, in ξ^0, and the dissipative nature of (5.4.12). From the mathematical point of view, a good ansatz should lead to separation of fast and slow variables in (5.4.11), (5.4.12), so that solving these equations is reduced to solving well-posed cell problems for the functions which form the kernel of $\mathbf{A}$. From the point of view of continuum mechanics, the choice is motivated by the following principle:

Instantaneous dissipation on the micro-scale leads to history-dependent dissipation on the macro-scale.

This is intuitively clear, since for small ε, the particles become more numerous, and the distance between them goes to zero, so that frictional dissipation on the surface of one particle is likely to influence the energy dissipation on the surface of several neighboring particles. With this in mind, consider the right-hand-side of (5.4.12). It can be written as

$$\left[-A\mathbf{e}(\xi^0)-B\mathbf{e}(\zeta^0)\right](t,\mathbf{x})\mathbf{n}(\mathbf{y})+\sum_j g_j(\mathbf{y})f_j(t,\mathbf{x}), \tag{5.4.16}$$

where the slow-variable functions f_j are products of powers of components of ξ^0, ζ^0. Evidently, the substitution should contain the same slow-variable functions as (5.4.16), otherwise the variables will not separate. Moreover, the history dependence should be incorporated. Therefore, we introduce the following ansatz:

$$\xi^1(t,\mathbf{x},\mathbf{y}) = K_1^{pq}(\mathbf{y})\mathbf{e}(\xi^0)_{pq} + \int_0^t K_2^{pq}(t-\tau,\mathbf{y})e(\zeta^0)(\tau,\mathbf{x})d\tau$$
$$+\sum_j \int_0^t M_j(t-\tau,\mathbf{y})f_j(\tau,\mathbf{x})d\tau, \qquad (5.4.17)$$

where the fast-variable functions K_1^{pq}, K_2^{pq} and M_j should be determined by solving cell problems. After the cell problems are solved, the effective equation for ξ^0, ζ^0 can be obtained from (5.4.8) by substituting (5.4.17) and applying the averaging operator $\langle\cdot\rangle$ defined by

$$\langle f\rangle(t,\mathbf{x}) = \frac{1}{|Y|}\int_Y f(t,\mathbf{x},\mathbf{y})d\mathbf{y}. \qquad (5.4.18)$$

Averaging (5.4.8), integrating by parts in the term $\nabla_y \sigma^1$ and using periodicity on the external boundary of the unit cell, we obtain the effective equation

$$\dot{\zeta^0} = \nabla_x \widehat{\mathbf{T}} + \widehat{\mathbf{S}} + \mathbf{F}, \qquad (5.4.19)$$

where

$$\begin{aligned}\widehat{\mathbf{T}} &= \langle \sigma^0 \rangle, \\ \widehat{\mathbf{S}} &= \langle \sigma^1 \mathbf{n} \rangle_\Gamma.\end{aligned} \qquad (5.4.20)$$

In the equation for $\widehat{\mathbf{S}}$, $\langle\cdot\rangle_\Gamma$ denotes the interface averaging operator

$$\langle f\rangle_\Gamma(t,\mathbf{x}) = \frac{1}{|Y|}\int_{\Gamma_Y} f(t,\mathbf{x},\mathbf{y})d\alpha(\mathbf{y}). \qquad (5.4.21)$$

While $\widehat{\mathbf{T}}$ can be identified with the effective stress tensor, the term $\widehat{\mathbf{S}}$ represents the density of a body force, which we call generalized drag force. The constitutive equations for $\widehat{\mathbf{T}}$ and $\widehat{\mathbf{S}}$ are given by (5.4.20). Combining (5.4.17) with (5.4.5) and (5.4.10), one can make two conclusions.

i) The effective equations are nonlinear and history-dependent. The effective stress contains the same nonlinear functions of ξ^0, ζ^0 as the contact conditions (5.4.9). The nonlinearities in the drag force are similar, but more complicated.

ii) The effective medium is not a simple material (see e.g. Truesdell (1997)), that is, both $\widehat{\mathbf{T}}$ and $\widehat{\mathbf{S}}$ depend not only on $\mathbf{e}(\xi^0), \mathbf{e}(\zeta^0)$, but also on ξ^0 and ζ^0.

It is important to derive explicit formulae for the effective equations, which we do in the following section for three different cases.

5.5 MODEL CASE I: LINEAR CONTACT CONDITIONS

First we consider the case

$$p(\xi^0) = 0, \quad F = 1, \quad G = const, \; G > 0, \qquad (5.5.1)$$

so the contact conditions on Γ_Y are

$$\sigma^0\mathbf{n} = -G\zeta^0_T, \qquad \sigma^1\mathbf{n} = -G\zeta^1_T. \tag{5.5.2}$$

The suggested ansatz is as follows.

$$\xi^1(t,\mathbf{x},\mathbf{y}) = K_1^{pq}(\mathbf{y})\mathbf{e}(\xi^0)_{pq}(t,\mathbf{x}) + \int_0^t K_2^{pq}(t-\tau,\mathbf{y})\mathbf{e}(\partial_t\xi^0)_{pq}(\tau,\mathbf{x})d\tau + \int_0^t M^p(t-\tau,\mathbf{y})\partial_t\xi^0_p(\tau,\mathbf{x})d\tau. \tag{5.5.3}$$

where the *scalars* $\xi^0_p, \mathbf{e}(\xi^0)_{pq}$ are the components of the vector ξ^0 and tensor $\mathbf{e}(\xi^0)$ respectively, and K_1^{pq}, K_2^{pq}, M^p are *vectors* to be determined. We substitute (5.5.3) and (5.5.2) into (5.4.13) and group the terms containing the same slow-variable functions. Factoring out the slow-variable terms, we obtain the cell problems for determination of the fast-variable functions.

5.5.1 CELL PROBLEMS

The problem for K_1^{pq} is obtained by collecting the terms containing $e(\xi^0)(t,\mathbf{x})_{pq}$.

$$\int_Y A(\mathbf{e}(K_1^{pq}) + I^{pq})(\mathbf{y}) \cdot \mathbf{e}(\mathbf{w})(\mathbf{y})d\mathbf{y} = 0, \tag{5.5.4}$$

where $I^{pq}_{kl} = \delta_{kp}\delta_{lq}$, so that $A_{ijkl}I^{pq}_{kl} = A_{ijpq}$.

The problem for $K_2^{pq}(0,\mathbf{y})$.

Collecting terms containing $\mathbf{e}(\partial_t\xi^0)_{pq}(t,\mathbf{x})$:

$$\int_Y B(\mathbf{e}(K_1^{pq}) + I^{pq})(\mathbf{y}) \cdot \mathbf{e}(\mathbf{w})(\mathbf{y})d\mathbf{y} + \int_Y B\mathbf{e}(K_2^{pq}(0,\mathbf{y})) \cdot \mathbf{e}(\mathbf{w})d\mathbf{y} = 0. \tag{5.5.5}$$

The problem for K_2^{pq}.

Collecting terms containing $e(\partial_t\xi_0)_{pq}(\tau,\mathbf{x})$.

$$-\int_Y Be(K_2^{pq})(0,\mathbf{y}) \cdot e(\mathbf{w})d\mathbf{y}d - \int_0^T \int_Y Be(K_2^{pq}) \cdot \partial_t e(\mathbf{w}(t,\mathbf{y}))d\mathbf{y}dt + \int_0^T \int_Y Ae(K_2^{pq}) \cdot e(\mathbf{w}(t,\mathbf{y}))d\mathbf{y}dt = 0, \tag{5.5.6}$$

The problem for $M^p(0,\mathbf{y})$.

Collecting terms containing $\partial_t\xi^0_p(t,\mathbf{x})$.

$$\int_Y B\mathbf{e}(M^p)(0,\mathbf{y}) \cdot \mathbf{e}(\mathbf{w})d\mathbf{y} + \int_{\Gamma_Y} Gw_{p,T}d\alpha = 0. \tag{5.5.7}$$

The problem for $M^p(t,\mathbf{y})$.

Collecting terms containing $\partial_t\xi^0_p(\tau,\mathbf{x})$.

$$-\int_Y \mathbf{Be}(M^p)(0,\mathbf{y})\cdot\mathbf{e}(\mathbf{w})d\mathbf{y}-\int_0^T\int_Y \mathbf{Be}(M^p)\cdot\partial_t\mathbf{e}(\mathbf{w}(t,\mathbf{y}))d\mathbf{y}dt+$$
$$\int_0^T\int_Y \mathbf{A}\mathbf{e}(M^p)\cdot\mathbf{e}(\mathbf{w})d\mathbf{y}dt=0. \quad (5.5.8)$$

The details on the derivation of the cell problems are presented in Appendix A.

5.5.2 AVERAGED EQUATIONS FOR MODEL I

To obtain the averaged equations, we integrate (5.4.8) over the periodicity cell and divide by $|Y|$ (see equation (5.4.19)). The contact condition (5.5.2) for σ_1 and the formula (5.5.3) we obtain

$$\widehat{\mathbf{S}}(\zeta^0)=-G\big(S_1^{pq}\mathbf{e}(\zeta^0)_{pq}+\int_0^t S_2^{pq}(t-\tau)e(\zeta^0)_{pq}(\tau,x)d\tau+$$
$$S_3^p\zeta_p^0+\int_0^t S_4^p(t-\tau)\zeta_p^0(\tau,x)d\tau\big), \quad (5.5.9)$$

where

$$S_1^{pq}=\langle K_{1,T}^{pq}+K_{2,T}^{pq}(0,\cdot)\rangle_\Gamma \qquad S_2^{pq}(t)=\langle K_{2,T}^{pq}(t,\cdot)\rangle_\Gamma$$
$$S_3^p=\langle M_T^p(0,\cdot)\rangle_\Gamma \qquad S_4^p(t)=\langle\partial_t M_T^p(t,\cdot)\rangle_\Gamma. \quad (5.5.10)$$

Equation (5.5.9) shows that the drag force is history-dependent, and depends linearly on the averaged velocity ζ^0 and the averaged strain rate $e(\zeta^0)$. The latter dependence shows that the drag force at point $\mathbf{x}$ is determined by the history of averaged velocities in a small neighborhood of $\mathbf{x}$, that is, the drag force is non-local in space as well as in time. Next we use (5.4.5) and (5.5.3) to calculate the effective stress:

$$\widehat{\mathbf{T}}=A(I+T_1)\mathbf{e}(\xi^0)+B(I+T_1)\mathbf{e}(\zeta^0)+(A+B\partial_t)\int_0^t T_2(t-\tau)\mathbf{e}(\zeta^0)(\tau,\cdot)d\tau+$$
$$(A+B\partial_t)\int_0^t T_3(t-\tau)\zeta^0(\tau,\cdot)d\tau, \quad (5.5.11)$$

where the components of $T_i, i=1,2,3$ are related to $K_{1,2}^{pq}, M^p$ as follows.

$$T_1^{klpq}=\langle\mathbf{e}(K_1^{pq})_{kl}\rangle, \quad (5.5.12)$$

$$T_2^{pqkl}=\langle\mathbf{e}(K_2^{pq})_{kl}(t-\tau,\cdot)\rangle, \quad (5.5.13)$$

$$T_3^{klp}=\langle\mathbf{e}(M^p)_{kl}(t-\tau,\cdot)\rangle, \quad (5.5.14)$$

The constitutive equation (5.5.11) shows that the effective material is not simple, since the last two terms in (5.5.11) are velocity-dependent.

5.6 MODEL II: QUADRATIC CONTACT CONDITIONS

In this case, we assume $p(\xi_n^0) = p\xi_n^0,\ F(\xi_n^0) = \xi_n^0,\ G(\zeta_T^0) = G$, where $p > 0, G > 0$ are constants. The contact conditions (5.4.9), (5.4.10) become

$$\underline{\underline{\sigma}}^0\mathbf{n} = -p\xi_n^0\mathbf{n} - G\xi_n^0\zeta_T^0, \tag{5.6.1}$$

and

$$\underline{\underline{\sigma}}^1\mathbf{n} = -p(\xi_n^1 - g_0)\mathbf{n} - G(\xi_n^1 - g_0)\zeta_T^0 - G\xi_n^0\zeta_T^1, \tag{5.6.2}$$

respectively. The fast-variable problem (5.4.13) becomes:

$$\int_0^T \int_{\Gamma_Y} p\xi_n^0\mathbf{n}\cdot\mathbf{w}d\alpha dt + \int_0^T \int_{\Gamma_Y} G\xi_n^0\zeta_T^0\cdot\mathbf{w}d\alpha dt$$
$$+\int_0^T \int_Y [A\mathbf{e}_y(\xi^1) + B\mathbf{e}_y(\zeta^1)]\mathbf{e}(\mathbf{w})d\mathbf{y}dt + \int_0^T \int_Y [A\mathbf{e}_x(\xi^0) + B\mathbf{e}_x(\zeta^0)]\mathbf{e}(\mathbf{w})d\mathbf{y}dt = 0. \tag{5.6.3}$$

To specify the ansatz (5.4.17), we note that the contact condition (5.6.1) contains terms linear in ξ^0 and quadratic terms of the form $\xi_p^0\zeta_q^0$ (components of the dyadic $\xi^0 \otimes \zeta^0$). Thus the slow-variable terms in the ansatz for ξ^1 should contain similar terms. Because the constitutive equations for the matrix contain $\mathbf{e}(\xi)$ and $\mathbf{e}(\zeta)$, the corresponding slow-variable terms must appear in the ansatz as well. Moreover, due to the dissipative nature of the contact conditions and the constitutive equations, we expect the effective medium to be history-dependent. Therefore we choose the following substitution.

$$\xi^1 = K_1^{pq}(\mathbf{y})\mathbf{e}(\xi^0)_{pq} + \int_0^t K_2^{pq}(t-\tau,\mathbf{y})\mathbf{e}(\zeta^0)_{pq}(\tau,\mathbf{x})d\tau +$$
$$\int_0^t \xi_p^0(\tau,\mathbf{x})M^{pq}(t-\tau,\mathbf{y})\partial_t\xi_q^0(\tau,\mathbf{x})d\tau + \int_0^t L^p(t-\tau,\mathbf{y})\xi_p^0(\tau,\mathbf{x})d\tau. \tag{5.6.4}$$

Substituting (5.6.4) into (5.6.3), and collecting slow-variable terms we get the following cell problems. **The problem for K_1^{pq}**. Collect the terms containing $e(\xi^0)_{pq}(t,\mathbf{x})$.

$$\int_Y A(I^{pq} + \mathbf{e}_y(K_1^{pq}))\cdot\mathbf{e}(\mathbf{w})d\mathbf{y} = 0. \tag{5.6.5}$$

The problem for $K_2^{pq}(0,\mathbf{y})$. Collect the terms containing $\mathbf{e}(\partial_t\xi^0)_{pq}(t,\mathbf{x})$.

$$\int_Y B(\mathbf{e}(K_2^{pq}(0,\cdot)) + \mathbf{e}(K_1^{pq}) + I^{pq})\cdot\mathbf{e}(\mathbf{w})d\mathbf{y} = 0. \tag{5.6.6}$$

The problem for $K_2^{pq}(t,\mathbf{y})$. Collect the terms containing $e(\partial_t\xi^0)_{pq}(\tau,\mathbf{x})$.

$$\int_0^T \int_Y (A\mathbf{e}_y(K_2^{pq}) + B\mathbf{e}_y(\partial_t K_2^{pq}))\cdot\mathbf{e}(\mathbf{w})d\mathbf{y}dt = 0. \tag{5.6.7}$$

The initial condition is given by the solution of the cell problem (5.6.6). **The problem for $M^{pq}(0,\mathbf{y})$.** Collect the terms containing $\xi_p^0\partial_t\xi_q^0(t,\mathbf{x})$.

$$\int_Y B\mathbf{e}_y(M^{pq}(0,\cdot))\cdot \mathbf{ew}) + \int_{\Gamma_Y} Gn_p w_{T,q} d\alpha = 0. \tag{5.6.8}$$

The problem for $M^{pq}(t,\mathbf{y})$. Collect the terms containing $\xi_p^0\partial_t\xi_q^0(\tau,\mathbf{x})$.

$$\int_0^T \int_Y (A\mathbf{e}_y(M^{pq}) + B\mathbf{e}_y(\partial_t M^{pq}))\cdot \mathbf{e}(\mathbf{w})d\mathbf{y} = 0 \tag{5.6.9}$$

The initial condition is given by the solution of (5.6.8).
The problem for $L^p(0,\mathbf{y})$. Collect the terms containing $\xi_p^0(t,\mathbf{x})$.

$$\int_Y B\mathbf{e}_y(L^p(0,\cdot))\cdot \mathbf{e}(\mathbf{w})d\mathbf{y} + \int_{\Gamma_Y} pn_p\mathbf{n}\cdot\mathbf{w}d\alpha = 0. \tag{5.6.10}$$

The problem for $L^p(t,\mathbf{y})$. Collect the terms containing $\xi_p^0(\tau,\mathbf{x})$.

$$\int_0^T \int_Y (A\mathbf{e}_y(L^p) + B\mathbf{e}_y(\partial_t L^p))\cdot \mathbf{e}(\mathbf{w})d\mathbf{y} = 0 \tag{5.6.11}$$

The initial condition is given by the solution of (5.6.10).

5.6.1 AVERAGED EQUATION FOR MODEL II

Constitutive equation for effective stress. We use (5.4.5) and (5.6.4) to express $\underline{\underline{\sigma}}^0$ in terms of ξ^0, ζ^0. Averaging out the fast variable we obtain

$$\begin{aligned}\widehat{\mathbf{T}} = A(T_1+I)\mathbf{e}(\xi^0) + B(T_1+I)\mathbf{e}(\zeta^0) + (A+B\partial_t)\int_0^t T_2(t-\tau)\mathbf{e}(\zeta^0)(\tau,\cdot)d\tau +\\ (A+B\partial_t)\int_0^t T_3(t-\tau)(\xi^0\otimes\zeta^0)(\tau,\cdot)d\tau + (A+B\partial_t)\int_0^t T_4(t-\tau)\zeta^0(\tau,\cdot)d\tau,\end{aligned} \tag{5.6.12}$$

where

$$T^1_{pqkl} = \langle \mathbf{e}(K_1^{pq})_{kl}\rangle, \qquad T^2_{pqkl} = \langle \mathbf{e}(K_2^{pq})_{kl}\rangle, \tag{5.6.13}$$

$$T^3_{pqkl} = \langle \mathbf{e}(M^{pq})_{kl}\rangle, \qquad T^4_{pkl} = \langle \int_Y \mathbf{e}(L^p)_{kl}\rangle. \tag{5.6.14}$$

The effective drag force. Using the contact conditions (5.6.2) we write

$$\widehat{\mathbf{S}} = \langle -p(\xi_n^1 - g^0)\mathbf{n} + G(\xi_n^1 - g^0)\zeta_T^0 + G\xi_n^0\zeta_T^1\rangle_\Gamma.$$

We make use of the formula $\zeta_T = (\mathbf{n}\times\zeta)\times\mathbf{n}$ for the "tangential projection" of a vector ζ. Component-wise,

$$\zeta_{T,l} = \varepsilon_{jkl}\varepsilon_{jus}n_k n_u \zeta_s = \zeta_s \mathcal{N}_{ls}. \tag{5.6.15}$$

Next, we make use of the formula (5.6.4) together with (5.6.15) to obtain

$$\langle -G(\xi_n^1\zeta_{T,l}^0+\xi_n^0\zeta_{T,l}^1\rangle_\Gamma = \partial_t\big[\xi_s^0\mathbf{e}(\xi^0)_{pq}W_{pqls}^1+$$
$$\int_0^t \xi_s^0(t,\cdot)e(\zeta^0)_{pq}(\tau,\cdot)W_{pqls}^2(t-\tau)d\tau+\int_0^t \xi_s^0(t,\cdot)\xi_p^0\zeta_q^0(\tau,\cdot)W_{pqls}^3(t-\tau)d\tau+$$
$$\int_0^t \xi_s^0(t,\cdot)\xi_p^0(\tau,\cdot)W_{pls}^4(t-\tau)d\tau\big], \quad (5.6.16)$$

where

$$W_{pqls}^1=\langle -G\mathcal{N}_{ls}n_mK_{1,m}^{pq}\rangle_\Gamma, \qquad W_{pqls}^2=\langle -G\mathcal{N}_{ls}n_mK_{2,m}^{pq}\rangle_\Gamma, \quad (5.6.17)$$

$$W_{pqls}^3=\langle -G\mathcal{N}_{ls}n_mM_m^{pq}\rangle_\Gamma, \qquad W_{pls}^4=\langle -G\mathcal{N}_{ls}n_mL_m^p\rangle_\Gamma, \quad (5.6.18)$$

Similarly,

$$\langle -p(\xi_n^1-g^0)n_l-Gg^0\zeta_{T,l}^0\rangle_\Gamma = V_l^1+\zeta_s^0V_{ls}^2+\mathbf{e}(\xi^0)_{pq}V_{pqls}^3+$$
$$\int_0^t \mathbf{e}(\zeta^0)_{pq}(\tau,\cdot)V_{pqls}^4(t-\tau)d\tau+\int_0^t \xi_p^0(\tau,\cdot)\zeta_q^0(\tau,\cdot)V_{pqls}^5(t-\tau)d\tau+$$
$$\int_0^t \xi_p^0(\tau,\cdot)V_{pls}^6(t-\tau)d\tau, \quad (5.6.19)$$

where

$$V_l^1=\langle pg^0n_l\rangle_\Gamma, \qquad V_{ls}^2=\langle Gg^0\mathcal{N}_{ls}\rangle_\Gamma,$$
$$V_{pql}^3=\langle -pK_{1,m}^{pq}n_mn_l\rangle_\Gamma, \qquad V_{pql}^4=\langle -pK_{2,m}^{pq}n_mn_l\rangle,$$
$$V_{pql}^5=\langle -pM_m^{pq}n_mn_l\rangle, \qquad V_{pl}^6=\langle -pL_m^pn_mn_l\rangle_\Gamma. \quad (5.6.20)$$

Combining (5.6.16) and (5.6.19) we obtain

$$\widehat{S}_l=V_l^1+\zeta_s^0V_{ls}^2+\mathbf{e}(\xi^0)_{pq}V_{pql}^3+\int_0^t \mathbf{e}(\zeta^0)_{pq}(\tau,\cdot)V_{pql}^4(t-\tau)d\tau+$$
$$\int_0^t \xi_p^0(\tau,\cdot)\zeta_q^0(\tau,\cdot)V_{pql}^5(t-\tau)d\tau+\int_0^t \xi_p^0(\tau,\cdot)V_{pl}^6(t-\tau)d\tau+$$
$$\partial_t\big[\xi_s^0\{\mathbf{e}(\xi^0)_{pq}W_{pqls}^1+\int_0^t \mathbf{e}(\zeta^0)_{pq}(\tau,\cdot)W_{pqls}^2(t-\tau)d\tau+$$
$$\int_0^t \xi_p^0\zeta_q^0(\tau,\cdot)W_{pqls}^3(t-\tau)d\tau+\int_0^t \xi_p^0(\tau,\cdot)W_{pls}^4(t-\tau)d\tau\}\big]. \quad (5.6.21)$$

5.7 MODEL III: POWER TYPE CONTACT CONDITION

5.7.1 CONTACT CONDITIONS, ANSATZ AND CELL PROBLEMS

In this case we take $p(z)=z^\gamma, F(z)=z^\beta$, where $0<\gamma<\beta$ are integers. We also linearize the δ-approximation:

$$\frac{\zeta_T}{(|\zeta_T|^2+\delta)^{1/2}}\approx R\zeta_T+\mathbf{r},$$

where R and r are the constant matrix and vector, respectively. The contact conditions are

$$\underline{\underline{\sigma}}^0\mathbf{n} = -(\xi_n^0)^a\mathbf{n} - (\xi_n^0)^b(R\zeta_T + \mathbf{r}), \tag{5.7.1}$$

$$\underline{\underline{\sigma}}^1\mathbf{n} = -a(\xi_n^0)^{a-1}(\xi_n^1 - g_0)\mathbf{n} - b(\xi_n^0)^{b-1}(\xi_n^1 - g_0)(R\zeta_T^0 + \mathbf{r}) - (\xi_n^0)^b R\zeta_T^1, \tag{5.7.2}$$

where $a > 1, b > a$ are integers. The right-hand-side of (5.7.1) is a sum of terms of two types:

$$C_a \xi^{0a_1}_{k_1} \xi^{0a_2}_{k_2} \xi^{0a_3}_{k_3}, \qquad C_{b,l} \xi^{0b_1}_{k_1} \xi^{0b_2}_{k_2} \xi^{0b_3}_{k_3} \zeta^0_{l}, \tag{5.7.3}$$

where the non-negative integers a_1, a_2, a_3 and b_1, b_2, b_3 satisfy $a_1 + a_2 + a_3 = a$ and $b_1 + b_2 + b_3 = b$, respectively, and (k_1, k_2, k_3) is a permutation of $(1,2,3)$. To keep track of numerous exponents and indices, it is convenient to use multiindex notation. In the multi-index notation, the expressions (5.7.3) are written as $C_a \xi^{0\alpha}_{\mathscr{K}}$, $C_b \xi^{0\beta}_{\mathscr{K}}$ with the multi-indices $\alpha = a_1a_2a_3$, $|\alpha| = a_1 + a_2 + a_3 = a, \beta = b_1b_2b_3$, $|\beta| = b_1 + b_2 + b_3 = b$, and $\mathscr{K} = k_1k_2k_3$. The whole contact condition (5.7.1) becomes

$$\underline{\underline{\sigma}}^0\mathbf{n} = \sum_{|\alpha|=0}^{a} C_\alpha \xi^{0\alpha}\mathbf{n} + \sum_{|\beta|\le b} C_\beta \xi^{0\beta}\zeta^0 + \sum_{|\beta|\le b} D_\beta \xi^{0\beta}(\mathbf{r}), \tag{5.7.4}$$

where $C_\alpha, C_\beta, D_\beta$ depend on (powers of) components of $\mathbf{n}$, components of R and binomial coefficients. The contact condition (5.7.2) is even more complicated. We represent it schematically as

$$(\sigma^1\mathbf{n})_l = \sum_{\mu,\nu,q} \xi^{0\mu}\zeta^{0\nu}\mathscr{A}^{ql}_{\mu\nu}(\mathbf{y})\xi^1_q + \sum_{\mu,\nu,q} \xi^{0\mu}\zeta^{0\nu}\mathscr{B}^{ql}_{\mu\nu}(\mathbf{y})\zeta^1_q + \sum_{\mu,\nu,q} \xi^{0\mu}\zeta^{0\nu}\mathscr{C}_{\mu\nu}(\mathbf{y})_l,$$
$$l = 1,2,3, \tag{5.7.5}$$

where multi-indices μ, ν are such that $|\mu| \le (b-1)$, $|\nu| \le 1$ and $q = 1,2,3$. The factors $\mathscr{A}, \mathscr{B}, \mathscr{C}$ depend on components of $\mathbf{n}$ R, $\mathbf{r}$ and binomial coefficients.

5.7.2 THE RELATION BETWEEN ξ^1 AND ξ^0

Motivated by (5.7.4), we choose the relation between ξ^1 and ξ^0 as follows.

$$\xi^1 = K_1^{pq}e(\xi^0)_{pq} + \int_0^t K_2^{pq}(t-\tau)e(\zeta^0)_{pq}(\tau)d\tau +$$
$$\sum \int_0^t M_\beta^{\mathscr{K},q}(t-\tau)\xi^{0\beta}_{\mathscr{K}}\zeta^0_q(\tau)d\tau + \sum \int_0^t L_\alpha^{\mathscr{K}}(t-\tau)\xi^{0\alpha}_{\mathscr{K}}(\tau)d\tau, \tag{5.7.6}$$

where the summation is over p, q, multiindices α, β and $\mathscr{K}$. In particular, $\mathscr{K}$ runs over all permutations of $1,2,3$. As before, the vectors $K_1^{pq}, K_{pq}^2, M_\alpha^{\mathscr{K},q}, L_\beta^{\mathscr{K}}$ are found by solving the cell problems. The latter are obtained by substituting (5.7.6) into the fast-variable problem (5.4.13) and collecting the terms containing the same slow-variable functions. The problems for K_1^{pq}, K_2^{pq} are just (5.6.5)-(5.6.7). The other cell problems have a structure similar to (5.6.9), (5.6.11). This structure can be described as follows. Let $X(t,\mathbf{y})$ denote any of the vectors $M_\alpha^{\mathscr{K},q}, L_\beta^{\mathscr{K}}$. The evolutionary cell problem for $e(X)$ is (the weak formulation of) a linear system of ordinary differential equations with respect to time, namely

$$A\mathbf{e}(X) + B\partial_t \underline{\underline{e}}(X) = 0, \tag{5.7.7}$$

where A, B are the constitutive tensors of the matrix. The initial condition $e(X(0,\mathbf{y}))$ for (5.7.7) is found by solving another cell problem. The latter is of the form (compare with (5.6.8), (5.6.10))

$$\nabla(B\mathbf{e}(X(0,\mathbf{y}))) = F_X, \tag{5.7.8}$$

where F_X is determined for each particular X by the contact conditions (5.7.1). Specifically, each X in (5.7.6) is multiplied by the corresponding (unique) slow-variable function $Sl_X(\xi^0)(t,x)$. The right-hand-side of (5.7.8) is determined by the coefficient at Sl_X in (5.7.4).

5.7.3 EFFECTIVE STRESS

Let us denote the solution operator of (5.7.7) by $e^{-B^{-1}At}$. Then

$$e(X(t,\mathbf{y})) = e^{-B^{-1}At} e(X(0,\mathbf{y})), \tag{5.7.9}$$

where $X(0,\mathbf{y})$ solves (5.7.8). The term corresponding to X in the expression (5.7.6) for ξ^1 will be of the form

$$\int_0^t X(t-\tau,\mathbf{y}) Sl_X(\xi^0)(\tau,x)d\tau. \tag{5.7.10}$$

Next, we determine the term corresponding to X in the expression (5.4.5) for σ^0. Applying the differential operator e_y to both sides of (5.7.10) and taking into account (5.7.9) we obtain

$$\int_0^t e^{B^{-1}A(t-\tau)} e_y(X(0,\mathbf{y})) Sl_X(\xi^0)(\tau,x)d\tau. \tag{5.7.11}$$

The contribution of the terms containing X to the effective stress tensor will be written as

$$(A+B\partial_t)\int_0^t \mathbf{e}^{B^{-1}A(t-\tau)} \langle \mathbf{e}_y(X(0,\mathbf{y}))\rangle Sl_X(\xi^0)(\tau,x)d\tau. \tag{5.7.12}$$

Using the formula (5.7.11) for each of the terms in the expression $\underline{\underline{\sigma}}^0 = A\mathbf{e}_x(\xi^0) + B\mathbf{e}_x(\zeta^0) + A\mathbf{e}_y(\xi^1) + B\mathbf{e}_y(\zeta^1)$, we obtain the constitutive equation for the effective stress

$$\widehat{T} = A(T^1+I)\mathbf{e}(\xi^0) + B(T^1+I)\mathbf{e}(\zeta^0) + (A+B\partial_t)\left[\int_0^t T^2(t-\tau)e(\xi^0)(\tau,\cdot)d\tau\right] + (A+B\partial_t)\left[\sum_{\mathscr{K},\beta,q}\int_0^t T_\beta^{\mathscr{K},q}(t-\tau)\xi^{0\beta}{}_{\mathscr{K}}\zeta^0{}_q(\tau,\cdot)d\tau + \sum_{\mathscr{K},\alpha}\int_0^t T_\alpha^{\mathscr{K}}(t-\tau)\xi^{0\alpha}{}_{\mathscr{K}}d\tau\right], \tag{5.7.13}$$

where

$$T^1_{pqkl} = \langle \mathbf{e}(K_1^{pq})_{kl} \rangle, \qquad T^2_{pqkl} = \langle \mathbf{e}(K_2^{pq})_{kl} \rangle,$$
$$T_\beta^{\mathscr{K},q} = \langle \mathbf{e}^{-B^{-1}A(t-\tau)} M_\beta^{\mathscr{K},q}(0,t-\tau)) \rangle,$$
$$T_\alpha^{\mathscr{K}} = \langle \mathbf{e}^{-B^{-1}A(t-\tau)} M_\alpha^{\mathscr{K}}(0,t-\tau)) \rangle. \tag{5.7.14}$$

5.7.4 EFFECTIVE DRAG FORCE

To obtain the effective drag force, we substitute ξ^1, ζ^1 determined by (5.7.6) into the condition (5.7.5) and average in $\mathbf{y}$ over Γ_Y. The resulting, rather complicated, equation is of the form

$$\widehat{\mathbf{S}} = \widehat{S}^1 + \widehat{S}^3 + \widehat{S}^3, \tag{5.7.15}$$

where $\widehat{S}^m, m = 1,2,3$ correspond to the three terms in (5.7.5), namely

$$\begin{aligned}\widehat{S}^1_l &= \sum_{\mu,\nu,s} \xi^{0\mu}\zeta^{0\nu}(t,\mathbf{x}) \Bigg[V^{1,l}_{\mu\nu}\mathbf{e}(\xi^0)(t,\mathbf{x}) + \int_0^t V^{2,l}_{\mu\nu}(t-\tau)\mathbf{e}(\zeta^0)(\tau,\mathbf{x})d\tau \\ &\quad + \sum_{|\beta|\le b,\mathscr{K}} \int_0^t V^{3,l}_{\mu\nu,\beta,\mathscr{K}}(t-\tau)\xi^{0\beta}{}_{\mathscr{K}}\zeta^0_q(\tau,x)d\tau \\ &\quad + \sum_{|\alpha|\le b,\mathscr{K}} \int_0^t V^{4,l}_{\mu\nu,\alpha,\mathscr{K}}(t-\tau)\xi^{0\alpha}{}_{\mathscr{K}}(\tau,x)d\tau \Bigg]\end{aligned},$$

$$\tag{5.7.16}$$

where

$$V^{1,lpq}_{\mu\nu} = \langle \mathscr{A}^{sl}_{\mu\nu} K^{pq}_{1,s} \rangle_\Gamma, \qquad V^{2,lpq}_{\mu\nu} = \langle \mathscr{A}^{sl}_{\mu\nu} K^{pq}_{2,s} \rangle_\Gamma,$$
$$V^{3,lq}_{\mu\nu,\beta,\mathscr{K}} = \langle \mathscr{A}^{sl}_{\mu\nu} M^{\mathscr{K}}_{\beta,s} \rangle_\Gamma, \qquad V^{4,l}_{\mu\nu,\beta,\mathscr{K}} = \langle \mathscr{A}^{sl}_{\mu\nu} L^{\mathscr{K}}_{\alpha,s} \rangle_\Gamma. \tag{5.7.17}$$

Similarly, $\widehat{S}^2$ corresponds to the second term in the right-hand-side of (5.7.5).

$$\begin{aligned}\widehat{S}^2_l &= \sum_{\mu,\nu,s} \xi^{0\mu}\zeta^{0\nu}(t,\mathbf{x})\partial_t \Bigg[W^{1,l}_{\mu\nu}e(\xi^0)(t,\mathbf{x}) + \int_0^t W^{2,l}_{\mu\nu}(t-\tau)e(\zeta^0)(\tau,\mathbf{x})d\tau \\ &\quad + \sum_{|\beta|\le b,\mathscr{K}} \int_0^t W^{3,l}_{\mu\nu,\beta,\mathscr{K}}(t-\tau)\xi^{0\beta}{}_{\mathscr{K}}\zeta^0_q(\tau,\mathbf{x})d\tau \\ &\quad + \sum_{|\alpha|\le b,\mathscr{K}} \int_0^t W^{4,l}_{\mu\nu,\alpha,\mathscr{K}}(t-\tau)\xi^{0\alpha}{}_{\mathscr{K}}(\tau,\mathbf{x})d\tau, \Bigg]\end{aligned}$$

$$\tag{5.7.18}$$

where

$$W^{1,lpq}_{\mu\nu} = \langle \mathscr{B}^{sl}_{\mu\nu} K^{pq}_{1,s} \rangle_\Gamma, \qquad W^{2,lpq}_{\mu\nu} = \langle \mathscr{B}^{sl}_{\mu\nu} K^{pq}_{2,s} \rangle_\Gamma,$$
$$W^{3,lq}_{\mu\nu,\beta,\mathscr{K}} = \langle \mathscr{B}^{sl}_{\mu\nu} M^{\mathscr{K}}_{\beta,s} \rangle_\Gamma, \qquad W^{4,l}_{\mu\nu,\beta,\mathscr{K}} = \langle \mathscr{B}^{sl}_{\mu\nu} L^{\mathscr{K}}_{\alpha,s} \rangle_\Gamma. \tag{5.7.19}$$

Finally, $\widehat{S}^3$ is simply

$$\widehat{S}^3_l = \sum_{\mu,\nu,q} \xi^{0\mu}\zeta^{0\nu}(t,\mathbf{x}) \langle \mathscr{C}_{\mu\nu,l} \rangle_\Gamma. \tag{5.7.20}$$

6 Acoustics in a Random Microstructure

6.1 INTRODUCTION

This chapter deals with a randomly fissured elastic material.[1] It is assumed that the associated random field is statistically homogeneous, with built-in scale separation. Although the underlying stochastic process does not necessarily have to be ergodic, we assume for simplicity of exposition that it is. This allows us to obtain an explicit and computationally easier auxiliary problem in a Representative Elementary Volume. We also derive the effective equations rigorously, using the notion of stochastic two-scale convergence in the mean, introduced in [43] and developed further in [229] and [42]. The effective equations model a one-phase viscoelastic medium with long-time history dependence.

We introduce the ε-problem as follows: Let (Ω,σ,μ) be a probability space; moreover, assume that an n-dimensional, dynamical system $\mathscr{T}$ is given on Ω, such that:

1. $\mathscr{T}(0)\omega = \mathbb{I}\omega = \omega$ on Ω and $\mathscr{T}(x_1+x_2) = \mathscr{T}(x_1)\mathscr{T}(x_2)$ for all $x_1,x_2 \in \mathbb{R}^n$.
2. $\forall x \in \mathbb{R}^n$ and $\forall E \in \Sigma$, $\mu(\mathscr{T}(x)(E)) = \mu(E)$, i.e. μ is an invariant measure for $\mathscr{T}$.
3. $\forall E \in \Sigma$ the set $\{(x,\omega) \in \mathbb{R}^n \times \Omega : \mathscr{T}(x)\omega \in E\}$ is an element of the σ-algebra $\mathscr{L} \times \Sigma$ on $\mathbb{R}^n \times \Omega$, where $\mathscr{L}$ is the usual Lebesgue σ-algebra on $\mathbb{R}^n$, i.e. $\mathscr{T}$ is a semiflow. [229, 43]

With the measurable dynamics, an associated n-parameter group of unitary operators may be introduced

$$(U(x)f)(\omega) := f(\mathscr{T}(x)\omega), \quad f \in L^2(\Omega) \tag{6.1.1}$$

on $L^2(\Omega) := L^2(\Omega,\Sigma,\mu)$. We suppose that $L^2(\Omega)$ is separable and the dynamical system $\{\mathscr{T}(x)\}$ is ergodic. Equation (6.1.1) expresses the fact that the material properties are statistically homogeneous. This assumption is more realistic than the assumption that the bone architecture is periodic.

In order to define the fluid part (blood-marrow) and the solid part (trabeculae), we fix a set $\mathscr{F} \in \Sigma$ such that $\mu(\mathscr{F}) > 0$ and $\mu(\Omega \setminus \mathscr{F}) > 0$. We define a random pore structure $F(\omega) \subset \mathbb{R}^n, \omega \in \Omega$ obtained from $\mathscr{F}$ by setting

$$F(\omega) = \{x \in \mathbb{R}^n : \mathscr{T}(x)\omega \in \mathscr{F}\} \tag{6.1.2}$$

[1] It is known that bone is not randomly configured; however, as will be mentioned in Chapter 12, it is possible to construct orthotropic randomly configured bone RVEs.

and assuming that $F(\omega)$ is open and connected almost surely (for almost all $\omega \in \Omega$). The random skeleton (trabeculae) is then constructed by setting

$$\mathscr{M} := \Omega \setminus \mathscr{F}, \quad M(\omega) = \mathbb{R}^n \setminus F(\omega) \text{ and}$$

$$M_\varepsilon(\omega) = \left\{x \in \mathbb{R}^n : \varepsilon^{-1}x \in M(\omega)\right\}. \tag{6.1.3}$$

Let G be a smooth bounded domain in $\mathbb{R}^n$ and after having chosen the random structure in $\mathbb{R}^n$, set

$$G_1^\varepsilon := \{x \in G : \mathrm{dist}(x, \partial G) \geq \varepsilon\}$$

The random pore system, which is assumed to be saturated with fluid, may now be introduced as

$$G_f^\varepsilon(\omega) = G \setminus \overline{M_\varepsilon(\omega) \cap G_1^\varepsilon}.$$

The random skeletal system in the domain $G \subset \mathbb{R}^n$ is given by

$$G_s^\varepsilon(\omega) = G \setminus \overline{G_f^\varepsilon(\omega)}$$

The solid and fluid stresses are given by

$$\sigma^{s,\varepsilon} = \mathbf{Ae}(\mathbf{u}) \text{ in } G_s^\varepsilon(\omega) \times (0,T) \tag{6.1.4}$$

$$\sigma^{f,\varepsilon} = -p^\varepsilon \mathbb{I} + 2\varepsilon^2 \eta \mathbf{e}(\partial_t \mathbf{u}) \text{ in } G_f^\varepsilon(\omega) \times (0,T), \tag{6.1.5}$$

where $\mathbf{A}$ is the elasticity tensor satisfying standard strong ellipticity conditions, and the parameters are the same that have been used in previous chapters. In the case of the blood-marrow mixture we consider the compressibility to be small and neglect the bulk viscosity term in the constitutive law. Hence, only the shear viscosity, ν, is considered. Also, in (6.1.4) we have used the small compressibility approximation $p^\varepsilon = -\rho_f c^2 \mathrm{div}\mathbf{u}$. For details of the derivation of this approximation we refer the reader to [363] . The asymptotic forms of the stresses are given by

$$\sigma^{s,\varepsilon} := \sigma^s\left(\mathscr{T}(\frac{x}{\varepsilon})\omega\right), \quad \sigma^{f,\varepsilon} := \sigma^f\left(\mathscr{T}(\frac{x}{\varepsilon})\omega\right) \tag{6.1.6}$$

then the asymptotic form of the equations of motion are

$$\rho_s \partial_t^2 \mathbf{u}^\varepsilon - \mathrm{div}\left(\mathbf{A}e(\mathbf{u}^\varepsilon)\right) = \mathbf{F}\rho_s \text{ in } G_s^\varepsilon(\omega) \times (0,T)$$

$$\rho_f \partial_t^2 \mathbf{u}^\varepsilon - \mathrm{div}\left(\sigma^{f,\varepsilon}\right) = \mathbf{F}\rho_f \text{ in } G_f^\varepsilon(\omega) \times (0,T); \tag{6.1.7}$$

moreover, we have that the interface conditions are the same as used in Chapter 3.

$$[\![\mathbf{u}^\varepsilon]\!] = 0, \; \sigma^{s,\varepsilon} \cdot \nu_s = \sigma^{f,\varepsilon} \cdot \nu_f \text{ on } \partial G_s^\varepsilon \cap G_f^\varepsilon, \tag{6.1.8}$$

where $[\![\cdot]\!]$ denotes the jump across the boundary, and ν_s, ν_f are the unit outward normals for G_s^ε and G_f^ε, respectively. On the outer boundary we impose zero boundary conditions. For simplicity of exposition, we impose zero boundary conditions on the outer boundary, and zero initial conditions:

$$\mathbf{u}^\varepsilon(0) = \partial_t \mathbf{u}^\varepsilon(0) = 0. \tag{6.1.9}$$

The variational formulation, for almost any realization, which corresponds to (6.1.7)–(6.1.9) is now given by:

Find $\mathbf{u}^\varepsilon \in H^1(0,T;H^1(G)^n)$ with $\dfrac{d^2\mathbf{u}^\varepsilon}{dt^2} \in L^2(0,T;L^2(G)^n)$ such that

$$\frac{d^2}{dt^2}\int_{G\times\Omega} \rho^\varepsilon \mathbf{u}^\varepsilon(t)\cdot\varphi\, dx\, d\mu + \frac{d}{dt}\int_{G_f^\varepsilon\times\Omega} 2\varepsilon^2\eta\mathbf{e}(\mathbf{u}^\varepsilon(t)) : e(\varphi)\, dx\, d\mu$$

$$+ \int_{G_s^\varepsilon\times\Omega} A\mathbf{e}(\mathbf{u}^\varepsilon(t)) : \mathbf{e}(\varphi)\, dx\, d\mu - \int_{G_f^\varepsilon\times\Omega} p^\varepsilon \operatorname{div}\varphi\, dx\, d\mu$$

$$= \int_{G\times\Omega} \rho^\varepsilon \mathbf{F}\cdot\varphi\, dx\, d\mu, \qquad \forall\varphi\in H^1(G\times\Omega)^n, \qquad (a.e.) \text{ in }]0,T[, \quad (6.1.10)$$

where $\rho^\varepsilon(x,\omega) = \rho_f\chi_{G_f^\varepsilon(\omega)}(x) + \rho_s\chi_{G_s^\varepsilon(\omega)}(x)$. For simplicity we suppose $\mathbf{F} \in H^2(0,T;L^2(G)^n)$.
After taking $\varphi = \frac{\partial u^\varepsilon}{\partial t}$ as the test function in (6.1.10), as in [135, 91], we obtain the same a priori estimates as in Chapter 3; moreover, the extension of the pressure field is the same as in Chapter 3, namely

$$\tilde{p}^\varepsilon(x,t) = \begin{cases} -\gamma \operatorname{div} u^\varepsilon(x,t) + \frac{\gamma}{|G|}\int_{G_f^\varepsilon} \operatorname{div} u^\varepsilon(x,t)\, dx, & x\in G_f^\varepsilon, \\ \frac{\gamma}{|G|}\int_{G_f^\varepsilon} \operatorname{div} u^\varepsilon(x,t)\, dx, & x\in G_s^\varepsilon, \end{cases}$$

where $\gamma = c^2\rho_f$, we get that $\|\tilde{p}^\varepsilon\|_{L^\infty(0,T;L^2(G_f^\varepsilon\cup G_s^\varepsilon))} \le C$.

6.2 STOCHASTIC TWO-SCALE LIMITS

In order to pass to the limit as ε goes to zero in (6.1.10), we use the notion of *stochastic two-scale convergence in the mean*, Bourgeat et al. [41, 43], Jikov et al. [229]. To this end, let us first recall some definitions. An element $\psi \in L^2(G\times\Omega)$ is said to be *admissible* if the function

$$\psi_{\mathcal{T}} : (x,\omega) \to \psi(x,\mathcal{T}(x)\omega), \quad (x,\omega)\in G\times\Omega,$$

defines an element of $L^2(G\times\Omega)$.

Definition 6.1. *The sequence $\{\mathbf{w}^\varepsilon\} \subset L^2(G\times\Omega)$ is said to converge stochastically two-scale in the mean (s.2-s.m.) to a limit $\mathbf{w} \in L^2(G\times\Omega)$ iff for any admissible $\psi \in L^2(G\times\Omega)$ we have*

$$\lim_{\varepsilon\to 0}\int_{G\times\Omega} \mathbf{w}^\varepsilon(x,\omega)\psi\left(x,\mathcal{T}(\frac{x}{\varepsilon})\omega\right) dx\, d\mu = \int_{G\times\Omega} \mathbf{w}(x,\omega)\psi(x,\omega)\, dx\, d\mu.$$

We recall some compactness properties of the stochastic two-scale convergence in the mean proved by Bourgeat et al. [43].

Lemma 6.1

Let $\{\mathbf{u}^\varepsilon\}$ be a bounded sequence in $L^2(G\times\Omega)$. Then there exists a subsequence of $\{\mathbf{u}^\varepsilon\}$ which stochastically two-scale converges in the mean to $\mathbf{u}\in L^2(Q\times\Omega)$. ■

Lemma 6.2

(i) Let $\mathbf{u}^\varepsilon$ and $\varepsilon\nabla_x\mathbf{u}^\varepsilon$ be bounded sequences in $L^2(G\times\Omega)$. Then there exists a function $\mathbf{u}\in L^2(G,H^1(\Omega))$ and a subsequence such that

$$\mathbf{u}^\varepsilon \overset{s.2-s.m}{\longrightarrow} \mathbf{u}$$

$$\varepsilon\nabla_x\mathbf{u}^\varepsilon \overset{s.2-s.m}{\longrightarrow} \nabla_\omega\mathbf{u}.$$

(ii) Let X be a norm-closed, convex subset of $H^1(G)$. And suppose that $\mathbf{u}^\varepsilon\in L^2(G\times\Omega)$ is a sequence satisfying the following conditions:

$\mathbf{u}^\varepsilon(\cdot,\omega)\in X$ for almost every $\omega\in\Omega$,
There exists an ε-independent constant C such that

$$\int_\Omega \|\mathbf{u}^\varepsilon(\cdot,\omega)\|^2_{H^1(G)}\,d\mu \le C.$$

Then there exist functions $\mathbf{u}\in H^1(G)$, $\mathbf{z}\in L^2(G\times\Omega)^n$ and a subsequence such that

$$\mathbf{u}^\varepsilon \overset{s.2-s.m}{\longrightarrow} \mathbf{u}$$

$$\nabla_x\mathbf{u}^\varepsilon \overset{s.2-s.m}{\longrightarrow} \nabla_x\mathbf{u}(x)+\mathbf{z}(x,\omega).$$

Moreover, $\mathbf{z}\in X$, where X is the closure of the space $\mathscr{V}^2_{\text{pot}}(\Omega)$ in $L^2(\Omega)^n$. ■

In the above, the corrector term $\mathbf{z}$ is a limit of some sequence of the potential functions and we can choose a potential function arbitrarily close to $\mathbf{z}$. As a consequence of our a priori estimates, we have the following convergence results:

$$\mathbf{u}^\varepsilon \overset{s.2-s.m}{\longrightarrow} \mathbf{u}^0(x,t)+\chi_{\mathscr{F}}(\omega)\mathbf{v}(x,\omega,t)$$

$$\chi_{G_s^\varepsilon}\nabla\mathbf{u}^\varepsilon \overset{s.2-s.m}{\longrightarrow} \chi_{\mathscr{M}}(\omega)[\nabla_x\mathbf{u}^0(x,t)+\nabla_\omega\mathbf{u}^1(x,\omega,t)]$$

$$\chi_{G_f^\varepsilon}\nabla\mathbf{u}^\varepsilon \overset{s.2-s.m}{\longrightarrow} \chi_{\mathscr{F}}(\omega)\nabla_\omega\mathbf{v}(x,\omega,t)$$

$$\tilde{p}^\varepsilon \overset{s.2-s.m}{\longrightarrow} \tilde{p}^0$$

with $\mathbf{v}=0$ in the solid part.

PASSING TO THE LIMIT $\varepsilon \to 0$

Consider the functions: $\varphi \in C^{\infty}(G)^n$, $\psi(x, \mathcal{T}(\frac{x}{\varepsilon})\omega) \in C^{\infty}(G \times \Omega)$, and $\zeta(x, \mathcal{T}(\frac{x}{\varepsilon})\omega) \in C^{\infty}(G \times \Omega)$, with $\zeta = 0$ on $\mathcal{M}$ and $\operatorname{div}_{\omega}\zeta = 0$ in $\mathcal{F}$. If $\varphi(x)$, $\varepsilon\psi\left(x, \frac{x}{\varepsilon}\right)$ or $\zeta\left(x, \frac{x}{\varepsilon}\right)$ are taken as the test functions in the variational formulation(6.1.10), after passing to the limit as $\varepsilon \to 0$ we obtain

$$\frac{d^2}{dt^2}\left\{\bar{\rho}\int_G \mathbf{u}^0(t)\varphi\,dx + \int_G \left(\int_\Omega \rho_f \mathbf{v}\,d\mu\right)\varphi\,dx\right\}$$
$$+\int_{G\times\Omega} \chi_{\mathcal{M}} A(\mathbf{e}_x(\mathbf{u}^0)+\mathbf{e}_\omega(\mathbf{u}^1)) : e_x(\varphi) - \int_G \left(\int_\Omega \tilde{p}^0\,d\mu\right)\operatorname{div}_x\varphi\,dx$$
$$= \int_G \bar{\rho}\mathbf{F}\varphi, \quad (6.2.1)$$

$$\int_{G\times\Omega} \chi_{\mathcal{M}} A(\mathbf{e}_x(\mathbf{u}^0)+\mathbf{e}_\omega(\mathbf{u}^1)) : e_\omega(\psi)\,dx\,d\mu - \int_{G\times\Omega} \tilde{p}^0 \operatorname{div}_\omega\psi = 0, \quad (6.2.2)$$

$$\frac{d^2}{dt^2}\int_{G\times\Omega} \chi_{\mathcal{F}}\rho_f(\mathbf{u}^0(t)+\mathbf{v}(t))\zeta + 2\eta\int_{G\times\Omega} \chi_{\mathcal{F}}\mathbf{e}_\omega\left(\frac{\partial \mathbf{v}}{\partial t}\right) : e_\omega(\zeta)\,dx\,d\mu$$
$$-\int_{G\times\Omega} \chi_{\mathcal{F}}\tilde{p}^0 \operatorname{div}_x\zeta\,dx\,d\mu = \int_{G\times\Omega} \chi_{\mathcal{F}}\rho_f\mathbf{F}\zeta\,dx\,d\mu, \quad (6.2.3)$$

where $\bar{\rho} = \rho_f\mu(\mathcal{F}) + \rho_s\mu(\mathcal{M})$, and $\mathbf{e}_x$ and $\mathbf{e}_\omega$ are the strain tensors with respect to x and to ω respectively (derivatives with respect to ω are taken in the sense of measure).

Therefore the limit functions $\{\mathbf{u}^0, \mathbf{u}^1, \mathbf{v}, \tilde{p}^0\}$, where the functions $\mathbf{u}^0, \mathbf{u}^1, \mathbf{v} \in H^3(0,T;(\Omega\times G)^n)$, $\tilde{p}^0 \in H^1(0,T;L^2(\Omega;L^2(\Omega))$, and $\operatorname{div}_x \int_\Omega \chi_{\mathcal{F}}(\omega)\mathbf{v}\,d\mu \in H^2(0,T;$ $L^2(G))$, satisfy the system (6.2.1)–(6.2.3) with zero initial conditions for $\mathbf{u}^0, \mathbf{u}^1, \frac{\partial u^1}{\partial t}, \mathbf{v}$, and $\frac{\partial \mathbf{v}}{\partial t}$.

6.3 PERIODIC APPROXIMATION

For simplicity we now use localized periodic approximation of effective coefficients. We consider a cube $S_\rho = [0,\rho]^3$ as the representative volume element (**RVE**) of the random medium and extend it periodically to the whole space $\mathbb{R}^3$. One can approximate effective coefficients of the original stochastic operator by the effective coefficients of such periodic operators. As shown by Bourgeat and Piatnitski [43, 39] as

$\rho \to \infty$ the periodic approximation operators converge to the effective operator obtained by homogenization of the original stochastic operator. We may treat the fluid stresses in a similar way:

$$\sigma_{\text{per}}^{f,\rho}(x,\omega) := -p^{\varepsilon}(\mathcal{T}_{x(\text{mod}S_\rho)}\omega)\, m\mathbb{I} + 2\varepsilon^2\eta\mathbf{e}\left(\partial_t\mathbf{u}(\mathcal{T}_{x(\text{mod}S_\rho)}\omega)\right)$$

and use these approximations whenever periodicity arguments are needed.
The formal description of construction of a periodically perforated domain G^{ε} is presented in detail in our earlier works. [135, 91, 149] Here we briefly recall that $\mathcal{Y} =]0,1[^n$, $n = 2,3$ is the unit cell in $\mathbb{R}^n$, $\mathcal{Y}_s$ (the solid part) is a closed subset of $\bar{\mathcal{Y}}$ and $\mathcal{Y}_f = \mathcal{Y}\backslash\mathcal{Y}_s$ (the fluid part). $G =]0,L[^n$ is covered with a regular mesh of size ε, each cell being a cube $\mathcal{Y}_i^{\varepsilon}$ homeomorphic to $\mathcal{Y}$, and we denote by G_s^{ε} and G_f^{ε} the solid and fluid parts of a porous medium G, respectively. Also, let $\chi_s^{\varepsilon}(x) := \chi_s\left(\frac{x}{\varepsilon}\right)$, and $\chi_f^{\varepsilon}(x) := \chi_f\left(\frac{x}{\varepsilon}\right)$.
We briefly summarize the main results and state the effective homogenized equations. For details of our work we refer the reader to [150].

Proposition 6.1. *The functions* $\mathbf{u}^0, \mathbf{v}, \mathbf{u}^1$, *and* $\tilde{p}^0$ *satisfy*

$$\text{div}_y\mathbf{v} = 0, \qquad \textit{in } G \times \mathcal{Y}_f \times]0,T[,$$

$$\tilde{p}^0(x,y,t) = \chi_s(y)B(t) + \chi_f(y)p^0(x,t), \qquad \textit{in } G \times \mathcal{Y} \times]0,T[,$$

where

$$B(t) = -\frac{\gamma}{|G|}\int_G\int_{\mathcal{Y}_s} \text{div}_y\mathbf{u}^1 dydx, \tag{6.3.1}$$

and

$$\frac{|\mathcal{Y}_f|}{\gamma}p^0(x,t) - \frac{|\mathcal{Y}_f|}{\gamma}B(t) + |\mathcal{Y}_f|\text{div}_x\mathbf{u}^0(x,t) \tag{6.3.2}$$
$$+\int_{\mathcal{Y}_f}\text{div}_x\mathbf{v}(x,y,t)dy - \int_{\mathcal{Y}_s}\text{div}_y\mathbf{u}^1(x,y,t)dy = 0,$$
$$\textit{in } G\times]0,T[.$$

Moreover,

$$B(t) = -\frac{|\mathcal{Y}_f|}{|G||\mathcal{Y}_s|}\int_G p^0(x,t)dx.$$

We derive the effective equations for $\mathbf{u}^0$ (solid displacement), p^0 (fluid pressure), and for

$$\mathbf{w} := \frac{1}{|\mathcal{Y}_f|}\int_{\mathcal{Y}_f}\mathbf{v}(x,y,t)dy,$$

which is the average displacement of the fluid relative to solid.
Starting from the strong form of the periodic equivalent of (6.2.2), we decompose $\mathbf{u}^1$, the corrector displacement in the solid part, and obtain

$$\mathbf{u}^1(x,y,t) = \sum_{i,j} (\mathbf{e}_x(\mathbf{u}^0))_{ij} \mathbf{W}^{ij}(y) + \left(p^0(x,t) + \frac{|\mathscr{Y}_f|}{|G||\mathscr{Y}_s|} \int_G p^0(x,t)dx \right) \mathbf{W}^0(y).$$

where the auxiliary functions $\mathbf{W}^{ij}$ and $\mathbf{W}^0$ can be calculated from the following cell problems:

$$\begin{cases} \operatorname{div}_y \left\{ A \left(\frac{e_i \otimes e_j + e_j \otimes e_i}{2} + e_y(\mathbf{W}^{ij}) \right) \right\} = 0 & \text{in } \mathscr{Y}_s, \\ A(\frac{e_i \otimes e_j + e_j \otimes e_i}{2} + e_y(\mathbf{W}^{ij}))\nu = 0 & \text{on } \partial\mathscr{Y}_s \backslash \partial\mathscr{Y}, \\ \int\limits_{G_s} \mathbf{W}^{ij}(y)\, dy = 0 & \end{cases} \tag{6.3.3}$$

and

$$\begin{cases} -\operatorname{div}_y \{ A\mathbf{e}_y(\mathbf{W}^0) \} = 0 & \text{in } \mathscr{Y}_s, \\ A\mathbf{e}_y(\mathbf{W}^0)\nu = -\nu & \text{on } \partial\mathscr{Y}_s \backslash \partial\mathscr{Y}, \\ \int\limits_{G_s} \mathbf{W}^0(y)\, dy = 0. & \end{cases} \tag{6.3.4}$$

On the other hand, using Proposition 3 and the periodic equivalents of the equations (6.2.3) and (6.2.1), we obtain the following effective equations of the Biot type.

Proposition 6.2. *The effective equations for* $\mathbf{u}^0$, $\mathbf{w}$ *and* p^0 *are*

$$\partial_{tt}(\rho^0 \mathbf{u}^0 + \rho_f \mathbf{w}) - \operatorname{div}(A^H \mathbf{e}_x(\mathbf{u}^0)) + B^H \nabla p^0 = \rho^0 \mathbf{F}, \tag{6.3.5}$$

$$\partial_t \mathbf{w} = \frac{1}{\rho_f} \int_0^t Y^H(t-\tau) \left(\rho_f \mathbf{F} - \nabla p^0 - \rho_f \partial_{tt} \mathbf{u}^0 \right)(x,\tau) d\tau, \tag{6.3.6}$$

and

$$p^0 = -L^H : e_x\left(\mathbf{u}^0\right) - M^H \operatorname{div}_x \mathbf{w}. \tag{6.3.7}$$

Here $\rho^0 = \rho_f + \frac{|\mathscr{Y}_s|}{|\mathscr{Y}_f|}\rho_s$, *and* A^H, B^H, M^H, L^H *are defined by* (6.3.8)

$$\begin{aligned} A^H_{ijkl} &= |\mathscr{Y}_f|^{-1} \int_{\mathscr{Y}_s} A_{ijmn} \left(\frac{\mathbf{e}_k \otimes \mathbf{e}_l + \mathbf{e}_l \otimes \mathbf{e}_k}{2} + e_y(\mathbf{W}^{kl}) \right)_{mn} dy, \\ B^H &= -|\mathscr{Y}_f|^{-1} \int_{\mathscr{Y}_s} A e_y(\mathbf{W}^0) dy + I, \\ L^H_{ij} &= \left(\int\limits_{\mathscr{Y}_s} \operatorname{div}_y \mathbf{W}^0\, dy - \frac{|\mathscr{Y}_f|}{\gamma} \right)^{-1} (\mathscr{C}^H_{ij} - |\mathscr{Y}_f|\delta_{ij},) \end{aligned} \tag{6.3.8}$$

$$M^H = -|\mathscr{Y}_f| \left(\int_{\mathscr{Y}_s} \operatorname{div}_y \mathbf{W}^0 \, dy - \frac{|\mathscr{Y}_f|}{\gamma} \right)^{-1},$$

where I is the identity matrix, and $\mathbf{W}^{ij}$ and $\mathbf{W}^0$ are obtained by solving (6.3.3) and (6.3.4). The kernel of the effective viscoelastic operator is a matrix Y^H given by

$$Y_{ik}^H = \frac{1}{|\mathscr{Y}_f|} \int_{\mathscr{Y}_f} V_i^k \left(y, \frac{\rho_f}{\eta}(t-\tau) \right) dy, \tag{6.3.9}$$

where $\mathbf{V}^k$ are the auxiliary vectors satisfying the cell problems

$$\begin{cases} \dfrac{\partial \mathbf{V}^k}{\partial t} - \Delta \mathbf{V}^k + \nabla \pi^k = 0, \\ \operatorname{div}_y \mathbf{V}^k = 0, \quad \mathbf{V}^k(y,0) = \mathbf{e}_k, \\ \mathbf{V}^k \,|_{\partial \mathscr{Y}_f \setminus \partial \mathscr{Y}} = 0, \quad \{\mathbf{V}^k, \pi^k\} \text{ 1-periodic in } y. \end{cases} \tag{6.3.10}$$

7 Non-Newtonian Interstitial Fluid

In this chapter we model the bone tissue as a periodic two-phase material composed of the porous viscoelastic solid matrix filled with a non-Newtonian fluid (representing the bone marrow). The model is made under the assumptions that the amplitude of the propagating waves is sufficiently small so that the material response of the matrix is linear, and that the interface between the phases is stationary. These assumptions do not exclude the possibility of large velocities, in which case the nonlinear effects in the fluid phase may be pronounced. It is natural to assume that the material response of the bone marrow is non-Newtonian; the actual bone marrow is a complicated bio-polymeric material consisting of collagen fibers and various blood and fatty (adipose) cells. In the chapter we use one of the simplest models of the non-Newtonian fluids: the Carreau law (see e.g. [32]). In this model, shear viscosity has a sub-linear, shear-thinning dependence on shear rate. Another feature of the acoustic model we present is a slightly compressible constitutive law, where pressure is proportional to the trace of the strain, or equivalently, to the divergence of the fluid displacement. Such models have long history, going back to work of Sanchez-Palencia [363].

Homogenization problems for the flow of non-Newtonian shear-thinning fluids in porous media were considered in [44], [45]. Here, we consider similar constitutive equations for the fluid phase, but the analytical difficulties that arise are quite different. The constitutive equations for our ε-problems are monotonic. This allows us to use the classical Minty-Browder argument for derivation of the effective constitutive equations. However, the proof is not straightforward due to the presence of inertial terms. In fact, the ε-problems we arrive at can be loosely classified as hyperbolic systems with dissipation, and homogenization results available for elliptic and parabolic problems with monotone operators cannot be directly applied.

7.1 THE SLIGHTLY COMPRESSIBLE POLYMER: MICROSCALE PROBLEM

In this section we assume that the solid phase Ω_s^ε of a periodically perforated domain is a viscoelastic material obeying the Kelvin-Voight constitutive equations

$$\sigma^{s,\varepsilon} = Ae(\mathbf{u}^\varepsilon) + Be(\mathbf{v}^\varepsilon), \quad e(\mathbf{u}^\varepsilon)_{ij} := \frac{1}{2}\left(\frac{\partial u_i^\varepsilon}{\partial x_j} + \frac{\partial u_j^\varepsilon}{\partial x_i}\right), \tag{7.1.1}$$

where $\mathbf{v}^\varepsilon$ denotes velocity, $\mathbf{u}^\varepsilon = \int_0^t \mathbf{v}^\varepsilon(\cdot,\tau)d\tau$ is the displacement, and $\sigma^{s,\varepsilon}$ is the stress tensor.

The equations of motion are given by

$$\rho_s \partial_t \mathbf{v}^\varepsilon - \operatorname{div}(Ae(\mathbf{u}^\varepsilon) + Be(\mathbf{v}^\varepsilon)) = \rho_s \mathbf{F} \tag{7.1.2}$$

in $\Omega_s^\varepsilon \times [0,T]$. In the above, $\mathbf{F}$ is the mass density of the external force (e.g. gravity) which we assume to be a constant vector.

For the slightly compressible case occurring in acoustics, the pressure depends on the density as follows:

$$\nabla p^\varepsilon = -c^2 \rho_f \operatorname{div} \mathbf{u}^\varepsilon. \tag{7.1.3}$$

For the details on the derivation of this equation see [363].

This implies that in the fluid part Ω_f^ε we can write

$$\rho_f \partial_t \mathbf{v}^\varepsilon - \operatorname{div}(\sigma^{f,\varepsilon}) = \rho_f \mathbf{F} \tag{7.1.4}$$

in $\Omega_f^\varepsilon \times [0,T]$, where

$$\sigma^{f,\varepsilon} := c^2 \rho_f \operatorname{div} \mathbf{u}^\varepsilon I + 2\mu \eta_r (e(\mathbf{v}^\varepsilon)) e(\mathbf{v}^\varepsilon). \tag{7.1.5}$$

To model the blood-marrow mixture inside the pores, we use a non-Newtonian (shear dependent) fluid model described by the Carreau law, which takes into account that polymers show a finite nonzero constant Newtonian viscosity at very low shear rates [3].

$$\begin{aligned} \eta_r(e(\mathbf{v})) &:= (\eta_0 - \eta_\infty)\left(1 + \lambda |e(\mathbf{v})|^2\right)^{\frac{r}{2}-1} + \eta_\infty \\ 1 < r < 2, &\quad \eta_0 > \eta_\infty \geq 0, \quad \lambda > 0. \end{aligned} \tag{7.1.6}$$

The parameter η_∞ is typically very small and through the remainder of the chapter we set $\eta_\infty = 0$. It should be noted that for $\eta_\infty > 0$, the term containing η_∞ is a standard linear Navier-Stokes type constitutive equation. Because the first term in (7.1.6) is sub-linear, the second term would dominate, with the resulting theory being very similar to the linear case. In particular, available a priori estimates would be L^2-based, rather than L^r-based. Therefore, the case when $\eta_\infty = 0$ is more difficult and mathematically more interesting.

The transition conditions between fluid and solid parts are given by the continuity of displacement

$$[\mathbf{u}^\varepsilon] = 0 \text{ on } \Gamma_\varepsilon \times]0,T[, \tag{7.1.7}$$

where $[\cdot]$ indicates the jump across the boundary, and the continuity of the contact force

$$\sigma^{s,\varepsilon} \cdot \nu = \sigma^{f,\varepsilon} \cdot \nu \text{ on } \Gamma_\varepsilon \times [0,T]. \tag{7.1.8}$$

At the exterior boundary we impose the following boundary conditions

$$\mathbf{v}^\varepsilon = \mathbf{u}^\varepsilon = 0. \tag{7.1.9}$$

The initial condition is

$$\mathbf{v}^\varepsilon(0,x) \quad = \quad \mathbf{v}_0, \quad \text{in } \Omega. \tag{7.1.10}$$

This implies that the initial displacement $\mathbf{u}^\varepsilon(0,x)$ is also zero. This simply means that that initial configuration plays the role of the reference configuration, and this reference configuration is not stressed.

Let θ^ε denote the characteristic function of Ω_s^ε. Note that $\theta^\varepsilon(x) = \theta(\frac{x}{\varepsilon})$ and $\theta(y)$ is the characteristic function of the solid part of the unit cell (periodically extended to the whole space).

Define the composite density ρ^ε and stress as follows:

$$\rho^\varepsilon = \rho_s \theta^\varepsilon + \rho_f(1-\theta^\varepsilon) \tag{7.1.11}$$

$$\sigma^\varepsilon = \sigma^{s,\varepsilon}\theta^\varepsilon + \sigma^{f,\varepsilon}(1-\theta^\varepsilon) \tag{7.1.12}$$

We shall also use the notation

$$\sigma^\varepsilon = \mathbf{T}^\varepsilon(e(\mathbf{v}^\varepsilon)),$$

where

$$\begin{aligned}\mathbf{T}^\varepsilon(\phi)\left(x,\frac{x}{\varepsilon},t\right) &= \theta^\varepsilon A\left(\int_0^t \phi\right) + \theta^\varepsilon B\phi \\ &\quad +(1-\theta^\varepsilon)c^2\rho_f \mathrm{tr}\left(\int_0^t \phi\right)I + (1-\theta^\varepsilon)2\mu\eta_r(\phi)\phi\end{aligned}$$

Also, define

$$\begin{aligned}\mathbf{T}^0(\phi)(x,y,t) &= \theta(y)A\left(\int_0^t \phi\right) + \theta(y)B\phi \\ &\quad +(1-\theta(y))c^2\rho_f \mathrm{tr}\left(\int_0^t \phi\right)I + (1-\theta(y))2\mu\eta_r(\phi)\phi\end{aligned}$$

for each smooth, symmetric matrix function $\phi(x,y,t)$. Here, $\mathrm{tr}\, M$ denotes trace of the matrix M.

Multiplying (11.1.4) and (11.1.6) by a smooth test function ϕ, equal to zero on $\partial\Omega$ and equal to zero for $t = T$, formally integrating by parts, and using boundary and interface conditions, we can write the weak formulation of the problem (7.1.1)-(7.1.10):

$$-\int_\Omega \rho^\varepsilon \mathbf{v}_0 \cdot \phi(x,0)dx - \int_0^T\int_\Omega \rho^\varepsilon \mathbf{v}^\varepsilon \cdot \partial_t\phi dxdt + \int_0^T\int_\Omega \sigma^\varepsilon : e(\phi)dxdt = \int_0^T\int_\Omega \rho^\varepsilon \mathbf{F}\cdot\phi, \tag{7.1.13}$$

for each $\phi \in C^\infty([0,T], C_0^\infty(\Omega)^n)$ such that $\phi(T,x) = 0$.

The spaces in which solutions are sought will be specified in the next section.

7.2 A PRIORI ESTIMATES

We assume that for each $\varepsilon > 0$, the problem (7.1.13) has a finite energy weak solution satisfying $\mathbf{v}^\varepsilon \in L^\infty(0,T;L^2(\Omega)^n)$, $\mathbf{u}^\varepsilon \in L^\infty(0,T;L^2(\Omega)^n)$, $e(\mathbf{v}^\varepsilon) \in L^r(0,T;L^r(\Omega)^n)$,

$e(\mathbf{u}^\varepsilon) \in L^\infty(0,T;L^r(\Omega)^n)$, where r is the constant in the power or Carreau law (7.1.6). Moreover, $e(\mathbf{v}^\varepsilon) \in L^2(0,T;L^2(\Omega_s)^n)$, $e(\mathbf{u}^\varepsilon) \in L^\infty(0,T;L^2(\Omega_s)^n)$. The above assumptions are reasonable. The following theorem shows that the above inclusions follow naturally from energy estimates. Therefore, the existence of the solution can be established using an approximation procedure, the energy estimates, and monotonicity of the principal part of the operator of the problem. In the next theorem we show that these weak solutions are uniformly bounded in ε.

Theorem 7.1

The finite energy weak solutions satisfy the following inequalities.

$$\| \mathbf{v}^\varepsilon \|_{L^\infty([0,T),L^2(\Omega)^n)} \leq C, \tag{7.2.1}$$

$$\| \mathbf{u}^\varepsilon \|_{L^\infty([0,T],L^2(\Omega)^n)} \leq C, \tag{7.2.2}$$

$$\|e(\mathbf{v}^\varepsilon)\|_{L^r(0,T;L^r(\Omega)^n)} \leq C, \tag{7.2.3}$$

$$\| e(\mathbf{v}^\varepsilon) \|_{L^2(0,T;L^2(\Omega_s)^n)} \leq C, \tag{7.2.4}$$

$$\|e(\mathbf{u}^\varepsilon)\|_{L^\infty(0,T;L^r(\Omega)^n)} \leq C, \tag{7.2.5}$$

$$\|e(\mathbf{u}^\varepsilon)\|_{L^\infty(0,T;L^2(\Omega_s^\varepsilon)^n)} \leq C, \tag{7.2.6}$$

$$\| \operatorname{div}\mathbf{u}^\varepsilon \|_{L^\infty(0,T;L^2(\Omega)^n)} \leq C, \tag{7.2.7}$$

$$\| \rho^\varepsilon \partial_t \mathbf{v}^\varepsilon \|_{L^r(0,T;W^{-1,r}(\Omega)^n))} \leq C, \tag{7.2.8}$$

$$\| \sigma^\varepsilon \|_{L^r(0,T;L^r(\Omega)^{n\times n}))} \leq C. \tag{7.2.9}$$

In the above, the number r is the power in the Carreau law (7.1.6), while C denotes a generic constant independent of ε. It should be noted that C may depend on T, Ω, ellipticity constants for the tensors A, B, viscosity of the fluid, parameters η_0, λ in the Carreau law, and also on ρ_s and ρ_f. ■

Note that $\mathbf{v}^\varepsilon$ alone is not a proper test function since $\mathbf{v}^\varepsilon(T)$ is not necessarily zero. For this reason we first introduce auxiliary functions ξ_τ for $\tau \in (0,T)$, and then use $\mathbf{v}^\varepsilon \xi_\tau$ as test functions.

For each fixed $\tau \in (0,T)$, let $\xi_\tau \in C^\infty([0,T])$ be such that

(i) $d_t \xi_\tau \leq 0$,

(ii) $\xi_\tau(t) = 1$ for $t \in [0,\tau]$, and $\xi_\tau(T) = 0$.

Note that $0 \leq \xi_\tau \leq 1$. Now, to obtain a priori estimates, we insert a test function $\mathbf{v}^\varepsilon \xi_\tau$ into the weak formulation (7.1.13). This yields

$$-\int_0^T \int_\Omega \rho^\varepsilon |\mathbf{v}^\varepsilon|^2 d_t \xi_\tau dx dt - \frac{1}{2} \int_0^T \int_\Omega \partial_t(\rho^\varepsilon |\mathbf{v}^\varepsilon|^2) \xi_\tau dx dt \tag{7.2.10}$$

$$+\varepsilon\int_0^T\int_\Omega \sigma^\varepsilon : e(\mathbf{v}^\varepsilon)\xi_\tau dxdt = \int_0^T\int_\Omega \rho^\varepsilon|\mathbf{v}_0|^2dx + \int_0^T\int_\Omega \rho^\varepsilon\mathbf{F}\cdot\mathbf{v}^\varepsilon\xi_\tau dxdt.$$

Integrating by parts with respect to t in the terms corresponding to the elastic part of the stress, we find

$$\int_0^T\int_\Omega \left(\theta^\varepsilon Ae(\mathbf{u}^\varepsilon) + (1-\theta^\varepsilon)c^2\rho_f\mathrm{div}\mathbf{u}^\varepsilon I\right) : e(\mathbf{v}^\varepsilon)\xi_\tau dxdt = \tag{7.2.11}$$

$$\frac{1}{2}\int_0^T\int_\Omega \theta^\varepsilon Ae(\mathbf{u}^\varepsilon) : e(\mathbf{u}^\varepsilon) + (1-\theta^\varepsilon)c^2\rho_f(\mathrm{div}\mathbf{u}^\varepsilon)^2\xi_\tau dx\Bigg|_{t=0}^{t=T}$$

$$-\ \frac{1}{2}\int_0^T\int_\Omega (\theta^\varepsilon Ae(\mathbf{u}^\varepsilon) : e(\mathbf{u}^\varepsilon) + (1-\theta^\varepsilon)c^2\rho_f(\mathrm{div}\mathbf{u}^\varepsilon)^2 d_t\xi_\tau dxdt$$

The first term in the right-hand side vanishes since $\xi_\tau(T) = 0$ and also because $\mathbf{u}^\varepsilon(0,x) = 0$, together with its first-order derivatives. Hence,

$$\int_0^T\int_\Omega \sigma^\varepsilon : e(\mathbf{v}^\varepsilon)\xi_\tau dxdt = \tag{7.2.12}$$

$$-\ \frac{1}{2}\int_0^T\int_\Omega (\theta^\varepsilon Ae(\mathbf{u}^\varepsilon) : e(\mathbf{u}^\varepsilon) + (1-\theta^\varepsilon)c^2\rho_f(\mathrm{div}\mathbf{u}^\varepsilon)^2 d_t\xi_\tau dxdt$$

$$+\ \int_0^T\int_\Omega \left(\theta^\varepsilon\mathbf{B} + 2\mu(1-\theta^\varepsilon)(\eta_0(1+\lambda|e(\mathbf{v}^\varepsilon)|^2)^{r/2-1}\right)|e(\mathbf{v}^\varepsilon)|^2\xi_\tau dxdt.$$

Each term in the right-hand side is non-negative by the definition of constitutive equations and also because $d_t\xi_\tau \le 0$. Therefore, we can consider the stress term in (7.2.10) (this is the last term in the left-hand side) separately from the other terms.

Estimating inertial terms. Omitting the non-negative stress term in the left-hand side of (7.2.10) yields

$$-\ \int_0^T\int_\Omega \rho^\varepsilon|\mathbf{v}^\varepsilon|^2 d_t\xi_\tau dxdt - \frac{1}{2}\int_0^T\int_\Omega \partial_t(\rho^\varepsilon|\mathbf{v}^\varepsilon|^2)\xi_\tau dxdt$$

$$\le\ \int_0^T\int_\Omega \rho^\varepsilon|\mathbf{v}_0|^2dx + \int_0^T\int_\Omega \rho^\varepsilon\mathbf{F}\cdot\mathbf{v}^\varepsilon\xi_\tau dxdt.$$

Integrating by parts with respect to time in the second term in the right-hand side, we find

$$-\frac{1}{2}\int_0^T\int_\Omega \rho^\varepsilon |\mathbf{v}^\varepsilon|^2 d_t\xi_\tau dxdt \leq \frac{1}{2}\int_0^T\int_\Omega \rho^\varepsilon |\mathbf{v}_0|^2 dx + \int_0^T\int_\Omega \rho^\varepsilon \mathbf{F}\cdot\mathbf{v}^\varepsilon \xi_\tau dxdt. \tag{7.2.13}$$

Now consider a sequence of functions $\xi_{\tau,n}$ such that, as $n \to \infty$,

$$\begin{aligned} &\xi_{\tau,n} \to \chi_{[0,\tau]} \text{ pointwise}, \\ &d_t\xi_{\tau,n} \to -\delta(t-\tau) \text{ in } \mathscr{D}'(0,T), \\ &\| d_t\xi_{\tau,n} \|_{L^1(0,T)} \leq 1 \end{aligned} \tag{7.2.14}$$

Here, $\chi_{[0,\tau]}$ is the characteristic function of $[0,\tau]$. To construct such a sequence take a compactly supported, smooth, even function $\kappa(t)$ such that $\kappa(t) \geq 0$, $\int \kappa dt = 1$, and the support of κ is the interval $[-a,a], a > 0$. Then set

$$\kappa_{\tau,n}(t) = n\kappa((t-\tau-a)n),$$

and define

$$\xi_{\tau,n} = 1 - \int_0^t \kappa_{\tau,n}(s)ds. \tag{7.2.15}$$

Inserting these functions into (7.2.13) and passing to the limit $n \to \infty$ we obtain

$$\left(\frac{1}{2}\int_0^T\int_\Omega |\mathbf{v}^\varepsilon|^2 dx\right)(\tau) \leq C + C|\mathbf{F}|\int_0^\tau \int_0^T\int_\Omega |\mathbf{v}^\varepsilon| dxdt. \tag{7.2.16}$$

with C independent of ε. Now we use the Gronwall inequality to deduce

$$\int_0^\tau \int_0^T\int_\Omega |\mathbf{v}^\varepsilon|^2 dxdt \leq \frac{1}{|\mathbf{F}|}(e^{2C|\mathbf{F}|\tau} - 1).$$

Since both sides are increasing functions of τ, we can differentiate both sides and obtain

$$\left(\int_0^T\int_\Omega |\mathbf{v}^\varepsilon|^2 dx\right)(\tau) \leq 2Ce^{2C|\mathbf{F}|T}. \tag{7.2.17}$$

This yields (7.2.1).

Next, write

$$|\mathbf{u}^\varepsilon|^2 = t^2\left|\frac{1}{t}\int_0^t \mathbf{v}^\varepsilon(\tau,\cdot)d\tau\right|^2 \leq t\int_0^t |\mathbf{v}^\varepsilon(\tau,\cdot)|^2 d\tau$$

where we used Jensen's inequality. Integrating both sides over Ω we obtain (7.2.2).

Gradient estimates. Integrating by parts in the inertial terms as in the derivation of (7.2.13) and noting that $d_t\xi_\tau \le 0$, we deduce

$$\int_0^T\int_\Omega \sigma^\varepsilon : e(\mathbf{v}^\varepsilon)\xi_\tau dxdt \le \frac{1}{2}\int_0^T\int_\Omega \rho^\varepsilon|\mathbf{v}_0|^2 dx + \int_0^T\int_\Omega \rho^\varepsilon \mathbf{F}\cdot\mathbf{v}^\varepsilon\xi_\tau dxdt. \tag{7.2.18}$$

Therefore, using the ellipticity of the solid viscosity tensor $\mathbf{B}$, we deduce

$$c\int_0^T\int_\Omega \left(\theta^\varepsilon\mathbf{B} + 2\mu(1-\theta^\varepsilon)(\eta_0(1+\lambda|e(\mathbf{v}^\varepsilon)|^2)^{r/2-1}\right)|e(\mathbf{v}^\varepsilon)|^2\xi_\tau dxdt \tag{7.2.19}$$
$$\le \frac{1}{2}\int_0^T\int_\Omega \rho^\varepsilon|\mathbf{v}_0|^2 dx + \int_0^T\int_\Omega \rho^\varepsilon\mathbf{F}\cdot\mathbf{v}^\varepsilon\xi_\tau,$$

and observing that

$$(1+\lambda|e(\mathbf{v}^\varepsilon)|^2)^{r/2-1} \le 1,$$

we obtain

$$\int_0^T\int_\Omega (1+\lambda|e(\mathbf{v}^\varepsilon)|^2)^{r/2-1}|e(\mathbf{v}^\varepsilon)|^2\xi_\tau dxdt$$
$$\le c\left(\frac{1}{2}\int_0^T\int_\Omega \rho^\varepsilon|\mathbf{v}_0|^2 dx + \int_0^T\int_\Omega \rho^\varepsilon\mathbf{F}\cdot\mathbf{v}^\varepsilon\xi_\tau dxdt\right)$$

with c depending only on μ, η_0, and the ellipticity constant for $\mathbf{B}$. The right-hand side is dominated by

$$C\left(\|\mathbf{v}_0\|^2_{L^2(\Omega)^n} + \int_0^T\int_\Omega |\mathbf{v}^\varepsilon|dxdt\right) \le C_1 + C_1(|\Omega|T)^{1/2}\left(\int_0^T\int_\Omega |\mathbf{v}^\varepsilon|^2 dxdt\right)^{1/2}.$$

Combining the last two inequalities and using (7.2.1) in the right-hand side, we find

$$\int_0^T\int_\Omega (1+\lambda|e(\mathbf{v}^\varepsilon)|^2)^{r/2-1}|e(\mathbf{v}^\varepsilon)|^2\xi_\tau dxdt \le C$$

with C independent of ε. Since the constant in the right-hand side is independent of τ, we can consider a sequence of functions $\xi_{\tau,j}$ that converges to 1 pointwise on $(0,T)$, as $j\to\infty$. Passing to the limit yields

$$\int (1+\lambda|e(\mathbf{v}^\varepsilon)|^2)^{r/2-1}|e(\mathbf{v}^\varepsilon)|^2 dxdt \le C \tag{7.2.20}$$

Since $1 < r < 2$, we have that $0 < 1 - \frac{r}{2} < 1$, and using a numerical inequality $(a+b)^p \leq a^p + b^p$ where $a, b \geq 0$ and $p \in (0,1]$, we obtain

$$(1+\lambda|e(\mathbf{v}^\varepsilon)|^2)^{\frac{r}{2}-1} > \frac{1}{1+\lambda^{\frac{2-r}{r}}|e(\mathbf{v}^\varepsilon)|^{2-r}}$$

This implies that

$$\int_0^T \int_\Omega \mu\eta_0(1+\lambda|e(\mathbf{v}^\varepsilon)|^2)^{\frac{r}{2}-1}|e(\mathbf{v}^\varepsilon)|^2\,dxdt \geq \int_0^T \int_\Omega \mu\eta_0 \frac{|e(\mathbf{v}^\varepsilon)|^2}{1+\lambda^{\frac{2-r}{r}}|e(\mathbf{v}^\varepsilon)|^{2-r}}dxdt \tag{7.2.21}$$

We now apply Jensen's inequality to the following convex function:

$$J(x) = \frac{x^{\frac{2}{r}}}{1+(\lambda x)^{\frac{2-r}{r}}}$$

with $x = |e(\mathbf{v}^\varepsilon)|^r$.

Let $S_T = \Omega \times (0,T)$, and let $|S_T|$ denote the Lebesgue measure of S_T. Now, Jensen's inequality

$$\frac{1}{|S_T|}\int_0^T \int_\Omega J(|e(\mathbf{v}^\varepsilon)|^r)\,dxdt \geq J\left(\frac{1}{|S_T|}\int_0^T \int_\Omega |e(\mathbf{v}^\varepsilon)|^r\,dxdt\right)$$

yields

$$\begin{aligned}\frac{1}{|S_T|}\int_0^T \int_\Omega \frac{|e(\mathbf{v}^\varepsilon)|^2}{1+\lambda^{\frac{2-r}{r}}|e(\mathbf{v}^\varepsilon)|^{2-r}} &\geq \frac{1}{|S_T|^{\frac{2}{r}}}\cdot\frac{\left(\int_0^T\int_\Omega |e(\mathbf{v}^\varepsilon)|^r\right)^{\frac{2}{r}}}{1+\lambda^{\frac{2-r}{r}}|S_T|^{\frac{r-2}{r}}\left(\int_0^T\int_\Omega |e(\mathbf{v}^\varepsilon)|^r\right)^{\frac{2-r}{r}}}\\ &= \frac{1}{|S_T|^{\frac{2}{r}}}\cdot\frac{\|e(\mathbf{v}^\varepsilon)\|^2_{L^r(S_T)}}{1+\lambda^{\frac{2-r}{r}}|S_T|^{\frac{r-2}{r}}\|e(\mathbf{v}^\varepsilon)\|^{2-r}_{L^r(S_T)}}\end{aligned}$$

Hence,

$$\int_0^T \int_\Omega \frac{|e(\mathbf{v}^\varepsilon)|^2}{1+\lambda^{\frac{2-r}{r}}|e(\mathbf{v}^\varepsilon)|^{2-r}} \geq |S_T|^{\frac{r-2}{r}}\cdot\frac{\|e(\mathbf{v}^\varepsilon)\|^2_{L^r(S_T)}}{1+\lambda^{\frac{2-r}{r}}|S_T|^{\frac{r-2}{r}}\|e(\mathbf{v}^\varepsilon)\|^{2-r}_{L^r(S_T)}}$$

And,

$$\int_0^T \int_\Omega \mu\eta_0(1+\lambda|e(\mathbf{v}^\varepsilon)|^2)^{\frac{r}{2}-1}|e(\mathbf{v}^\varepsilon)|^2 dxd\tau \geq C\frac{\|e(\mathbf{v}^\varepsilon)\|^2_{L^r(S_T)}}{1+\lambda^{\frac{2-r}{r}}|S_T|^{\frac{r-2}{r}}\|e(\mathbf{v}^\varepsilon)\|^{2-r}_{L^r(S_T)}}$$

Combining (7.2.20) with (7.2.21) results in

$$\|e(\mathbf{v}^\varepsilon)\|^2_{L^r(S_T)} \le C(1+\lambda^{\frac{2-r}{r}}|S_T|^{\frac{r-2}{r}}\|e(\mathbf{v}^\varepsilon)\|^{2-r}_{L^r(S_T)}) \tag{7.2.22}$$

For $1<r<2$, using Young's inequality

$$ab \le \frac{a^p}{p}+\frac{b^q}{q}, \quad \frac{1}{p}+\frac{1}{q}=1$$

with $p=\frac{2}{2-r}, q=\frac{2}{r}$, implies that

$$\|e(\mathbf{v}^\varepsilon)\|^2_{L^r(0,T;L^r(\Omega)^n)} \le C\left(1+\delta^{\frac{2}{2-r}}\|e(\mathbf{v}^\varepsilon)\|^2_{L^r(0,T;L^r(\Omega)^n)}+\left(\frac{\lambda}{|S_T|}\right)^{\frac{2(2-r)}{r^2}}(\frac{1}{\delta})^{2/r}\right)$$

Setting $\delta=\left(\frac{1}{2C}\right)^{\frac{2-r}{2}}$ and rearranging yields (7.2.3).

Next, return to (7.2.19) and note that L^2-estimates can be obtained for the restriction of $e(\mathbf{v}^\varepsilon)$ to the solid domain Ω^ε_s. Estimating the right-hand side of (7.2.19) as above and noting that the contribution of the fluid domain is non-negative yields (7.2.4).

From (7.2.3) and (7.2.4) we can obtain estimates on the gradient of displacement. First, we use Jensen's inequality to write

$$\begin{aligned} |e(\mathbf{u}^\varepsilon)|^r(t,\cdot) &= t^r\left|\frac{1}{t}\int_0^t e(\mathbf{v}^\varepsilon)(\tau,\cdot)d\tau\right|^r \\ &\le t^{r-1}\int_0^t |e(\mathbf{v}^\varepsilon)|^r(\tau,\cdot)d\tau \\ &\le T^{r-1}\int_0^T |e(\mathbf{v}^\varepsilon)|^r(\tau,\cdot)d\tau \end{aligned}$$

for almost all $t\in(0,T)$. Integrating both sides over Ω yields

$$\|e(\mathbf{u}^\varepsilon)\|^r_{L^\infty(0,T;L^r(\Omega))} = \left(\int_0^T\int_\Omega |e(\mathbf{u}^\varepsilon)|^r dx\right)(t) \le T^{r-1}\int_0^T\int_\Omega |e(\mathbf{v}^\varepsilon)|^r dxdt.$$

The right-hand side is bounded by C independent of ε because of (7.2.3), and we obtain (7.2.5). Similarly, (7.2.4) yields (7.2.6). This estimate provides a higher spatial integrability index ($2>r$), but only on the solid subdomain.

Next, returning to (7.2.18), we consider the elastic part of the stress and write

$$\begin{aligned} &-\frac{1}{2}\int(\theta^\varepsilon Ae(\mathbf{u}^\varepsilon):e(\mathbf{u}^\varepsilon)+(1-\theta^\varepsilon)c^2\rho_f(\mathrm{div}\mathbf{u}^\varepsilon)^2)d_t\xi_\tau dxdt \quad (7.2.23)\\ &\le \frac{1}{2}\int_0^T\int_\Omega \rho^\varepsilon|\mathbf{v}_0|^2dx+\int_0^T\int_\Omega \rho^\varepsilon\mathbf{F}\cdot\mathbf{v}^\varepsilon\xi_\tau dxdt \\ &\le C \end{aligned}$$

with C independent of ε. Since the integrand is in $L^1(\Omega)$ for almost all t, we can consider a sequence $\xi_{\tau,n}$ as above and pass to the limit $n \to \infty$ to obtain (7.2.7) in addition to (7.2.6), with C independent of ε. Because of the degeneration in the elastic part of the stress, the L^2-estimate is available on the whole domain Ω only for the trace of $e(\mathbf{u}^\varepsilon)$ (equal to $\mathrm{div}\mathbf{u}^\varepsilon$), but not for off-diagonal components.

Bounds on the stress and time derivative. We begin by noting that

$$\begin{aligned}\sigma^\varepsilon &= \left(\theta^\varepsilon A e(\mathbf{u}^\varepsilon) + c^2\rho_f(1-\theta^\varepsilon)\mathrm{div}\mathbf{u}^\varepsilon I\right)\\ &+ \left(\theta^\varepsilon B e(\mathbf{v}^\varepsilon) + (1-\theta^\varepsilon)2\mu\eta_r(|e(\mathbf{v}^\varepsilon)|)e(\mathbf{v}^\varepsilon)\right)\\ &= \sigma^{1,\varepsilon}(e(\mathbf{u}^\varepsilon)) + \sigma^{2,\varepsilon}(e(\mathbf{v}^\varepsilon)),\end{aligned}$$

and

$$|\sigma^{1,\varepsilon}|^2 \le C_1|e(\mathbf{u}^\varepsilon)|^2 + C_2|\mathrm{div}\mathbf{v}^\varepsilon|^2,$$

with C_1, C_2 independent of ε. Next, since $\eta_r(|e(\mathbf{v}^\varepsilon)|)^r \le 1$ we have

$$|\sigma^{2,\varepsilon}|^r \le C_3|e(\mathbf{v}^\varepsilon)|^r,$$

with C_3 independent of ε. Integrating the first inequality over Ω, the second over $(0,T)\times\Omega$, and using already obtained bounds on the gradients of $\mathbf{u}^\varepsilon, \mathbf{v}^\varepsilon$ we deduce

$$\| \sigma^{1,\varepsilon} \|_{L^\infty(0,T;L^2(\Omega)^{n\times n}))} \le C \tag{7.2.24}$$

$$\| \sigma^{2,\varepsilon} \|_{L^r(0,T;L^r(\Omega)^{n\times n}))} \le C. \tag{7.2.25}$$

Since $r \le 2$, this implies (7.2.9). Therefore,

$$|\langle \mathrm{div}\sigma^\varepsilon, \phi\rangle| \le \int_0^T\int_\Omega |\sigma^\varepsilon||\nabla\phi|dxdt \le \| \sigma^\varepsilon \|_{L^r(0,T;L^r(\Omega)^n)} \| \nabla\phi \|_{L^{r'}(0,T;L^{r'}(\Omega)^n))}, \tag{7.2.26}$$

where $1/r' + 1/r = 1$. In view of zero boundary trace of ϕ, application of Poincare inequality yields

$$|\langle \mathrm{div}\sigma^\varepsilon, \phi\rangle| \le C \| \phi \|_{L^{r'}(0,T;W^{1,r'}(\Omega)^n))}.$$

Thus

$$\| \mathrm{div}\sigma^\varepsilon \|_{L^r(0,T;W^{-1,r}(\Omega)^n))} \le C \tag{7.2.27}$$

Next, recall that

$$\partial_t(\rho^\varepsilon\mathbf{v}^\varepsilon) = \rho^\varepsilon\partial_t\mathbf{v}^\varepsilon = \mathrm{div}\sigma^\varepsilon + \rho^\varepsilon\mathbf{F},$$

where $\rho^\varepsilon \in L^\infty((0,T)\times\Omega)$ and $\mathbf{F}$ is constant. Therefore, using (7.2.27) to estimate the right-hand side, we deduce (7.2.8).

7.3 TWO-SCALE SYSTEM

In order to prove the main convergence results of this chapter, we use the notion of *two-scale convergence*. We refer the reader to Chapter 1 for a review of some basic results related to the two-scale convergence with time dependence. Using a priori estimates from Theorem 7.1 and sequential compactness of two-scale convergence, we obtain for almost all $t \in [0,T]$ existence of subsequences $\{\mathbf{v}^\varepsilon\}$, $\{\sigma^\varepsilon\}$ (not relabeled), and functions $\overline{\mathbf{v}}(t,x), \mathbf{w}(t,x,y), \sigma^0(t,x,y)$ such that

$$\mathbf{v}^\varepsilon \to \overline{\mathbf{v}}(t,x) \text{ in the two-scale sense} \tag{7.3.1}$$

$$e(\mathbf{v}^\varepsilon) \to e_x(\overline{\mathbf{v}})(t,x) + e_y(\mathbf{w})(t,x,y) \text{ in the two-scale sense} \tag{7.3.2}$$

$$\sigma^\varepsilon \to \sigma^0 \text{ in the two-scale sense} \tag{7.3.3}$$

where

$$\overline{\mathbf{v}} \in L^r([0,T]; W^{1,r}(\Omega)^n))$$
$$\mathbf{w} \in L^r([0,T]; L^r(\Omega, W^{1,r}_{\text{per}}(Y)/\mathbb{R})^n)$$

and

$$\sigma^0 \in L^r([0,T]; L^r(\Omega \times Y/\mathbb{R})^{n\times n}).$$

Next, we pass to the limit as $\varepsilon \to 0$ in (7.1.13) using a generic test function $\phi(x,t)$. Let

$$\overline{\rho} = \int_Y (\rho_s \theta(y) + \rho_f (1 - \theta(y)))dy. \tag{7.3.4}$$

Noting the well-known fact that $\rho^\varepsilon \to \overline{\rho}$ weak-$\star$ in $L^\infty(\Omega)$ allows us to pass to the limit in the first term in the left-hand side of (7.1.13), and also in the right-hand side. To pass to the limit in the second term in the left-hand side, we use Lemma 9.2), which yields $\rho^\varepsilon \mathbf{v}^\varepsilon \to \overline{\rho \mathbf{v}}$ in the sense of distributions on $(0,T)\times\Omega$. In the term containing σ^ε we use another well-known fact ([315],[8],[412]) that two-scale convergence $\sigma^\varepsilon \to \sigma^0$ implies weak convergence of σ^ε to $\int_Y \sigma^0(x,y,t)dy$. Combining the results we obtain

$$-\int_\Omega \overline{\rho}\mathbf{v}_0 \cdot \phi(x,0)dx - \int_0^T \int_\Omega \overline{\rho \mathbf{v}} \cdot \partial_t \phi dxdt + \int_0^T \int_\Omega \left(\int_Y \sigma^0(x,y,t)dy\right) : e(\phi)dxdt$$
$$= \int_0^T \int_\Omega \overline{\rho}\mathbf{F}\cdot\phi, \tag{7.3.5}$$

which is a weak formulation of

$$\bar{\rho}\partial_t \overline{\mathbf{v}} - \operatorname{div}_x \int_Y \sigma^0(x,y,t)dy = \bar{\rho}\mathbf{F}, \tag{7.3.6}$$

with the initial condition $\overline{\mathbf{v}}(x,0) = \mathbf{v}_0$. Next we use a test function $\phi^\varepsilon(x,t) = \varepsilon\phi(x, \frac{x}{\varepsilon}, t)$ in the (7.1.13) and pass to the limit. The result is

$$0 = \int_Y \sigma^0 : e(\phi(y,t))dy = -\int_Y \operatorname{div}_y \sigma^0 \cdot \phi(y,t)dy, \ \forall \phi(y,t), \tag{7.3.7}$$

which is a weak formulation of the equation

$$\mathrm{div}_y \sigma^0(x,y,t) = 0. \tag{7.3.8}$$

To obtain (7.3.7), we note that, thanks to the bounds on ρ^ε and $\mathbf{v}^\varepsilon$ (see Theorem 7.1),

$$\left(\int_\Omega \rho^\varepsilon \mathbf{v}_0 \cdot \varepsilon \phi^\varepsilon dx \right)(0) \ \to \ 0,$$

$$\int_0^T \int_\Omega \rho^\varepsilon \mathbf{v}^\varepsilon \cdot \varepsilon \partial_t \phi^\varepsilon dxdt \ \to \ 0$$

as well as

$$\int_0^T \int_\Omega \rho^\varepsilon \mathbf{F} \cdot \varepsilon \phi^\varepsilon dxdt \ \to \ 0$$

Noting that $\varepsilon e(\phi^\varepsilon) = e_y(\phi^\varepsilon) + \varepsilon e_x(\phi^\varepsilon)$ and using bounds on σ^ε from Theorem 7.1, after passing to the limit in (7.1.13) we obtain

$$\begin{aligned} &\lim_{\varepsilon \to 0} \int_0^T \int_\Omega \sigma^\varepsilon : \varepsilon e(\phi^\varepsilon) dxdt = \lim_{\varepsilon \to 0} \int_0^T \int_\Omega \sigma^\varepsilon : e_y(\phi) dxdt \\ &= \int_0^T \int_\Omega \int_Y \sigma^0(x,y,t) : e_y(\phi)(x,y,t) dydxdt = 0. \end{aligned}$$

Finally, if we choose a special form of the test function, namely $\phi = \phi_1(x,t)\phi_2(y,t)$, we get that $e_y(\phi) = \phi_1(x,t) e_y(\phi_2(y,t))$ which gives us

$$\int_0^T \int_\Omega \phi_1(x,t) \left(\int_Y \sigma^0 : e_y(\phi_2(y,t)) dy \right) dxdt = 0, \ \forall \phi_1$$

and (7.3.7) follows.

In the next section we use monotonicity arguments to derive the equations for the effective stress.

7.4 DESCRIPTION OF THE EFFECTIVE STRESS

In this section we obtain a closed-form effective system from equations (7.3.5) and (7.3.7). We specify an effective constitutive equation describing the dependence of the limiting stress tensor σ^0 on the gradients of $\overline{\mathbf{v}}$ and $\mathbf{w}$. First, we prove a lemma that establishes monotonicity properties of the operators associated with σ^ε.

Lemma 7.1

Let $\mathbf{T}^\varepsilon$ be as in (7.1.13) and $\xi_\tau(t)$ be a function defined in (7.2.10). Then, for each pair of symmetric matrices $\phi, \psi \in C^\infty(\Omega \times Y \times [0,T])^{n\times n}$, we have

$$\int_0^T \int_\Omega (\mathbf{T}^\varepsilon(\phi) - \mathbf{T}^\varepsilon(\psi)) : (\phi - \psi)\,\xi_\tau dxdt \geq 0 \tag{7.4.1}$$

■

Proof. Write $\mathbf{T}^\varepsilon = \mathbf{T}_1^\varepsilon + \mathbf{T}_2^\varepsilon$, where

$$\mathbf{T}_1^\varepsilon(\phi) = \theta^\varepsilon A\left(\int_0^t \phi\right) + \theta^\varepsilon B\phi + (1-\theta^\varepsilon)c^2\rho_f \mathrm{tr}\left(\int_0^t \phi\right) I,$$

and

$$\mathbf{T}_2^\varepsilon(\phi) = (1-\theta^\varepsilon)2\mu\eta_r(\phi)\phi. \tag{7.4.2}$$

With the above definitions, we can split the left-hand side of (7.4.1) in the following way

$$\begin{aligned}
&\int_0^T \int_\Omega (\mathbf{T}^\varepsilon(\phi) - \mathbf{T}^\varepsilon(\psi)) : (\phi - \psi)\,\xi_\tau dxdt \qquad (7.4.3)\\
&= \int_0^T \int_\Omega (\mathbf{T}_1^\varepsilon(\phi) - \mathbf{T}_1^\varepsilon(\psi)) : (\phi - \psi)\,\xi_\tau dxdt\\
&+ \int_0^T \int_\Omega (\mathbf{T}_2^\varepsilon(\phi) - \mathbf{T}_2^\varepsilon(\psi)) : (\phi - \psi)\,\xi_\tau dxdt.
\end{aligned}$$

Let us denote $\delta = \phi - \psi$. Since $\mathbf{T}_1^\varepsilon$ is a linear mapping,

$$\int_0^T \int_\Omega (\mathbf{T}_1^\varepsilon(\phi) - \mathbf{T}_1^\varepsilon(\psi)) : (\phi - \psi)\,\xi_\tau dxdt = \int_0^T \int_\Omega (\mathbf{T}_1^\varepsilon(\phi - \psi)) : (\phi - \psi)\,\xi_\tau dxdt.$$

Substituting expression for $\mathbf{T}_1^\varepsilon$, we get:

$$\int_0^T \int_\Omega \left[\theta^\varepsilon A\left(\int_0^t \delta\right) + \theta^\varepsilon B\delta + (1-\theta^\varepsilon)c^2\rho_f \mathrm{tr}\left(\int_0^t \delta\right) I\right] : (\delta)\xi_\tau dxdt$$

$$= \frac{1}{2}\int_0^T\int_\Omega \partial_t \left[\theta^\varepsilon A\left(\int_0^t \delta\right) : \left(\int_0^t \delta\right)\xi_\tau\right] dxdt$$

$$+ \int_0^T\int_\Omega \theta^\varepsilon B\delta : \delta\xi_\tau dxdt + \frac{1}{2}\int_0^T\int_\Omega \partial_t \left[(1-\theta^\varepsilon)c^2\rho_f \mathrm{tr}\left(\int_0^t \delta\right)^2 \xi_\tau\right] dxdt$$

$$- \frac{1}{2}\int_0^T\int_\Omega \theta^\varepsilon A\left(\int_0^t \delta\right) : \left(\int_0^t \delta\right) d_t\xi_\tau dxdt$$

$$- \frac{1}{2}\int_0^T\int_\Omega (1-\theta^\varepsilon)c^2\rho_f \mathrm{tr}\left(\int_0^t \delta\right)^2 d_t\xi_\tau dxdt$$

The terms containing ∂_t vanish, because at the lower limit $t = 0$ we have $\int_0^t \delta = 0$, and at the upper limit $t = T$ we have $\xi_\tau(T) = 0$. The remaining three terms in the right-hand side are all non-negative since A, B are positive definite, $\xi_\tau \geq 0$, and $d_t\xi_\tau \leq 0$. Thus

$$\int_0^T\int_\Omega (\mathbf{T}_1^\varepsilon(\phi) - \mathbf{T}_1^\varepsilon(\psi)) : (\phi - \psi)\,\xi_\tau dxdt \geq 0,$$

i.e. the first term in the right-hand side of (7.4.3) is non-negative.

Next consider the other term in the right-hand side of (7.4.3). Recall equation (7.1.6) that defines η_r. Using this equation we can write

$$\eta_r(|\phi|)\phi = \nabla_\phi F(|\phi|),$$

for each smooth symmetric matrix function ϕ. Here

$$F(s) = \frac{1}{r}(\eta_0 - \eta_\infty)(1+\lambda s^2)^{r/2} + \frac{1}{2}\eta_\infty s^2.$$

We claim that $F(|\phi|)$ is a convex function of ϕ. First, observe that $F(s) : \mathbf{R}^+ \to \mathbf{R}^+$ is an increasing and convex function. This can be checked directly by differentiating. Therefore, using the triangle inequality and the fact that F increases, we have that

$$F(|\kappa\phi_1 + (1-\kappa)\phi_2|) \leq F(\kappa|\phi_1| + (1-\kappa)|\phi_2|),$$

for each $\kappa \in [0,1]$. On the other hand, by convexity of F, for each $\kappa \in [0,1]$,

$$F(\kappa|\phi_1| + (1-\kappa)|\phi_2|) \leq \kappa F(|\phi_1|) + (1-\kappa)F(|\phi_2|).$$

And we can conclude that $F(|\phi|)$ is a convex function of ϕ.

Next, to derive (7.4.1), we note that

$$(\mathbf{T}_2^\varepsilon(\phi) - \mathbf{T}_2^\varepsilon(\psi)) : (\phi - \psi) = \left(\nabla_\phi F(|\phi|) - \nabla_\psi F(|\psi|)\right) : (\phi - \psi) \geq 0$$

by convexity of $F(|\phi|)$. Since $\xi_\tau \geq 0$,

$$\int_0^T \int_\Omega (\mathbf{T}_2^\varepsilon(\phi) - \mathbf{T}_2^\varepsilon(\psi)) : (\phi - \psi)\, \xi_\tau dxdt \geq 0,$$

i.e. the second term in the right-hand side of (7.4.3) is non-negative. This completes the proof of (7.4.1). □

Theorem 7.2

The following holds:

$$\sigma^0 = \mathbf{T}^0(e_x(\overline{\mathbf{v}}) + e_y(\mathbf{w})).$$

■

Proof. Let $s > 0$ be a fixed number, and let $\phi \in \mathscr{D}(\Omega; C_{per}^\infty(Y))^{n\times n}$ be a symmetric matrix, and $\phi_1 \in \mathscr{D}(\Omega; C_{per}^\infty(Y))^n$. We define a test function

$$\mathbf{z}^\varepsilon\left(x, \frac{x}{\varepsilon}, t\right) = \left(e(\overline{\mathbf{v}}(x,t) + \varepsilon\phi_1(x, \frac{x}{\varepsilon})) + s\phi(x, \frac{x}{\varepsilon})\right) \xi_\tau(t) \tag{7.4.4}$$

where $\xi_\tau(t)$ is a function defined in Section 7.2.

Clearly, the above-defined $\mathbf{z}^\varepsilon$ two-scale converges to

$$\mathbf{z}^0 = (e_x(\overline{\mathbf{v}}) + e_y(\phi_1) + s\phi)\, \xi_\tau.$$

By Lemma 7.1

$$\int_0^T \int_\Omega (\mathbf{T}^\varepsilon(e(\mathbf{v}^\varepsilon)) - \mathbf{T}^\varepsilon(\mathbf{z}^\varepsilon)) : (e(\mathbf{v}^\varepsilon) - \mathbf{z}^\varepsilon)\xi_\tau dxdt \geq 0 \tag{7.4.5}$$

Write the left-hand side as a sum of four terms:

$$\int_0^T \int_\Omega (\mathbf{T}^\varepsilon(e(\mathbf{v}^\varepsilon)) - \mathbf{T}^\varepsilon(\mathbf{z}^\varepsilon)) : (e(\mathbf{v}^\varepsilon) - \mathbf{z}^\varepsilon)\xi_\tau dxdt = I_1 + I_2 + I_3 + I_4,$$

where

$$I_1 = \int_0^T \int_\Omega \mathbf{T}^\varepsilon(e(\mathbf{v}^\varepsilon)) : e(\mathbf{v}^\varepsilon)\xi_\tau dxdt, \qquad I_2 = -\int_0^T \int_\Omega \mathbf{T}^\varepsilon(e(\mathbf{v}^\varepsilon)) : \mathbf{z}^\varepsilon \xi_\tau dxdt,$$

$$I_3 = -\int_0^T \int_\Omega \mathbf{T}^\varepsilon(\mathbf{z}^\varepsilon) : e(\mathbf{v}^\varepsilon)\xi_\tau dxdt, \qquad I_4 = \int_0^T \int_\Omega \mathbf{T}^\varepsilon(\mathbf{z}^\varepsilon) : \mathbf{z}^\varepsilon \xi_\tau dxdt.$$

Since $I_1 = \int_0^T \int_\Omega \sigma^\varepsilon : e(\mathbf{v}^\varepsilon)\xi_\tau dxdt$, we can use (7.2.10), integrate by parts in t and rearrange the terms to obtain

$$I_1 = \int_0^T \int_\Omega \rho^\varepsilon \mathbf{F} \cdot \mathbf{v}^\varepsilon \xi_\tau dxdt + \frac{1}{2}\int_\Omega \rho^\varepsilon \mathbf{v}_0 \cdot \mathbf{v}_0 dx \tag{7.4.6}$$

$$+ \quad \frac{1}{2}\int_0^T \int_\Omega \rho^\varepsilon \mathbf{v}^\varepsilon \cdot \mathbf{v}^\varepsilon d_t \xi_\tau dx dt.$$

To pass to the limit $\varepsilon \to 0$ in (7.4.6), we use the following lemma ([274], Lemma 5.1).

Lemma 7.2

Let g^n, h^n converge weakly to g, h, respectively in $L^{p_1}(0,T;L^{p_2}(\Omega))$ and $L^{q_1}(0,T;L^{q_2}(\Omega))$, where $1 \le p_1, p_2 \le \infty$,

$$\frac{1}{p_1} + \frac{1}{q_1} = \frac{1}{p_2} + \frac{1}{q_2} = 1.$$

We assume in addition that $\partial_t g^n$ is bounded in $L^1(0,T;W^{-m,1}(\Omega))$ for some $m \ge 0$ independent of n and

$$\| h^n - h^n(\cdot + \xi, t) \|_{L^{q_1}(0,T;L^{q_2}(\Omega))} \to 0$$

as $|\xi| \to 0$, uniformly in n.

Then $g^n h^n$ converges to gh in the sense of distributions on $\Omega \times (0,T)$. ■

In the first term in the right-hand side of (7.4.6) we apply this lemma with $g^\varepsilon = \mathbf{F}\rho^\varepsilon$ and $h^\varepsilon = \mathbf{v}^\varepsilon$. This is legitimate since $\partial_t \rho^\varepsilon = 0$, $\rho^\varepsilon \to \overline{\rho}$ weak-$\star$ in $L^\infty(\Omega \times (0,T))$ and thus in all $L^p(\Omega \times (0,T))$ with $1 \le p < \infty$. Also, $\mathbf{v}^\varepsilon \to \overline{\mathbf{v}}$ in $L^\infty(0,T;L^2(\Omega))^n$ and has uniform in ε Sobolev regularity by Theorem 7.1 and first Korn inequality (recall that $\mathbf{v}^\varepsilon = 0$ on $\partial\Omega$). Application of lemma 9.2 yields

$$\rho^\varepsilon \mathbf{F} \cdot \mathbf{v}^\varepsilon \to \overline{\rho} \mathbf{F} \cdot \overline{\mathbf{v}} \tag{7.4.7}$$

in $\mathscr{D}'((0,T) \times \Omega)$ and also weakly in $L^p(0,T;L^2(\Omega))$ for each $p < \infty$.

The second term in the right-hand side of (7.4.6) trivially passes to the limit, and in the third term we apply lemma 9.2 with $g^\varepsilon = \rho^\varepsilon \mathbf{v}^\varepsilon$ and $h^\varepsilon = \mathbf{v}^\varepsilon$. The assumptions of the lemma are verified, because of the convergence just obtained for $\rho^\varepsilon \mathbf{v}^\varepsilon$ (recall that $\mathbf{F}$ is a constant vector that can be arbitrarily chosen). Also, the time derivative of $\rho^\varepsilon \mathbf{v}^\varepsilon$ satisfies the estimate (7.2.8), and thus is bounded in $L^1(0,T;W^{-1,1}(\Omega))$ independent of ε. The lemma yields

$$\rho^\varepsilon \mathbf{v}^\varepsilon \cdot \mathbf{v}^\varepsilon \to \overline{\rho}\, \overline{\mathbf{v}} \cdot \overline{\mathbf{v}} \tag{7.4.8}$$

in $\mathscr{D}'((0,T) \times \Omega)$ and also weakly in $L^p(0,T;L^2(\Omega))$ for each $p < \infty$. Collecting the results (7.4.7), (7.4.8) we obtain

$$\begin{aligned} \lim_{\varepsilon \to 0} I_1 &= \int_0^T \int_\Omega \overline{\rho} \mathbf{F} \cdot \overline{\mathbf{v}} \xi_\tau dx dt + \frac{1}{2}\int_\Omega \overline{\rho} \mathbf{v}_0 \cdot \mathbf{v}_0 dx \\ &+ \frac{1}{2}\int_0^T \int_\Omega \overline{\rho}\, \overline{\mathbf{v}} \cdot \overline{\mathbf{v}} d_t \xi_\tau dx dt. \end{aligned} \tag{7.4.9}$$

Since $\mathbf{T}^\varepsilon(\mathbf{v}^\varepsilon) \to \sigma^0$ in the two-scale sense, and $\mathbf{z}^\varepsilon$ strongly two-scale converges to $e_x(\overline{\mathbf{v}}) + e_y(\phi_1) + s\phi$, the integral of their product passes to the limit:

$$\lim_{\varepsilon\to 0} I_2 = -\int_0^T \int_\Omega \int_Y \sigma^0 : (e_x(\overline{\mathbf{v}}) + e_y(\phi_1) + s\phi)\xi_\tau dxdydt.$$

Next we observe that both $\mathbf{z}^\varepsilon$ and $\mathbf{T}^\varepsilon(\mathbf{z}^\varepsilon)$ are admissible test functions for two-scale convergence. The use of the discontinuous characteristic function $\theta(y)$ creates a bit of difficulty with checking this, but that difficulty can be dealt with using the same argument as in [8]. Therefore,

$$\lim_{\varepsilon\to 0} I_3 = -\int_0^T \int_\Omega \int_Y \mathbf{T}^0(e_x(\overline{\mathbf{v}}) + e_y(\phi_1) + s\phi) : (e_x(\overline{\mathbf{v}}) + e_y(\mathbf{w}))\xi_\tau dxdydt$$

$$\lim_{\varepsilon\to 0} I_4 = \int_0^T \int_\Omega \int_Y \mathbf{T}^0(e_x(\overline{\mathbf{v}}) + e_y(\phi_1) + s\phi) : (e_x(\overline{\mathbf{v}}) + e_y(\phi_1) + s\phi)\xi_\tau dxdydt.$$

Next, we consider a sequence of functions $\phi_{1,k}$ strongly converges to $\mathbf{w}(x,y)$ in the space $L^r(0,T;L^r(\Omega, L^r_{per}(Y)))$ as $k \to \infty$. Passing to the limit we observe that

$$\lim_{k\to\infty}\lim_{\varepsilon\to 0}(I_3 + I_4) = \int_0^T \int_\Omega \int_Y \mathbf{T}^0(e_x(\overline{\mathbf{v}}) + e_y(\mathbf{w}) + s\phi) : s\phi\xi_\tau dxdydt \tag{7.4.10}$$

and

$$\lim_{k\to\infty}\lim_{\varepsilon\to 0} I_2 = -\int_0^T \int_\Omega \int_Y \sigma^0 : (e_x(\overline{\mathbf{v}}) + e_y(\mathbf{w}) + s\phi)\xi_\tau dxdydt. \tag{7.4.11}$$

Now, combining (7.4.9)-(7.4.11) from (7.4.5) we obtain

$$T_1 + T_2 + T_3 \geq 0, \tag{7.4.12}$$

where

$$\begin{aligned} T_1 &= \int_0^T \int_\Omega \overline{\rho}\mathbf{F}\cdot\overline{\mathbf{v}}\xi_\tau dxdt + \frac{1}{2}\int_\Omega \overline{\rho}\mathbf{v}_0\cdot\mathbf{v}_0 dx \\ &+ \frac{1}{2}\int_0^T \int_\Omega \overline{\rho}\,\overline{\mathbf{v}}\cdot\overline{\mathbf{v}}d_t\xi_\tau dxdt - \int_0^T \int_\Omega \int_Y \sigma^0 : e_x(\overline{\mathbf{v}})\xi_\tau dxdydt, \end{aligned} \tag{7.4.13}$$

$$T_2 = -\int_0^T \int_\Omega \int_Y \sigma^0 : e_y(\mathbf{w})\xi_\tau dxdydt, \tag{7.4.14}$$

$$T_3 = \int_0^T \int_\Omega \int_Y \mathbf{T}^0(e_x(\overline{\mathbf{v}}) + e_y(\mathbf{w}) + s\phi) : s\phi\xi_\tau dxdydt. \tag{7.4.15}$$

After integrating the terms in T_1 by parts with respect to t and rearranging, we find

$$T_1 = \int_0^T \int_\Omega \overline{\rho}\mathbf{F}\cdot\overline{\mathbf{v}}\xi_\tau dxdt + \int_\Omega \overline{\rho}\mathbf{v}_0\cdot\mathbf{v}_0 dx \tag{7.4.16}$$

$$+ \quad \int_0^T \int_\Omega \overline{\rho}\, \overline{\mathbf{v}} \cdot \partial_t(\overline{\mathbf{v}}\xi_\tau)dxdt - \int_0^T \int_\Omega \int_Y \sigma^0 : e_x(\overline{\mathbf{v}})\xi_\tau dxdydt.$$

Comparison with (7.1.13) shows that this is the left-hand side in the weak formulation of the effective equation

$$-\overline{\rho}\partial_t\overline{\mathbf{v}} + \mathrm{div}_x\left(\int_Y \sigma^0 dy\right) - \overline{\rho}\mathbf{F} = 0$$

with a particular test function equal to $\overline{\mathbf{v}}\xi_\tau$, and the initial condition $\overline{\mathbf{v}}(\cdot,0) = \mathbf{v}_0$. Therefore, T_1 is zero. Similarly, T_2 is the left-hand side in the weak formulation of the corrector problem

$$\mathrm{div}_y\sigma^0 = 0$$

with the test function $\mathbf{w}$, and thus $T_2 = 0$ as well. This reduces (7.4.12) to

$$T_3 \geq 0.$$

After division by $s > 0$, and letting $s \to 0$, we obtain

$$\int_0^T \int_\Omega \int_Y \left[\mathbf{T}^0(e_x(\overline{\mathbf{v}}) + e_y(\mathbf{w})) - \sigma^0\right] : \phi(x,y)\xi_\tau dxdydt \geq 0 \tag{7.4.17}$$

for any test function $\phi(x,y)$ and any $\tau \in (0,T)$. This means that the left-hand side of (7.4.17) must be zero. Considering a sequence of functions $\xi_{\tau,k}$ converging to the characteristic function $\chi_{(0,\tau)}$ yields

$$\int_0^\tau \int_\Omega \int_Y \left[\mathbf{T}^0(e_x(\overline{\mathbf{v}}) + e_y(\mathbf{w})) - \sigma^0\right] : \phi(x,y)dxdy = 0$$

for almost all $\tau \in (0,T)$, and every test function ϕ. This yields

$$\sigma^0 = \mathbf{T}^0(e_x(\overline{\mathbf{v}}) + e_y(\mathbf{w}))$$

almost everywhere on $\Omega \times Y \times (0,T)$. □

7.5 EFFECTIVE EQUATIONS

We now have all the ingredients needed to specify the effective equations. They are of the two-velocity type, meaning that the equation for the effective velocity $\overline{\mathbf{v}}$ is coupled to the equation for the corrector velocity $\mathbf{w}$. The constitutive equation for the effective stress is provided by Theorem 7.2. Combining this theorem with (7.3.6), (7.3.8) we obtain

$$\overline{\rho}\partial_t\overline{\mathbf{v}} - \mathrm{div}\int_Y \mathbf{T}^0(e_x(\overline{\mathbf{v}}) + e_y(\mathbf{w}))dy = \overline{\rho}\mathbf{F}, \tag{7.5.1}$$

$$\mathrm{div}_y\left(\mathbf{T}^0(e_x(\overline{\mathbf{v}}) + e_y(\mathbf{w}))\right) = 0. \tag{7.5.2}$$

Recall that $\mathbf{T}^0$ is given by (7.1.13). The first equation governs the evolution of $\overline{\mathbf{v}}$, while the second is the corrector equation for $\mathbf{w}$. The initial conditions for (7.5.1) are $\overline{\mathbf{v}}(x,0) = \mathbf{v}_0$ and the natural condition $\overline{\mathbf{u}}(x,0) = 0$, where $\overline{\mathbf{u}} = \int_0^t \overline{\mathbf{v}}(\tau,\cdot)d\tau$ is the effective displacement. The corrector equation is also an evolution equation, but its type is different. This is better seen if we rewrite $\mathbf{T}^0(e_x(\overline{\mathbf{v}}) + e_y(\mathbf{w}))$ in terms of the corrector displacement

$$\mathbf{w}^d(x,y,t) = \int_0^t \mathbf{w}(,x,y,\tau)d\tau$$

and the effective displacement. Using (7.1.13) we obtain

$$\begin{aligned}
\mathbf{T}^0(e_x(\overline{\mathbf{v}}) + e_y(\mathbf{w})) &= \theta(y)A(e(\overline{\mathbf{u}}) + e_y(\mathbf{w}^d)) + \theta(y)B(e(\overline{\mathbf{v}}) + e_y(\mathbf{w}_t^d)) \\
&+ (1-\theta(y))c^2\rho_f(\mathrm{div}\overline{\mathbf{u}} + \mathrm{div}_y\mathbf{w}^d) \\
&+ (1-\theta(y))2\mu\eta_r(e(\overline{\mathbf{v}}) + e_y(\mathbf{w}_t^d))(e(\overline{\mathbf{v}}) + e_y(\mathbf{w}_t^d)),
\end{aligned} \tag{7.5.3}$$

where $\mathbf{w}_t^d$ denotes $\mathbf{w}$. Equation (7.5.2) is a first-order in time evolution equation for $\mathbf{w}^d$ with the natural initial condition $\mathbf{w}^d(x,y,0) = 0$. On the other hand, equation (7.5.1), considered as an equation for the effective displacement, is second-order in time.

8 Multiscale FEM for the Modeling of Cancellous Bone

The present chapter deals with the application of the multiscale finite element method (FEM) to the modeling of cancellous bone as an alternative to Biot's model, whereby the main intention is to decrease the extent of the necessary laboratory tests. The multiscale FEM is an up-to-date numerical method intensively used in many fields for modeling heterogeneous materials [219, 243]. At the beginning, the chapter gives a brief explanation of the multiscale concept (Section 8.1) and thereafter focuses on the modeling of the representative volume element (RVE) and on the calculation of the effective material parameters including an analysis of their change with respect to increasing porosity (Section 8.2). Section 8.3 concentrates on the macroscopic calculations, which is illustrated by the simulation of ultrasonic testing and a study of the attenuation dependency on material parameters and excitation frequency. An alternative, simplified RVE, along with a comprehensive comparison with the experimental results, is presented in Section 8.4, whereas the bone anisotropy is studied in Section 8.5. The model proposed in Section 8.6 is based on the fact that the reflection might have a significant contribution to the attenuation of the acoustic waves propagating through the cancellous bone. The numerical implementation of the mentioned effect is realized by the development of a new representative volume element that includes an infinitesimally thin transient layer on the contact surface of the bone and the marrow. This layer serves to model the amplitude transformation of the incident wave by the transition through media with different acoustic impedances and to take into account the energy loss due to the reflection. Section 8.7 deals with the application of the inverse homogenization method [242, 244] to the determination of geometrical properties of cancellous bone. The approach represents a combination of an extended version of the Levenberg-Marquardt method with the multiscale FEM. The extension of the method is concerned with the selection strategy for distinguishing the global minimum from the plethora of local minima. Figures in this chapter are printed with permission from Springer Nature and World Scientific Publishing. Originally, they have been published in works [219, 222, 241, 242].

8.1 CONCEPT OF THE MULTISCALE FEM

Multiscale FEM belongs to the group of homogenization methods which are applicable to statistically uniform materials [223, 243]. These materials possess a representative volume element (RVE) whose analysis yields effective material parameters, under the limitation that the ratio of the characteristic lengths of RVE and the simulated body has to tend to zero. For this reason, terms macro- and microscale are commonly used in this context. As both scales are analyzed simultaneously, the standard notation distinguishes between quantities related to the different scales by introducing an overbar symbol. Thus position vector, displacements vector, strain tensor, stress tensor, and potential are respectively denoted by

$$\bar{\mathbf{x}},\ \bar{\mathbf{u}},\ \bar{\varepsilon},\ \bar{\sigma},\ \bar{\psi}=\bar{\psi}(\bar{\varepsilon},\bar{\mathbf{x}}) \qquad \text{at the macrocontinuum,} \tag{8.1.1}$$

$$\mathbf{x},\ \mathbf{u},\ \varepsilon,\ \sigma,\ \psi=\psi(\varepsilon,\mathbf{x}) \qquad \text{at the microcontinuum.} \tag{8.1.2}$$

The method is based on the principle of volume averaging, leading to the definition of the macrostress tensor in the form

$$\bar{\sigma} = \frac{1}{V}\int_{\mathscr{B}} \sigma dV$$

where the integration is carried out over the RVE $\mathscr{B}$, with the volume V. The well-posedness of the problem on the microscale still requires the equality of macrowork with the volume average of microwork

$$\bar{\sigma} : \bar{\varepsilon} = \frac{1}{V}\int_{\mathscr{B}} \sigma : \varepsilon \, dV \tag{8.1.3}$$

which is known as the Hill-Mandel macrohomogeneity condition [191, 197]. Expression (8.1.3) is satisfied by three types of boundary conditions at the microlevel: static, kinematic, and periodic ones. It is also worth mentioning that within the scope of cancellous bone modeling and especially in approaches where the accent is put on a realistic representation of geometry [339], the consideration of a sample smaller than the present RVE is often needed. Such an analysis yields "apparent" instead of effective material parameters, and apart from the static, kinematic, and periodic boundary conditions, the mixed boundary condition guarantee that Hill's condition is fulfilled [185, 333]. As we assume in our model that without loss of generality the microstructure consists of periodically repeated RVEs, we will work in the following analysis exclusively with periodic boundary conditions. In this case the microdeformation is assumed to be dependent on the macrostrain tensor $\bar{\varepsilon}$ and on the microfluctuation $\tilde{\mathbf{u}}$

$$\mathbf{u} = \bar{\varepsilon}\mathbf{x} + \tilde{\mathbf{u}}. \tag{8.1.4}$$

Now microfluctuations have to be periodic and traction antiperiodic on the periodic boundary of the RVE

$$\begin{aligned} \tilde{\mathbf{u}}^{+} &= \tilde{\mathbf{u}}^{-}, \\ \mathbf{t}^{+} &= -\,\mathbf{t}^{-} \end{aligned}$$

and the additive decomposition is characteristic of the microstrain tensor

$$\varepsilon = \bar{\varepsilon} + \tilde{\varepsilon}, \qquad \tilde{\varepsilon} = \frac{1}{2}(\nabla\tilde{\mathbf{u}} + \tilde{\mathbf{u}}\nabla) = \nabla^{s}\tilde{\mathbf{u}}. \tag{8.1.5}$$

Such a decomposition permits us to split the problem of simulation of a heterogeneous body into two parts, each consisting of one boundary value problem (BVP).

The first BVP relates to the simulation of the homogenized macroscopic body and the second one to the analysis of the RVE. For solving these two BVPs, any standard method can be applied and FEM is chosen here for this purpose. Moreover, using the described theory and standard program FEAP [415], a new multiscale FE program is written. Its main difference in comparison with a standard FE program code is that calculations at the microlevel replace the missing effective constitutive law at the macroscale (Fig. 8.1.1).

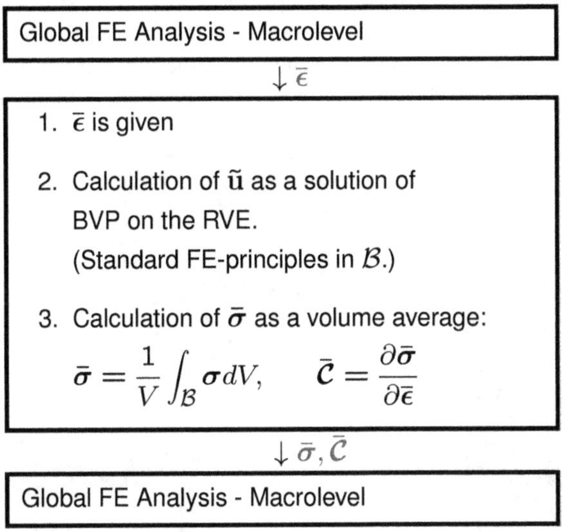

FIGURE 8.1.1 Connection of scales in the program code.

The simplified flow chart (Fig. 8.1.1) shows that macroscale calculations provide the macrostrain tensor $\bar{\varepsilon}$ which has to be understood as a priori given at the microlevel. Solution of the BVP at this level results in distributions of the microfluctuations $\tilde{\mathbf{u}}$ and of the microstresses σ whose volume average represents the sought counterpart on the macrolevel $\bar{\sigma}$. The calculations at this level also supply the elasticity tensor $\bar{\mathscr{C}}$. In contrast to the case of finite deformations, this tensor has to be calculated only once as it is independent from the deformation state.

8.2 MICROSCALE: THE RVE PROPOSAL AND EFFECTIVE PROPERTIES

8.2.1 MODELING OF THE RVE FOR CANCELLOUS BONE

Due to the complex geometry of cancellous bone, different types of the RVE can be proposed for its modeling. Moreover, it is well known that the bone structure changes dependent on many factors and that the deterioration of trabecular bone is characterized by a conversion from bone plates to bone rods. Consequently, the terms "rod-like" and "plate-like" are frequently used for a subjective classification of cancellous bone. For an objective quantification, a morphometric parameter called the structure model index (SMI) is introduced by [195]. This index shows the amount of plates and rods composing the structure. For the ideal plate and rod structures, the SMI values are 0 and 3 respectively. In our model, we treat the solid frame as a system of thin walls whose thickness and width can be varied. In limit cases, such an RVE can contain complete solid facets or thin columns in its edges. The RVE is assumed to have a cubic form with the side length a. The other parameters that are needed to completely determine the RVE, are the thickness of the wall d and width of the wall b. Figure 8.2.1 shows an example of the real microstructure of cancellous bone affected by osteoporosis, and the proposal for the corresponding RVE. In the part of the figure that shows the real structure of the bone, it can be seen that most of the solid walls are resorbed and only the edge columns remain. To calculate the

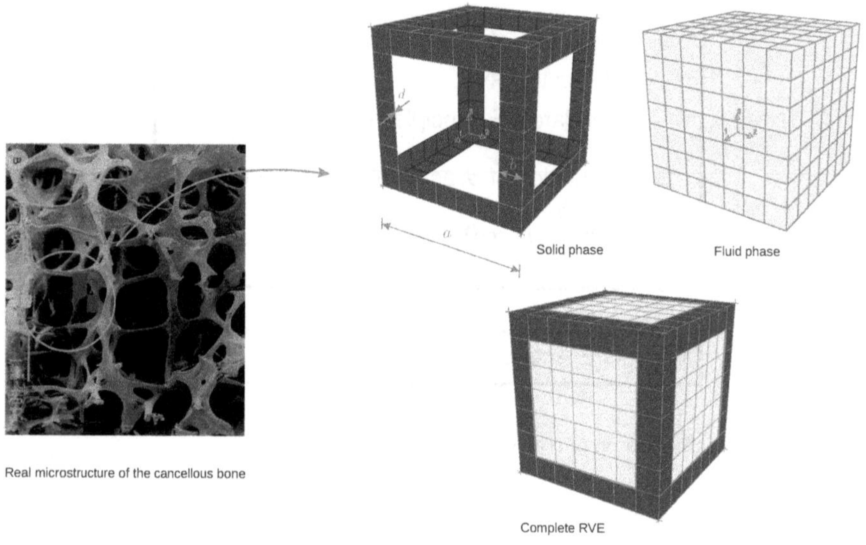

FIGURE 8.2.1 Real bone microstructure and corresponding RVE.

effective material parameters, a dynamic analysis of the proposed RVE is necessary, and periodic excitation is preferable because of its simplicity. In such a case, the load and induced deformations are harmonic functions in time

$$\mathbf{p}(\mathbf{x},t) = \mathbf{p}(\mathbf{x})e^{i\omega t}, \qquad \mathbf{u}(\mathbf{x},t) = \mathbf{u}(\mathbf{x})e^{i\omega t}, \tag{8.2.1}$$

where $\mathbf{p}$ and $\mathbf{u}$ may be complex-valued, ω is the frequency of excitation and i represents the imaginary unit. The application of (8.2.1) yields the simplified form of the equation of motion

$$\rho\ddot{\mathbf{u}}(\mathbf{x},t) - \nabla\cdot\sigma(\mathbf{x},t) = \rho\mathbf{b}(\mathbf{x},t) \qquad \rightarrow \qquad -\omega^2\rho\mathbf{u}(\mathbf{x}) - \nabla\cdot\sigma(\mathbf{x}) = \rho\mathbf{b}(\mathbf{x})$$

depending on density ρ and body forces $\mathbf{b}$. To describe the problem completely, the constitutive laws of the fluid and solid phases still need to be stated precisely. Regarding the bone material, a linear analysis is typical so that for the solid part a linear relation between stresses and strains (8.2.3) and for the marrow the constitutive law of barotropic fluids (8.2.5) are assumed. The state of deformations in the solid part Ω_s can now be described by the system

$$-\omega^2 \rho_s \mathbf{u} - \nabla \cdot \sigma_s = \rho_s \mathbf{b}(\mathbf{x}), \tag{8.2.2}$$

$$\sigma_s = \mathscr{C} : \varepsilon \tag{8.2.3}$$

and in the fluid part Ω_f, by:

$$-\omega^2 \rho_f \mathbf{u} - \nabla \cdot \sigma_f = \rho_f \mathbf{b}(\mathbf{x}), \tag{8.2.4}$$

$$\sigma_f = c^2 \rho_f \nabla \cdot \mathbf{u} \mathbf{I} + 2i\omega\eta\, \varepsilon + i\omega\xi\, \nabla \cdot \mathbf{u} \mathbf{I}. \tag{8.2.5}$$

Here $\mathscr{C}$ is the elasticity tensor of the solid phase, c the sound velocity of the marrow so that $c = \sqrt{K/\rho_f}$, K is the bulk modulus, and η and ξ are the viscosity coefficients. The indices s and f are used to distinguish the phases. Furthermore, the coupling condition requires that there is no deformation jump orthogonal to the interface of phases

$$u_s^{\perp} = u_f^{\perp} \quad \text{on} \quad \Gamma = \Omega_s \cap \Omega_f. \tag{8.2.6}$$

Recall that all the expressions (8.2.2)-(8.2.6) are defined in the complex domain. The material parameters of the solid phase, bulk modulus K_s and shear modulus μ_s are also complex-valued and they can be written in the form

$$K_s = K_s^R + iK_s^I, \qquad \mu_s = \mu_s^R + i\mu_s^I \tag{8.2.7}$$

where the imaginary parts are related to the real ones according to

$$K_s^I = \frac{\delta}{\pi} K_s^R, \qquad \mu_s^I = \frac{\delta}{\pi} \mu_s^R, \tag{8.2.8}$$

and δ denotes the logarithmic decrement amounting to 0.1 for underwater acoustics [48, 50, 117]. Remaining material parameters appearing in the constitutive laws defined above, are listed in Table 8.1.

8.2.2 MODELING OF THE SOLID PHASE

Due to the geometric properties, we are going to model the solid phase by using the shell elements [220]. But as the shell elements already implemented in the program FEAP [415] are not applicable for simulations in the complex domain, further adaption has been necessary. In particular we focused on the extension of an element convenient for simulation of flat and shallow shells. The formulation of this element

TABLE 8.1
Material parameters for cancellous bone according to Williams and Johnson [399].

Parameter	Symbol	Value	Unity
Pore fluid density	ρ_f	950	kg m^{-3}
Fluid bulk modulus	K_f	2.00×10^9	Pa
Pore fluid viscosity	η	1.5	Ns m^{-2}
Frame material density	ρ_s	1960	kg m^{-3}
Bulk modulus of the solid phase (real part)	K_s^R	2.04×10^{10}	Pa
Shear modulus of the solid phase (real part)	E_s^R	0.833×10^{10}	Pa
Young's modulus of the solid phase (real part)	μ_s^R	2.2×10^{10}	Pa
Poisson's ratio	ν	0.32	-

is based on the superposition of the linear theory for a plate loaded in its plane and a plate loaded by bending, whose potentials are

$$\Pi_p^e = \Pi_p^{e,int} + \Pi_p^{e,ext} = \frac{1}{2}\int_{\mathscr{A}^e} \mathbf{u}^T \cdot \mathbf{L}_p^T \cdot \mathscr{C}_p \cdot \mathbf{L}_p \cdot \mathbf{u}\, da + \Pi_p^{e,ext}, \tag{8.2.9}$$

$$\Pi_b^e = \Pi_b^{e,int} + \Pi_b^{e,ext} = \frac{1}{2}\int_{\mathscr{A}^e} \theta^T \cdot \mathbf{L}_b^T \cdot \mathscr{C}_b \cdot \mathbf{L}_b \cdot \theta\, da + \Pi_b^{e,ext}. \tag{8.2.10}$$

Hereafter, subscript p is taken to denote the plate loaded in its plane and subscript b for the bending of the plate. $\mathbf{L}_p$ and $\mathbf{L}_b$ are linear operators and $\mathscr{C}_p$ and $\mathscr{C}_b$ material tensors whose precise definition can be found in the literature [415]. The integration is carried out over the middle area of the element $\mathscr{A}^e$. As expression (8.2.9) depends on the displacements $\mathbf{u}_p = \{u\ v\}^T$ and (8.2.10) on the rotations $\theta = \{\theta_x\ \theta_y\}^T$, an FE approximation

$$\mathbf{u}^m = \mathbf{N}^m \cdot \hat{\mathbf{a}}^{me}, \tag{8.2.11}$$

$$\mathbf{u}^m = \{\ u \quad v \quad \theta_x \quad \theta_y\ \}^T, \qquad \hat{\mathbf{a}}_i^m = \{\ \hat{u}_i \quad \hat{v}_i \quad \hat{\theta}_{zi} \quad \hat{w}_i \quad \hat{\theta}_{xi} \quad \hat{\theta}_{yi}\ \}^T \tag{8.2.12}$$

leads after substitution into the first variation of (8.2.9) and (8.2.10) to the weak form of the problem

$$\delta\Pi^e = (\delta\hat{\mathbf{a}}^{me})^T \cdot \left(\int_{\mathscr{A}^e} (\mathbf{N}^m)^T \cdot \mathbf{L}^T \cdot \mathscr{C} \cdot \mathbf{L} \cdot \mathbf{N}^m\, da \right) \cdot \hat{\mathbf{a}}^{me} + \delta\Pi^{e,ext}, \tag{8.2.13}$$

$$= (\delta\hat{\mathbf{a}}^{me})^T \cdot \left(\int_{\mathscr{A}^e} (\mathbf{B}^m)^T \cdot \mathscr{C} \cdot \mathbf{B}^m\, da \right) \cdot \hat{\mathbf{a}}^{me} + \delta\Pi^{e,ext}. \tag{8.2.14}$$

In expressions (8.2.11) and (8.2.12), $\mathbf{N}^m$ represents the matrix of the shape functions, and superscript m indicates that a modified approximation is used ($\hat{\theta}_{zi}$ and $\hat{w}_i$ are used in the approximation of the vector $\mathbf{u}$). The symbol $\hat{\mathbf{a}}^{me}$ relates to the vector of the element DOFs and $\hat{\mathbf{a}}_i^{\mathrm{m}}$ to the vector of the nodal DOFs. The matrices appearing in (8.2.14) are composed as follows

$$\mathbf{L} = \begin{bmatrix} \mathbf{L}_{\mathrm{p}} & 0 \\ 0 & \mathbf{L}_{\mathrm{b}} \end{bmatrix}, \qquad \mathbf{B}^{\mathrm{m}} = \mathbf{L}\cdot\mathbf{N}^{\mathrm{m}}, \qquad \mathscr{C} = \begin{bmatrix} \mathscr{C}_{\mathrm{p}} & 0 \\ 0 & \mathscr{C}_{\mathrm{b}} \end{bmatrix} \tag{8.2.15}$$

Note that the described element [381, 415] is also adapted to structures with small curvature by introducing correction factors in which case the formulation, in contrast to (8.2.15), becomes coupled. The weak form of the problem (8.2.14) finally results in the discretized equilibrium equation of an element dependent on the stiffness matrix $\mathbf{K}^{\mathrm{e}}$ and the vector of nodal forces $\mathbf{f}^{\mathrm{e}}$

$$\mathbf{K}^{\mathrm{e}}\cdot\hat{\mathbf{a}}^{\mathrm{me}} = \mathbf{f}^{\mathrm{e}}. \tag{8.2.16}$$

Furthermore, the application of the element in the dynamic case requires the minimization of the Largrangian $L = E_k - \Pi$

$$\delta\int_t L dt = \delta\int_t (E_k - \Pi)\, dt = 0 \tag{8.2.17}$$

where integration of the time t is considered, Π is the total potential including (8.2.9) and (8.2.10), and E_k represents the kinetic energy

$$E_{\mathrm{k}} = \frac{1}{2}\int_{\Omega} \rho\, \dot{\mathbf{u}}^T\cdot\dot{\mathbf{u}}\, dv, \tag{8.2.18}$$

with the first variation in the form

$$\delta\int_t E_{\mathrm{k}}(\dot{\mathbf{u}})\, dt = -\int_t\int_{\Omega} \rho\, \delta\mathbf{u}^T\cdot\ddot{\mathbf{u}}\, dv\, dt = \omega^2\int_t\int_{\Omega} \rho\, \delta\mathbf{u}^T\cdot\mathbf{u}\, dv\, dt. \tag{8.2.19}$$

The kinetic energy is now approximated by using the same set of DOFs as in Eqs. (8.2.11) and (8.2.12), however, here, the contribution of rotations is neglected at the global level

$$\mathbf{u}_{\mathrm{k}} = \mathbf{N}_{\mathrm{k}}\cdot\hat{\mathbf{a}}^{\mathrm{me}}, \qquad \mathbf{u}_{\mathrm{k}} = \left\{ u \quad v \quad w \right\}^{\mathrm{T}}, \qquad \hat{\mathbf{a}}_i^{\mathrm{m}} = \left\{ \hat{u}_i \quad \hat{v}_i \quad \hat{\theta}_{zi} \quad \hat{w}_i \quad \hat{\theta}_{xi} \quad \hat{\theta}_{yi} \right\}^{\mathrm{T}}. \tag{8.2.20}$$

Substitution of (8.2.20) in (8.2.19) and using (8.2.18) now leads to the discretized equation of motion including the mass matrix $\mathbf{M}^{\mathrm{e}}$

$$\left(-\omega^2\mathbf{M}^{\mathrm{e}} + \mathbf{K}^{\mathrm{e}}\right)\cdot\hat{\mathbf{a}}^{\mathrm{me}} = \mathbf{f}^{\mathrm{e}}, \qquad \mathbf{M}^{\mathrm{e}} = \int_{\Omega^{\mathrm{e}}} \rho\, \mathbf{N}^{\mathrm{T}}\cdot\mathbf{N} dv.$$

The final extension from the real to the complex domain includes steps similar to those previously described. As here the displacements and the rotations possess the imaginary counterparts, the following approximations have to be used instead of (8.2.11), (8.2.12) and (8.2.18)

$$\mathbf{u}_{\mathrm{c}}^{\mathrm{m}} = \left\{ (\mathbf{u}^{\mathrm{mR}})^{\mathrm{T}} \quad i(\mathbf{u}^{\mathrm{mI}})^{\mathrm{T}} \right\}^{\mathrm{T}}, \qquad \mathbf{u}_{\mathrm{c}} = \left\{ (\mathbf{u}^{\mathrm{R}})^{\mathrm{T}} \quad i(\mathbf{u}^{\mathrm{I}})^{\mathrm{T}} \right\}^{\mathrm{T}}, \tag{8.2.21}$$

$$\hat{\mathbf{a}}_{\mathrm{c}}^{\mathrm{me}} = \left\{ (\hat{\mathbf{a}}^{\mathrm{eR}})^{\mathrm{T}} \quad i(\hat{\mathbf{a}}^{\mathrm{eI}})^{\mathrm{T}} \right\}^{\mathrm{T}}, \tag{8.2.22}$$

$$\mathbf{u}_{\mathrm{c}}^{\mathrm{m}} = \mathbf{N}_{\mathrm{c}}^{\mathrm{m}} \cdot \hat{\mathbf{a}}_{\mathrm{c}}^{\mathrm{me}}, \qquad \mathbf{N}_{\mathrm{c}}^{\mathrm{m}} = \begin{bmatrix} \mathbf{N}^{\mathrm{m}} & 0 \\ 0 & \mathbf{N}^{\mathrm{m}} \end{bmatrix}, \tag{8.2.23}$$

$$\mathbf{u}_{\mathrm{c}} = \mathbf{N}_{\mathrm{c}} \cdot \hat{\mathbf{a}}_{\mathrm{c}}^{\mathrm{me}}, \qquad \mathbf{N}_{\mathrm{c}} = \begin{bmatrix} \mathbf{N} & 0 \\ 0 & \mathbf{N} \end{bmatrix}. \tag{8.2.24}$$

Here the real and imaginary DOFs of an element are grouped separately, which is convenient for further derivations. The index c denotes the complex quantities. Due to the complex material parameters (8.2.7) the elasticity tensor also has complex form with the real and imaginary submatrices dependent on the real and imaginary parts of Young's modulus and Poisson's number

$$\mathscr{C}_{\mathrm{c}} = \begin{bmatrix} \mathscr{C}^{\mathrm{R}} & i\mathscr{C}^{\mathrm{I}} \\ i\mathscr{C}^{\mathrm{I}} & \mathscr{C}^{\mathrm{R}} \end{bmatrix}, \qquad \mathscr{C}^{\mathrm{R}} = \mathscr{C}(E^{\mathrm{R}}, \nu), \qquad \mathscr{C}^{\mathrm{I}} = \mathscr{C}(E^{\mathrm{I}}, \nu). \tag{8.2.25}$$

Equation (8.2.7) and standard relations between material parameters may be used to show that $E^{\mathrm{I}} = \frac{\delta}{\pi} E^{\mathrm{R}}$ and that the imaginary part of Poisson's ratio is equal to zero. The implementation of (8.2.23), (8.2.24) and (8.2.25) in complex counterparts of (8.2.13) and (8.2.19) now gives the complex form of the equation of motion

$$\left(-\omega^2 \mathbf{M}_{\mathrm{c}}^{\mathrm{e}} + \mathbf{K}_{\mathrm{c}}^{\mathrm{e}}\right) \cdot \hat{\mathbf{a}}_{\mathrm{c}}^{\mathrm{me}} = \mathbf{f}_{\mathrm{c}}^{\mathrm{e}} \tag{8.2.26}$$

dependent on the following quantities

$$\mathbf{M}_{\mathrm{c}}^{\mathrm{e}} = \int_{\Omega^{\mathrm{e}}} \rho \mathbf{N}_{\mathrm{c}}^{\mathrm{T}} \cdot \mathbf{N}_{\mathrm{c}} \, dv = \begin{bmatrix} \mathbf{M}^{\mathrm{e}} & 0 \\ 0 & \mathbf{M}^{\mathrm{e}} \end{bmatrix}, \quad \mathbf{K}_{\mathrm{c}}^{\mathrm{e}} = \int_{\mathscr{A}^{\mathrm{e}}} (\mathbf{B}_{\mathrm{c}}^{\mathrm{m}})^T \cdot \mathscr{C} \cdot \mathbf{B}_{\mathrm{c}}^{\mathrm{m}} \, da = \begin{bmatrix} \mathbf{K}^{\mathrm{eR}} & i\mathbf{K}^{\mathrm{eI}} \\ i\mathbf{K}^{\mathrm{eI}} & \mathbf{K}^{\mathrm{eR}} \end{bmatrix},$$

$$\mathbf{B}_{\mathrm{c}}^{\mathrm{m}} = \mathbf{L}_{\mathrm{c}} \cdot \mathbf{N}_{\mathrm{c}}^{\mathrm{m}} = \begin{bmatrix} \mathbf{B}^{\mathrm{m}} & 0 \\ 0 & \mathbf{B}^{\mathrm{m}} \end{bmatrix}, \qquad \mathbf{L}_{\mathrm{c}} = \begin{bmatrix} \mathbf{L} & 0 \\ 0 & \mathbf{L} \end{bmatrix}. \tag{8.2.27}$$

Note that the imaginary parts of the elasticity tensor in the present model are used to introduce the attenuation effects, while, in the engineering practice, Rayleigh damping is applied for it. According to this approach, an additive decomposition of the damping matrix into two parts proportional to the mass matrix and stiffness matrix is assumed [21, 415].

8.2.3 MODELING OF THE FLUID PHASE

An extension to the complex domain is also necessary for the 8-node cubic element chosen to simulate the marrow part. As in this case the derivation procedure is much simpler, it can be started directly using the complex form of the potential characteristic for this element

$$L = \frac{1}{2} \int_{\Omega} \rho \, \dot{\mathbf{u}}_{\mathrm{c}}^T \cdot \dot{\mathbf{u}}_{\mathrm{c}} \, dv - \frac{1}{2} \int_{\Omega} \varepsilon_{\mathrm{c}} \cdot \mathscr{C}_{\mathrm{c}} \cdot \varepsilon_{\mathrm{c}} \, dv - \Pi^{\mathrm{ext}}. \tag{8.2.28}$$

Here, the vector of the complex displacements

$$\mathbf{u}_c = \left\{(\mathbf{u}^R)^T \; i(\mathbf{u}^I)^T\right\}^T = \left\{u^R \quad v^R \quad w^R \quad iu^I \quad iv^I \quad iw^I\right\}^T \tag{8.2.29}$$

has to be approximated in the following way:

$$\mathbf{u}_c = \mathbf{N}_c \cdot \hat{\mathbf{a}}_c^e, \qquad \hat{\mathbf{a}}_c^e = \left\{ (\hat{\mathbf{a}}^{eR})^T \quad i(\hat{\mathbf{a}}^{eI})^T \right\}^T, \tag{8.2.30}$$

$$\hat{\mathbf{a}}_i^R = \left\{\hat{u}^R \quad \hat{v}^R \quad \hat{w}^R\right\}^T, \qquad \hat{\mathbf{a}}_i^I = \left\{\hat{u}^I \quad \hat{v}^I \quad \hat{w}^I\right\}^T, \qquad \mathbf{N}_c = \begin{bmatrix} \mathbf{N} & 0 \\ 0 & \mathbf{N} \end{bmatrix} \tag{8.2.31}$$

and $\mathbf{N}$ represents the matrix consisting of the shape functions typical for the standard 8-node element of the program FEAP [21, 415]. By using approximation (8.2.30) and the complex elasticity tensor based on (8.2.5), the variation of (8.2.28) becomes

$$-\omega^2(\delta\hat{\mathbf{a}}_c^e)^T \cdot \left(\int_{\Omega^e} \rho \mathbf{N}_c^T \cdot \mathbf{N}_c \, dv\right) \cdot \hat{\mathbf{a}}_c^e + (\delta\hat{\mathbf{a}}_c^e)^T \cdot \left(\int_{\Omega^e} \mathbf{N}_c^T \cdot \mathbf{L}_c^T \cdot \mathscr{C}_c \cdot \mathbf{L}_c \cdot \mathbf{N}_c \, dv\right) \cdot \hat{\mathbf{a}}_c^e + \delta\Pi^{e,ext} = 0$$

yielding the equation of motion

$$(-\omega^2 \mathbf{M}_c^e + \mathbf{K}_c^e) \cdot \hat{\mathbf{a}}_c^{me} = \mathbf{f}_c^e \tag{8.2.32}$$

Although the same notation is used as in the case of shell element (8.2.26), all the quantities appearing here are defined in a different manner, corresponding to the solid element.

8.2.4 SUMMARY OF THE EQUATIONS DEFINING THE BVP ON THE MICROLEVEL

Let us for convenience summarize the description of the RVE

$$(-\omega^2 \mathbf{M}_s + \mathbf{K}_s) \cdot \tilde{\mathbf{a}}_s = \mathbf{f}_s(\bar{\varepsilon}) \quad \text{in} \quad \Omega_s, \tag{8.2.33}$$

$$\sigma_s = \mathscr{C} : \varepsilon \quad \text{in} \quad \Omega_s, \tag{8.2.34}$$

$$(-\omega^2 \mathbf{M}_f + \mathbf{K}_f) \cdot \tilde{\mathbf{a}}_f = \mathbf{f}_f(\bar{\varepsilon}) \quad \text{in} \quad \Omega_f, \tag{8.2.35}$$

$$\sigma_f = c^2 \rho_f \nabla \cdot \mathbf{u} \mathbf{I} + 2i\omega\eta \, \varepsilon + i\omega\xi \, \nabla \cdot \mathbf{u} \mathbf{I} \quad \text{in} \quad \Omega_f, \tag{8.2.36}$$

$$[\tilde{u}_\perp] = 0 \quad \text{on} \quad \Gamma = \Omega_s \cup \Omega_f, \tag{8.2.37}$$

$$\tilde{\mathbf{u}}^+ = \tilde{\mathbf{u}}^- \quad \text{on} \quad \partial\Omega. \tag{8.2.38}$$

Here again, the indices s and f are used for different phases. In the equations of motion (8.2.33) and (8.2.35) the volume forces are neglected but the influence of macrodeformation is introduced. On opposite faces of the RVE, microfluctuations must be periodic due to the Hill-Mandel macrohomogeneity condition (8.2.38). As defined in Section 8.2.1, equations (8.2.34), (8.2.36) are constitutive laws and (8.2.37) is a coupling condition.

8.2.5 EFFECTIVE ELASTICITY TENSOR: OUTPUT FROM THE MICROSCALE

The final result of the microscale calculations is the effective elasticity tensor. In the theory of linear hyperelasticity, this tensor is defined as the second derivative of the potential functional $\psi(\bar{\varepsilon})$ with respect to the infinitesimal strain tensor or equivalently, the first derivative of the Cauchy stress tensor with respect to the same tensor

$$\bar{\mathscr{C}} = \frac{\partial^2 \psi(\bar{\varepsilon})}{\partial \bar{\varepsilon}^2} = \frac{\partial \bar{\sigma}}{\partial \bar{\varepsilon}}. \tag{8.2.39}$$

In our case, due to the heterogeneity of the material, an analytic expression for the stress is not available, so the numerical interpretation of (8.2.39) is necessary. Bearing in mind the vector notation for stresses and strains, typical for FE

$$\bar{\sigma} = \{(\sigma^{\mathrm{R}})^{\mathrm{T}} \quad i(\bar{\sigma}^{\mathrm{I}})^{\mathrm{T}}\}^{\mathrm{T}}, \tag{8.2.40}$$

$$(\sigma^{\mathrm{R}})^{\mathrm{T}} = \{\sigma_{11}^{\mathrm{R}}\ \sigma_{22}^{\mathrm{R}}\ \sigma_{33}^{\mathrm{R}}\ \sigma_{12}^{\mathrm{R}}\ \sigma_{23}^{\mathrm{R}}\ \sigma_{13}^{\mathrm{R}}\}, \qquad (\sigma^{\mathrm{I}})^{\mathrm{T}} = \{\sigma_{11}^{\mathrm{I}}\ \sigma_{22}^{\mathrm{I}}\ \sigma_{33}^{\mathrm{I}}\ \sigma_{12}^{\mathrm{I}}\ \sigma_{23}^{\mathrm{I}}\ \sigma_{13}^{\mathrm{I}}\} \tag{8.2.41}$$

$$\bar{\varepsilon} = \{(\varepsilon^{\mathrm{R}})^{\mathrm{T}} \quad i(\bar{\varepsilon}^{\mathrm{I}})^{\mathrm{T}}\}^{\mathrm{T}}, \tag{8.2.42}$$

$$(\varepsilon^{\mathrm{R}})^{\mathrm{T}} = \{\varepsilon_{11}^{\mathrm{R}}\ \varepsilon_{22}^{\mathrm{R}}\ \varepsilon_{33}^{\mathrm{R}}\ 2\varepsilon_{12}^{\mathrm{R}}\ 2\varepsilon_{23}^{\mathrm{R}}\ 2\varepsilon_{13}^{\mathrm{R}}\} \qquad (\varepsilon^{\mathrm{I}})^{\mathrm{T}} = \{\varepsilon_{11}^{\mathrm{I}}\ \varepsilon_{22}^{\mathrm{I}}\ \varepsilon_{33}^{\mathrm{I}}\ 2\varepsilon_{12}^{\mathrm{I}}\ 2\varepsilon_{23}^{\mathrm{I}}\ 2\varepsilon_{13}^{\mathrm{I}}\} \tag{8.2.43}$$

the numerical interpretation of (8.2.39) can be written

$$\bar{\mathscr{C}}_{ij} \approx \frac{1}{\varepsilon}[\bar{\sigma}_i(\bar{\varepsilon}^{\varepsilon j}) - \bar{\sigma}_i(\bar{\varepsilon})] \tag{8.2.44}$$

where $\bar{\varepsilon}^{\varepsilon j}$ is the perturbated strain tensor so that $\bar{\varepsilon}_i^{\varepsilon j} = \bar{\varepsilon}_i$ for $i \neq j$ and $\bar{\varepsilon}_i^{\varepsilon j} = \bar{\varepsilon}_i + \varepsilon$ for $i = j$, $i, j = 1, \ldots 12$. Symbol ε denotes a small perturbation. The whole elasticity tensor $\bar{\mathscr{C}}$ is now of the dimension 12×12 and it consists of two real and two imaginary submatrices (form analogous to (8.2.25)).

Repeating calculations based on (8.2.44) for different geometries of the RVE, it is possible to study the change of material strength caused by the process of osteoporosis. In the following, three groups of tests are performed. In each of the groups one of the geometry parameters defined in Section 8.2.1 is varied while the other ones are kept constant. The diagrams in Figure 8.2.2 show the change of two terms of the elasticity tensors, $\mathscr{C}_{11}$ and $\mathscr{C}_{12}$, with respect to the porosity. The increasing porosity is simulated by changing the wall width b in the range from 0.2 to 0.125 mm, whereby the thickness of the solid wall d amounts to 0.1 mm as a first case and 0.05 mm as a second one. Side length a is 1 mm in both cases. Except the parameters given in Table 8.1 the remaining values used are logarithmic decrements corresponding to the sound motion through the bone $\delta = 0.1$, viscosity coefficient $\xi = 0$, and excitation frequency 100 kHz. The change of terms $\mathscr{C}_{17}$ and $\mathscr{C}_{18}$ belonging to the imaginary part are shown in Figure 8.2.3.

Since experimental investigations have shown that the disappearing of the complete solid walls has the strongest influence on the decrease in material strength, such a process was simulated, too. According to the results presented in the work of

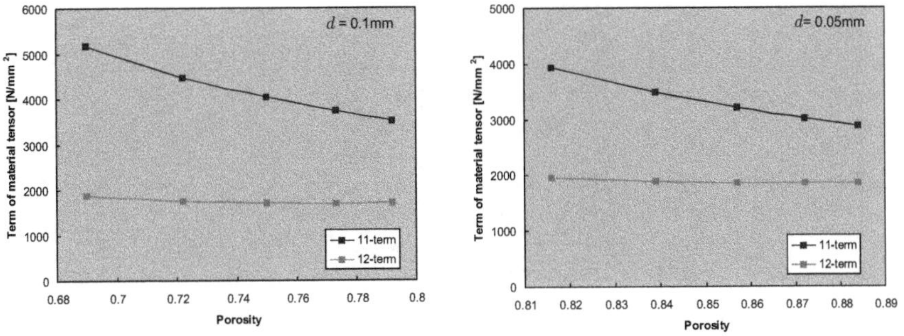

FIGURE 8.2.2 Change in the real components $\mathscr{C}_{11}$ and $\mathscr{C}_{12}$ of the elasticity tensor with increasing porosity. The solid wall thickness d takes the values 0.1 and 0.05 mm.

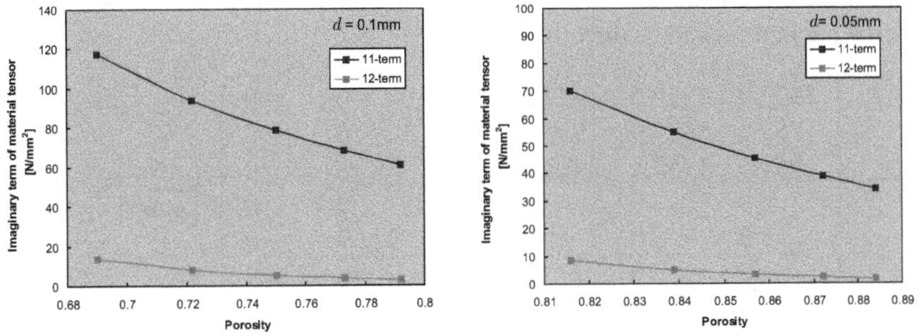

FIGURE 8.2.3 Change in the imaginary components of the elasticity tensor $\mathscr{C}_{71}$ and $\mathscr{C}_{81}$ with increasing porosity for different thicknesses of the solid wall d.

[294], osteoporosis causes an increase of the spacings of the solid walls from 0.471 to 2.2 mm which is simulated by increasing the side length of the cube-shaped RVE, thereby the wall thickness is kept at $d = 0.05$ mm and wall width at $b = a/6$. The results are presented in Figure 8.2.4 where the abscissas are side length and porosity respectively. The latter diagram shows that with increasing porosity, both terms of the elasticity tensor tend to the value 2000 N/mm^2, which corresponds to the limiting case of pure marrow.

8.2.6 EFFECTIVE MATERIAL PARAMETERS

The calculated effective elasticity tensors further permit the evaluation of the effective material parameters, which will be presented in the following example. For the RVE determined by the parameters $a = 1$ mm, $d = 0.05$ mm, $b = 0.25$ mm, the real part of the elasticity tensor is given by (8.2.45) and it corresponds to the cubic material symmetry, in which case the compliance tensor has a form shown in (8.2.46)

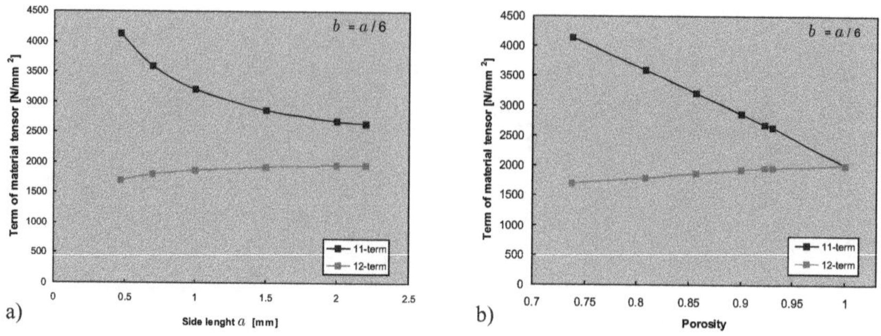

FIGURE 8.2.4 Change in components $\mathscr{C}_{11}$ and $\mathscr{C}_{12}$ with increasing spans a and corresponding porosity. Width of solid wall $b = a/6$.

with the shear modulus G. Obviously, calculating the inverse matrix of (8.2.45), the material parameters can be directly evaluated:

$$\mathscr{C}^R = \begin{bmatrix} 3927.9 & 1959.5 & 1959.5 & 0 & 0 & 0 \\ 1959.5 & 3927.9 & 1959.5 & 0 & 0 & 0 \\ 1959.5 & 1959.5 & 3927.9 & 0 & 0 & 0 \\ 0 & 0 & 0 & 272.9 & 0 & 0 \\ 0 & 0 & 0 & 0 & 272.9 & 0 \\ 0 & 0 & 0 & 0 & 0 & 272.9 \end{bmatrix} \left[\frac{\text{N}}{\text{mm}^2}\right], \quad (8.2.45)$$

$$\mathbf{S} = \begin{bmatrix} \frac{1}{E} & -\frac{\nu}{E} & -\frac{\nu}{E} & 0 & 0 & 0 \\ -\frac{\nu}{E} & \frac{1}{E} & -\frac{\nu}{E} & 0 & 0 & 0 \\ -\frac{\nu}{E} & -\frac{\nu}{E} & \frac{1}{E} & 0 & 0 & 0 \\ 0 & 0 & 0 & \frac{1}{G} & 0 & 0 \\ 0 & 0 & 0 & 0 & \frac{1}{G} & 0 \\ 0 & 0 & 0 & 0 & 0 & \frac{1}{G} \end{bmatrix}. \quad (8.2.46)$$

A further topic of interest is to follow the influence of osteoporosis on effective material behavior. To this end, the effective material parameters are calculated for the effective elasticity tensors mentioned in the previous section. In Fig. 8.2.5, the results corresponding to the RVEs with wall width b in the interval 0.25–0.125 mm and wall thickness $d = 0.1$ mm and $d = 0.05$ mm are considered. The side length $a = 1$ mm is kept constant.

Dependence of Young's modulus and shear modulus on porosity is a smooth, monotonically decreasing function, while Poisson's ratio shows opposite behavior: it increases with increasing porosity. Young's modulus takes the values in the interval 1417.17–4163.74 N/mm^2, the shear modulus 32.55–477.82 N/mm^2, and Poisson's ratio 0.266–0.392 for the density 1067–1263 kg/m^3.

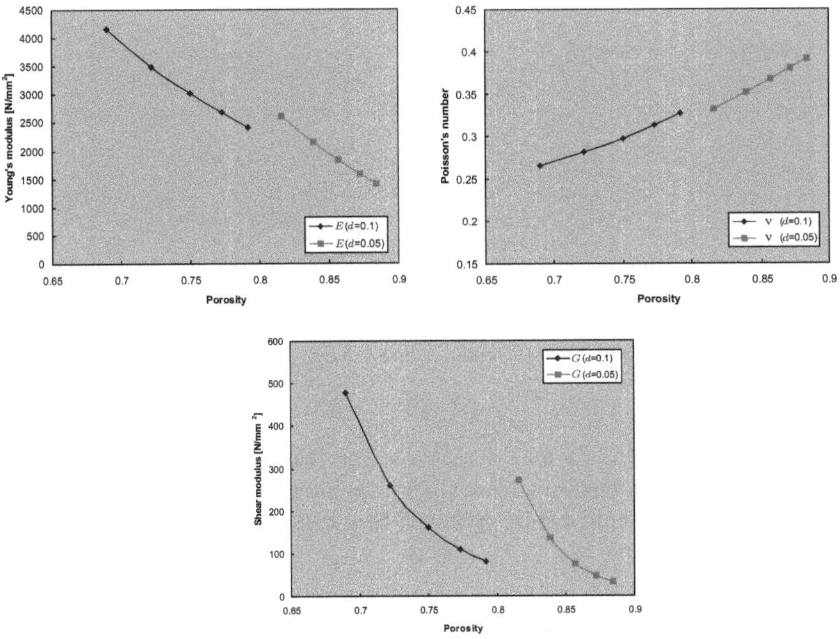

FIGURE 8.2.5 Change in the real part of effective material parameters versus porosity.

8.2.7 ANALYSIS OF THE DRY SKELETON

The standard literature [15, 202, 351, 399, 420] mostly investigates the effective elasticity parameters of the pure solid phase. For the purpose of comparison, the calculations described in Section 8.2.4 are repeated for the RVE without marrow core (Fig. 8.2.1). In this case, in the system (8.2.33)–(8.2.38), equations related to the solid phase and Hill macrohomogeneity condition stay active, while the equations related to the liquid phase and coupling condition have to be left out. The final results at the microlevel are again effective elasticity tensors and material parameters.

Figure 8.3.1 shows that in the case of wall thickness of 0.1 mm, the effective Young's modulus takes the values 1883–3576 N/mm^2 while for a wall thickness of 0.05 mm, the values are between 1050 and 2114 N/mm^2. These results agree well with the results obtained by Ashman and Rho [15], who found, using ultrasonic tests, that the structural elasticity modulus of cancellous bone has values in the interval 985–2110 N/mm^2.

The shear modulus (results not shown here) takes values in the range 81.17–468.97 N/mm^2 in the first case ($d = 0.1$ mm) and 31.93–263.34 N/mm^2 in the second case ($d = 0.05$ mm). A comparison with the results obtained by an analysis of biphasic material (Fig. 8.2.5) shows that the presence of a liquid phase does not significantly influence the values of this material parameter. This of course can be justified by the fact that the shear resistance of the fluid can be neglected. The situation in the case of Poisson's ratio is quite opposite (Fig. 8.3.1). This parameter takes values

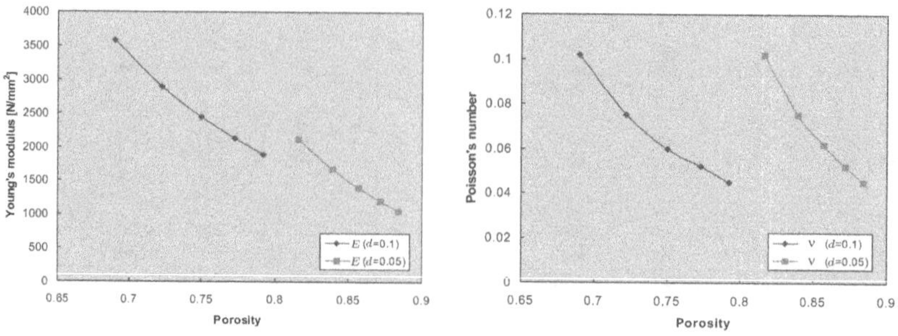

FIGURE 8.3.1 Change in Young's modulus and Poisson's ratio over porosity for the dry skeleton.

in the range 0.045–0.102 for the dry skeleton. The values are falling with increasing porosity while for the biphasic material they are increasing. The great influence of marrow on Poisson's ratio can be explained by the nearly incompressible nature of this material.

8.3 MACROSCALE: SIMULATION OF THE ULTRASONIC TEST

8.3.1 ULTRASONIC ATTENUATION TEST

The main topic of the present section is the application of the method developed for the simulation of the behavior of the whole bone or of its parts. This task belongs to the domain of macroscale calculations and a simulation of the ultrasonic attenuation test is chosen for an illustration.

The setup of such a test carried out by Hosokawa-Otani [203] is shown in Figure 8.3.2. Here a transmitter and hydrophone are submerged in distilled water at 23 ± 0.5°C and the bone specimen is placed between them. The chosen frequency bandwidth of excitation waves is 0.5–5 MHz. The test uses samples measuring 20–30 mm, with two different thicknesses $d_1 = 9$ mm and $d_2 = 7$ mm. The samples are chosen to represent the different types of cancellous bone whose densities vary in the range of 1120–1200 kg/m^3. Before proceeding to the experiments, the samples are saturated with water in order to remove air bubbles formed in the process of preparation. Using such a test, wave speed v and attenuation α can be calculated according to the standard expressions [203]:

$$v = \frac{\Delta d\, v_0}{\Delta d - (\Delta\phi/\omega)v_0} (=) \frac{\text{mm}}{\text{s}}, \tag{8.3.1}$$

$$\alpha = \frac{\ln \Delta V}{\Delta d} (=) \frac{\text{Np}}{\text{mm}} \tag{8.3.2}$$

where Δd represents the difference in thickness $\Delta d = d_1 - d_2$, v_0 is the propagation speed in water, $\Delta\phi$ is the phase difference, ω the frequency of initial signals and ΔV is the ratio of amplitude spectra for two different thickness $\Delta V = \frac{V2}{V1}$.

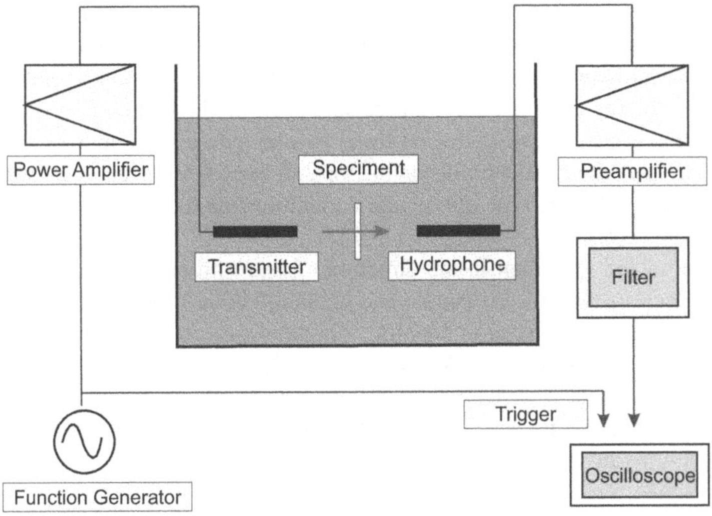

FIGURE 8.3.2 Laboratory test of A. Hosokawa and T. Otani [203].

8.3.2 FEM MODEL OF THE ULTRASONIC TEST

In the FE modeling of the described test, we start from the fact that the applied sound excitation is given by a harmonic function in time $\bar{\mathbf{p}}(\mathbf{x},t) = \bar{\mathbf{p}}(\mathbf{x})e^{i\omega t}$, causing periodic displacements $\bar{\mathbf{u}}(\mathbf{x},t) = \bar{\mathbf{u}}(\mathbf{x})e^{i\omega t}$. The problem which has to be solved is then summarized as follows:

$$-\omega^2\bar{\rho}\,\bar{\mathbf{u}}(\mathbf{x}) - \bar{\nabla}\cdot\bar{\sigma}(\mathbf{x}) = \mathbf{0}, \qquad (8.3.3)$$

$$\bar{\sigma}(\mathbf{x}) = \bar{\mathscr{C}} : \bar{\varepsilon}(\mathbf{x}), \qquad (8.3.4)$$

$$\bar{\mathbf{u}}(\mathbf{x}) = \bar{\mathbf{u}}^*(\mathbf{x}) \quad \text{on} \quad \partial\bar{\mathscr{B}}_u, \qquad \bar{\sigma}(\mathbf{x})\cdot\bar{\mathbf{n}}(\mathbf{x}) = \bar{\mathbf{p}}(\mathbf{x}) \quad \text{on} \quad \partial\bar{\mathscr{B}}_p. \qquad (8.3.5)$$

Here in the equation of motion (8.3.3), volume forces are neglected, and the only external load is due to the sound pressure acting on the boundary part $\partial\bar{\mathscr{B}}_p$, and $\bar{\mathbf{u}}^*(\mathbf{x})$ are displacements given on the Dirichlet boundary part $\partial\bar{\mathscr{B}}_u$. The constitutive law (8.3.4) depends on the effective elasticity tensor $\bar{\mathscr{C}}$ obtained from the microscale. As stated at the beginning, the overbar denotes quantities related to the macrolevel.

Concerning the geometrical properties of the FE model, the following remarks are necessary. First, as the water in the original test is used only as a transmitter whose attenuation can be neglected, in the FE model, consideration has to be given only to the behavior of the sample. Second, as the sound wave is longitudinal, the whole simulation may be considered as a 2D problem of wave propagation through the thin slice of the sample (Fig. 8.3.3a). Moreover, the displacements in x_2 direction at all points have to be suppressed, while the results will show that displacements in the x_3 direction are of an order smaller than those in the x_1 direction, which fits in with the nature of sound waves.

The dimension of the sample transversal to the direction of wave propagation is assumed to be 50 mm, which in any case is greater than the wavelength of the excitation sound waves (see Section 8.3.3). Two kinds of boundary conditions are simulated at the top and the bottom of the specimen. In the first case, it is presumed that all displacements at all points on these two boundaries are constrained. In the second case, only the middle points on the top and bottom boundaries are supported. The results show that the type of boundary conditions on these two boundaries does not influence the results.

The size of the specimen in the direction of wave propagation is chosen as 30 mm. The discretization of the sample and applied load are shown in Figure 8.3.3a. Here the number of elements is 100×50. The thickness in the direction "2" is 0.5 mm. The sound pressure $p = 8$ kPa acts on the left boundary of the sample. The diameter of the pulser is assumed to be 10 mm. The last few parameters (thickness of the specimen, pressure and diameter of the transmitter) are chosen arbitrarily as they do not have any significant influence on the simulation.

8.3.3 TEST EXAMPLE

The velocity of wave propagation through the solid bodies is dependent on the material properties, but also on the shape of the body and especially on its dimensions. In particular there are two kinds of sound velocity to be distinguished: the wave velocity through the bounded medium, also known as the bar velocity (c_b), and the velocity through the unbounded medium (c_u). Here, the term "bounded" means that cross-sectional dimensions of the sample, transverse to the direction of wave motion, have to be smaller than the wavelength. The mentioned velocities are defined by the following expressions [15, 291]

$$c_b = \sqrt{\frac{E}{\rho}}, \tag{8.3.6}$$

$$c_u = \sqrt{\frac{\lambda + 2\mu}{\rho}}. \tag{8.3.7}$$

where λ and μ are the Lamé parameters. The observation is that the velocity in an unbounded medium becomes greater than in the bounded one. In the scope of biomechanics, Eq. (8.3.7) appears often in the form $\mathscr{C}_{11} = \rho c_u^2$, where $\mathscr{C}_{11}$ represents the 11-component of the elasticity tensor [399].

Using the results from Section 8.2.6 and the expressions (8.3.6) and (8.3.7), the wave velocities through the cancellous bone can be calculated and used as a check of the FEM simulations. This is illustrated by an example where consideration is given to the wave propagation through the homogenized bone with an RVE geometry determined by the parameters $a = 1$ mm, $d = 0.1$ mm and $b = 0.25$ mm. The model shown in Figure 8.3.3a corresponds to the case of an unbounded medium, so that the expected wave velocity amounts to $c_u = 2021.69$ m/s. For an arbitrarily chosen excitation frequency $f = 0.6$ MHz, the expected wavelength is 3.37 mm. The results of the FEM simulation shown in Figure 8.3.3b endorse such expectations as the resulting wavelength λ is approximately 3.24 mm (9.25 wavelengths on the length of sample which amounts to 30 mm). The wavelength is connected to velocity c and frequency f by the relation $\lambda = c/f$. An additional check looks at the amplitude of particle oscillations u related to wave pressure p by using expression

$$u = \frac{p}{2\pi f c \rho}. \tag{8.3.8}$$

This relation can be derived using the definition of the wave impedance and belongs to the basics of acoustic theory [180]. By using (8.3.8), it can be shown that the approximate particle amplitude for the example discussed in this section has the value $8.32 \cdot 10^{-7}$mm, which agrees with the displacement values shown in Figure 8.3.3b where the mean value of the slightly attenuated amplitudes amounts to $8.335 \cdot 10^{-7}$mm.

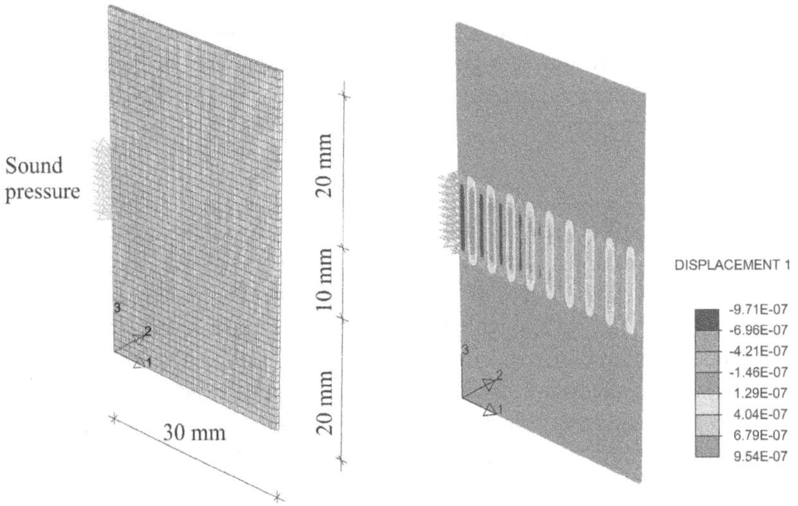

FIGURE 8.3.3 a) Model for FEM simulation of wave propagation through cancellous bone. b) Wave propagation through cancellous bone with an RVE geometry determined by the parameters $a = 1$ mm, $d = 0.1$ mm, $b = 0.25$ mm. Assumed excitation frequency 0.6 MHz.

8.3.4 WAVE ATTENUATION

The last group of simulations considers the analysis of wave attenuation. Here the propagation of waves of different frequencies through samples with different material parameters is simulated in order to check the experimentally obtained result that increasing excitation frequency and material density cause increasing attenuation.

Firstly, consideration is given to the influence of increasing excitation frequency on bone behavior. To this end the type of material microstructure in the simulations is fixed, and sound excitation at different frequencies is applied. As the influence of attenuation is more noticeable in the case of higher frequencies, excitation is simulated in the domain 0.9–1.7 MHz. The microstructure is chosen according to the geometry of the RVE determined by the parameters $a = 1$ mm, $b = 0.25$ mm and $d = 0.05$ mm ($\rho = 1136$ kg/m^3). The results of the simulations are shown in Figure 8.3.4a where the stronger attenuation obviously corresponds to higher frequencies.

The study of the relationship between attenuation and density is more complex than the study of the influence of excitation frequency. This can be expected, because the RVE geometry presented in Section 8.2.1 is determined by three parameters (wall thickness d, wall width b and side length a). Correspondingly, three different types of tests can be carried out. In each group of tests, two of the geometrical parameters have to be kept constant and the remaining one is varied. For illustration, the results obtained by changing the wall width are shown in Figure 8.3.4b. The results are obtained by investigating materials with an RVE such that $a = 1$ mm, $d = 0.05$ mm and the width of wall b takes the values 0.25–0.125 mm. Figure 8.3.4b shows that decreasing density causes decreasing attenuation.

From Figure 8.3.4, the conclusion can be drawn that the numerical simulations endorse experimental results: the greater excitation frequency and sample density correspond to the greater attenuation. The numerical values of the attenuation are obtained by using expression

$$\alpha = \frac{\ln \Delta V'}{\Delta d'} \tag{8.3.9}$$

where, in contrast to Eq. (8.3.2), $\Delta V'$ represents the ratio of the amplitudes at two points of the same sample, at the same horizontal level, but with the distance $\Delta d'$ amounting to the integer number of wave lengths. The obtained values amount, on average, to 0.115–0.225 Np/cm, which is approximately the same as if they were obtained by superposition of the attenuation due to the solid frame and fluid core separately. However, the values are smaller than those obtained experimentally. This can be explained by the missing reflection effects, an aspect considered in Section 8.6.

The final question to be considered here is the distinction of the fast and the slow waves. The existence of these two kinds of waves was predicted by Biot [28, 29] and experimentally observed by [203]. But the question of and when these two waves appear still is an open issue and in some experimental studies only a single wave is observed [106, 318, 376, 396]. In their work, Hosokawa and Otani [205] show that the appearance of a second wave is dependent on the inclination of the incident wave to the trabeculae alignment. If the wave propagates perpendicular to the direction of the trabeculae, the slow and fast wave overlap completely in time and the second wave cannot be observed. In the case that the propagation direction of the wave coincides with the longitudinal direction of the trabeculae, the second wave will appear. Another approach documents the existence of mixed modes, where both waves overlap in time [12, 22, 210, 337]. The results presented in this chapter show that at the current stage, the homogenization technique allows simulation solely of a single longitudinal wave, which can be regarded as a mixed-mode wave.

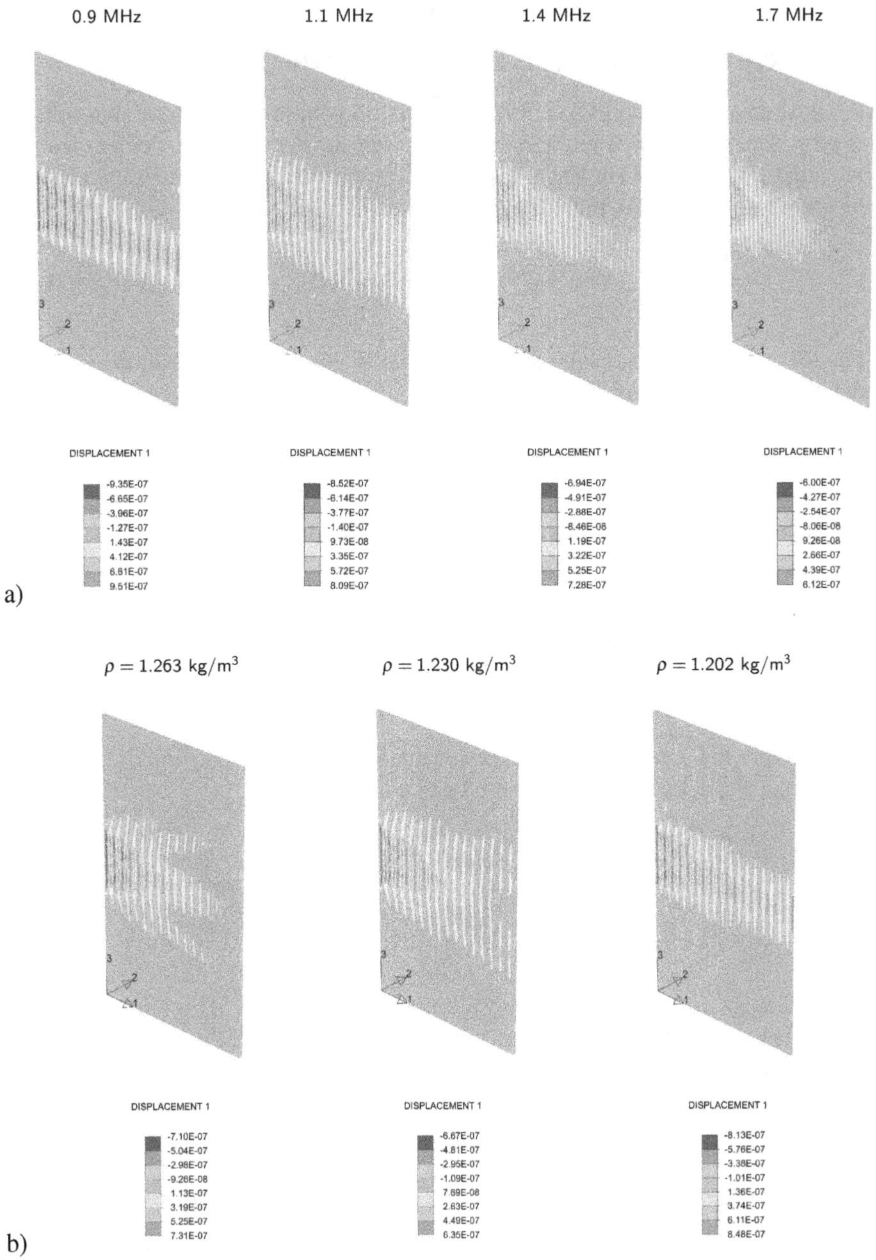

FIGURE 8.3.4 a) Wave propagation of sound waves of different frequencies through the bone whose geometry of RVE is determined by the parameters $a = 1$ mm, $b = 0.25$ mm and $d = 0.05$ mm. b) Wave propagation of the wave of frequency 1 MHz in the case that material density varies.

8.4 SIMPLIFIED VERSION OF THE RVE

8.4.1 RVE II: SOLID PHASE CONSISTING OF THIN COLUMNS

As an alternative possibility, an RVE with the solid phase consisting of thin columns is proposed. Here, for simulation of both phases, the 8-node brick element with the potential (8.2.28) is used. The phases are distinguished only through the elasticity tensor which is complex in both cases and has the form

$$\mathscr{C}_{\text{c}} = \begin{bmatrix} \mathscr{C}^{\text{R}} & i\mathscr{C}^{\text{I}} \\ i\mathscr{C}^{\text{I}} & \mathscr{C}^{\text{R}} \end{bmatrix}. \tag{8.4.1}$$

For the solid phase, the submatrices $\mathscr{C}^{\text{R}}$ and $\mathscr{C}^{\text{I}}$ have to be calculated using the definition for the standard elasticity tensor but introducing the value for the real and imaginary parts of the material parameters

$$\mathscr{C}^{\text{R}} = \mathscr{C}(K^{\text{R}}, \mu^{\text{R}}), \qquad \mathscr{C}^{\text{I}} = \mathscr{C}(K^{\text{I}}, \mu^{\text{I}}). \tag{8.4.2}$$

In the extended form, it can be written

$$\mathscr{C}^{\text{R}} = \begin{bmatrix} \kappa^{\text{R}} + \frac{1}{2}\mu^{\text{R}} & \kappa^{\text{R}} - \frac{3}{2}\mu^{\text{R}} & \kappa^{\text{R}} - \frac{3}{2}\mu^{\text{R}} & 0 & 0 & 0 \\ \kappa^{\text{R}} - \frac{3}{2}\mu^{\text{R}} & \kappa^{\text{R}} + \frac{1}{2}\mu^{\text{R}} & \kappa^{\text{R}} - \frac{3}{2}\mu^{\text{R}} & 0 & 0 & 0 \\ \kappa^{\text{R}} - \frac{3}{2}\mu^{\text{R}} & \kappa^{\text{R}} - \frac{3}{2}\mu^{\text{R}} & \kappa^{\text{R}} + \frac{1}{2}\mu^{\text{R}} & 0 & 0 & 0 \\ 0 & 0 & 0 & \mu^{\text{R}} & 0 & 0 \\ 0 & 0 & 0 & 0 & \mu^{\text{R}} & 0 \\ 0 & 0 & 0 & 0 & 0 & \mu^{\text{R}} \end{bmatrix}, \quad \mathscr{C}^{\text{I}} = \begin{bmatrix} \kappa^{\text{I}} + \frac{1}{2}\mu^{\text{I}} & \kappa^{\text{I}} - \frac{3}{2}\mu^{\text{I}} & \kappa^{\text{I}} - \frac{3}{2}\mu^{\text{I}} & 0 & 0 & 0 \\ \kappa^{\text{I}} - \frac{3}{2}\mu^{\text{I}} & \kappa^{\text{I}} + \frac{1}{2}\mu^{\text{I}} & \kappa^{\text{I}} - \frac{3}{2}\mu^{\text{I}} & 0 & 0 & 0 \\ \kappa^{\text{I}} - \frac{3}{2}\mu^{\text{I}} & \kappa^{\text{I}} - \frac{3}{2}\mu^{\text{I}} & \kappa^{\text{I}} + \frac{1}{2}\mu^{\text{I}} & 0 & 0 & 0 \\ 0 & 0 & 0 & \mu^{\text{I}} & 0 & 0 \\ 0 & 0 & 0 & 0 & \mu^{\text{I}} & 0 \\ 0 & 0 & 0 & 0 & 0 & \mu^{\text{I}} \end{bmatrix}. \tag{8.4.3}$$

On the other hand, using the constitutive law (8.2.5), the real part and the imaginary part of the elasticity tensor for the fluid phase are

$$\mathscr{C}^{\text{R}} = \begin{bmatrix} c^2\rho & c^2\rho & c^2\rho & 0 & 0 & 0 \\ c^2\rho & c^2\rho & c^2\rho & 0 & 0 & 0 \\ c^2\rho & c^2\rho & c^2\rho & 0 & 0 & 0 \\ 0 & 0 & 0 & 0 & 0 & 0 \\ 0 & 0 & 0 & 0 & 0 & 0 \\ 0 & 0 & 0 & 0 & 0 & 0 \end{bmatrix}, \quad \mathscr{C}^{\text{I}} = \begin{bmatrix} 2\omega\eta + \omega\xi & \omega\xi & \omega\xi & 0 & 0 & 0 \\ \omega\xi & 2\omega\eta + \omega\xi & \omega\xi & 0 & 0 & 0 \\ \omega\xi & \omega\xi & 2\omega\eta + \omega\xi & 0 & 0 & 0 \\ 0 & 0 & 0 & \omega\eta & 0 & 0 \\ 0 & 0 & 0 & 0 & \omega\eta & 0 \\ 0 & 0 & 0 & 0 & 0 & \omega\eta \end{bmatrix}. \tag{8.4.4}$$

The geometry of the new type of RVE (hereafter called RVE II) is shown in Figure 8.4.1 where a denotes the side length and b is the thickness of the solid column. The RVE shown in Fig. 8.2.1, applying shell elements, will be hereafter called RVE I. The coupling conditions are mutual for both RVEs: there is no jump of the displacements perpendicularly to the contact surface.

8.4.2 NUMERICAL VALUES OF THE EFFECTIVE MATERIAL PARAMETERS

For the calculation of the effective material parameters, we first consider RVE I whereby the increasing porosity is simulated by decreasing the wall width b from 0.25 to 0.125 mm. The side length of the RVE is kept constant $a = 1$ mm, and the simulations are repeated for two different thicknesses of the solid wall $d = 0.1$ mm and $d = 0.05$ mm. As illustration, the results obtained for Young's modulus are presented in Figure 8.4.2a. This diagram shows that increasing porosity causes decreasing of

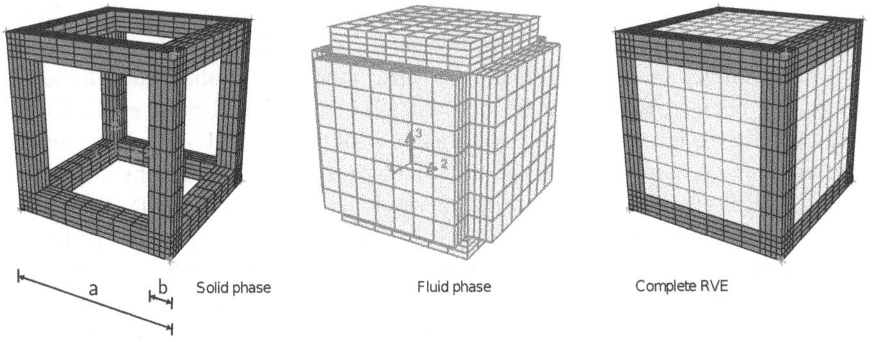

FIGURE 8.4.1 Geometry of RVE II.

Young's modulus and the obtained values are in the range 890.47–2649.22 MPa for porosity 0.69–0.884. In the diagram, two graphs can be recognized and each of them corresponds to one of the wall thicknesses (0.1 mm or 0.05 mm). This indicates that not only the porosity, but also the form of solid walls influences the effective values. Furthermore, it means that the RVEs with the same porosity but different thickness and width of solid walls might have different effective values.

A similar analysis can be performed by using the RVE II. Here the geometry of the RVE is described by two parameters: side length a and column thickness b, and the influence of both of them on the effective material parameters is studied. The results for Young's modulus are shown in Figure 8.4.2b. In the first group of tests (E_a), the column thickness is taken to be in the range 0.1–0.2 mm but the side length is kept at $a = 1$ mm. In the second group of tests (E_b) the column thickness is constant $b = 0.2$ mm, and the side length a is variable: 1.00–2.25 mm. It is interesting to note that for this kind of the RVE, both groups of tests yield the same dependency of the material parameters on the porosity. The resulting line is continuous without any jumps although obtained for two groups of tests.

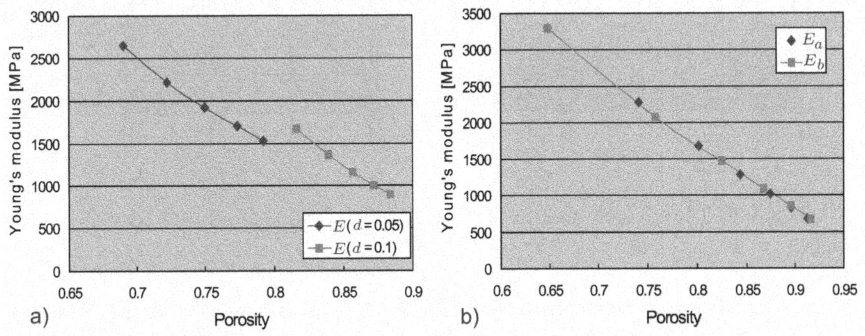

FIGURE 8.4.2 Young's modulus obtained by using a) RVE I, b) RVE II.

A comparison of the values for four effective material parameters obtained by the investigation of RVE I and RVE II is presented in Fig. 8.4.3. The diagrams 8.4.3a,b show that similar values for Young's and shear modulus are obtained in these two ways. The results obtained by the investigation of the RVE II behave as a kind of trend line for the results obtained by the analysis of the RVE I. For the bulk modulus and the Poisson ratio (Fig. 8.4.3c,d), the numerical values obtained by an analysis of the RVE II are slightly greater than those obtained in the case of RVE I, but the difference decreases with increasing porosity. The diagrams show that increasing porosity causes decreasing Young's shear and bulk modulus but increasing Poisson's ratio. Obviously, the effective material parameters are dependent on the porosity, but the assumption for the shape of the individual trabeculae does not contribute significantly to their change. In contrast to this, the disposition, the spacings and the orientation between the individual trabeculae are of great importance for the effective values, as will be shown in Section 8.5, which is concerned with the anisotropy of the cancellous bone.

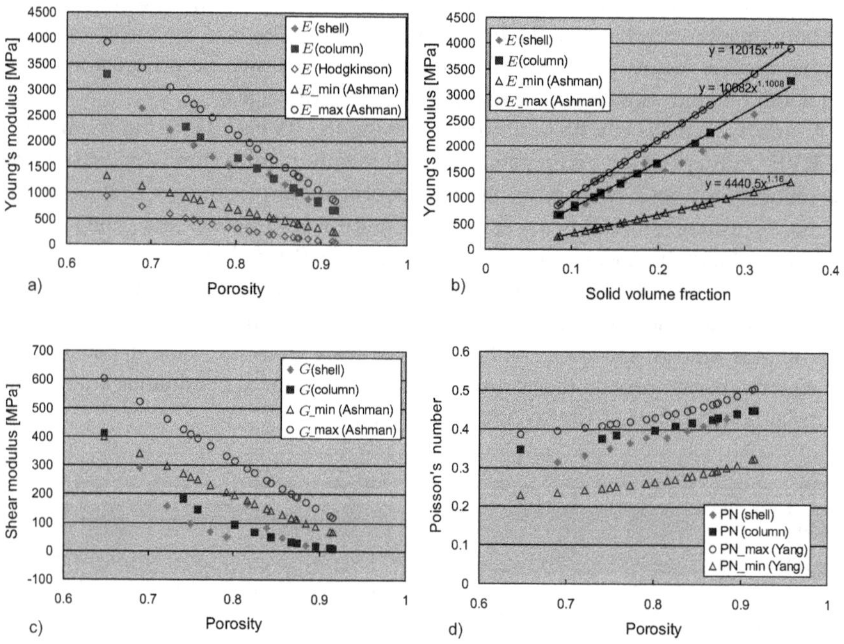

FIGURE 8.4.3 The effective material parameters and their comparison with the experimental values.

The effective material parameters presented in Fig. 8.4.3 lie in the following range: for RVE I and a porosity of 0.69–0.884, the Young modulus amounts to 890.47–2649.22 MPa, the shear modulus to 18.66–290.94 MPa and Poisson's ratio to 0.314–0.427. For RVE II and a porosity of 0.648–0.916, the Young modulus amounts to 679.42–3292.92 MPa, the shear modulus to 8.37–410.10 MPa and Poisson's ratio to 0.347–0.450. Apart from the calculated values, Fig. 8.4.3 also shows a comparison with values based on experimental investigations.

The results for Young's modulus are compared with the solution of Hodgskinson and Currey [95, 200] and Ashman et al. [16, 33]. The solution of Hodgkinson and Currey proposes the following values

$$E = 0.003715\rho_{\mathrm{w}}^{1.96} \tag{8.4.5}$$

where the result is expressed in MPa, if the density is expressed in kg/m^3. The upper and lower bounds predicted by Ashman et al. have the form

$$E_1 = E_{\min} = 0.52\rho_{\mathrm{w}}^{1.16}\ [\mathrm{MPa}], \tag{8.4.6}$$

$$E_3 = E_{\max} = 2.84\rho_{\mathrm{w}}^{1.07}[\mathrm{MPa}]. \tag{8.4.7}$$

Note that within this contribution, the wet structural density is used for the evaluation of (8.4.5)–(8.4.7), which is approximately 25% higher than the dry structural density [33]. It is expressed in the following way

$$\rho_{\mathrm{w}} = 1.25\rho_{\mathrm{s}}V_{\mathrm{v}}$$

where V_{v} is the solid volume fraction.

Diagram 8.4.3a shows that the values obtained numerically, by using the homogenization technique, lie within the bounds predicted by Ashman et al., but they are higher than Hodgkinson's. Fig. 8.4.3b shows the bounds of Ashman as well as the potential trendline for the values obtained by using RVE II, which is determined by the equation $E = 10082V_{\mathrm{v}}^{1.1008}$. The trendline for the values obtained by using RVE I is not included in the diagram for reasons of a clearer form of presentation, but it is close to the trend line corresponding to RVE II and is determined by the relation $E = 8131.1V_{\mathrm{v}}^{1.0095}$. The model predicts nearly linear dependence of the Young modulus on the solid volume fraction which is similar to the predictions of Ashman's group. However, it differs from the solution proposed by Gibson [130] according to which a dependency on the square of the structural density and solid volume fraction for the cubic open cell model is proposed:

$$\frac{E}{E_{\mathrm{s}}} \propto \left(\frac{\rho}{\rho_{\mathrm{s}}}\right)^2.$$

The effective values of the shear modulus (Fig. 8.4.3c) are also compared with the bounds proposed by Ashman et al. [16]:

$$G_{12} = G_{\min} = 0.08\rho_{\mathrm{w}}^{1.26}\ [\mathrm{MPa}], \tag{8.4.8}$$

$$G_{23} = G_{\max} = 0.29\rho_{\mathrm{w}}^{1.13}\ [\mathrm{MPa}], \tag{8.4.9}$$

which are again evaluated for the wet structural density. The comparison shows that the model presented here yields quite low values for the shear models. This furthermore indicates that an alternative geometry of the RVE or an alternative set of the material parameters (Table 8.1) should be assumed in order to improve the numerical values.

Since the contribution of Ashman et al. [16] does not include data on the Poisson number, the data of Yang et al. [405] are used in order to verify this parameter

$$\nu_{12} = \nu_{\min} = 0.176 V_v^{-0.248}, \tag{8.4.10}$$

$$\nu_{13} = \nu_{\max} = 0.316 V_v^{-0.191}, \tag{8.4.11}$$

and Figure 8.4.3d shows that the obtained values lie within the permitted domain.

Apart from the effective material parameters, the effective elasticity tensor calculated at the microlevel also renders information on the acoustic wave velocities. To this end the expression $c_u = \sqrt{\mathscr{C}_{11}/\rho}$ is used (see Eq. (8.3.7)). The change of the wave velocity with respect to the solid volume fraction is shown in Fig. 8.4.4 where a nearly linear dependence can easily be recognized. A similar behavior is predicted in many contributions, e.g. in the work by Haïat et al. [177] where two cases are distinguished that depend on the direction of the wave propagation. In the first case, the wave propagates in the anteroposterior direction, the fast and slow wave overlaps in time, and the approximate increase of the velocity is 80 m/s for 10 percent of the solid volume fraction. In the second case, the wave propagates in the direction which coincides with the principal orientation of the trabeculae in the transversal plane. Here, there is no overlapping of the waves in time, and the increase of the velocity of the fast wave is about 200 m/s for 10 percent of the solid volume fraction. The velocities range from 1515–1715 m/s for RVE I and 1578–2000 m/s for RVE II. The values lie between the physical bounds which amount to 1451 m/s for the pure marrow and to 2614 m/s for the bounded solid phase and to 3128 m/s for the unbounded solid phase.

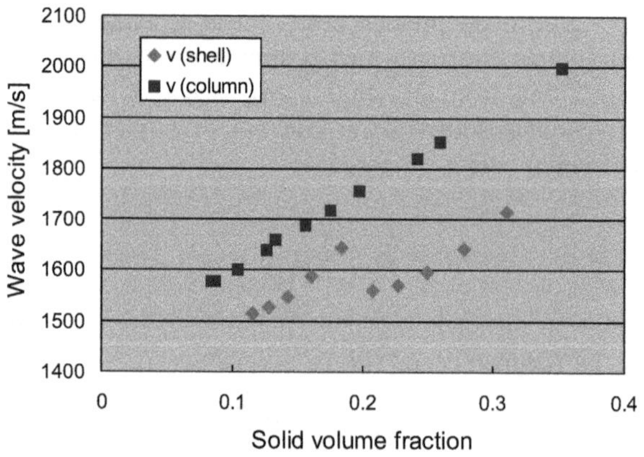

FIGURE 8.4.4 Dependence of the wave velocity on the solid volume fraction.

8.5 ANISOTROPY OF CANCELLOUS BONE

A number of experimental investigations show that cancellous bone has a very complex pattern along which the trabeculae are aligned and that this type of bone should be modeled as an anisotropic medium. The illustration considers an example of transversal anisotropy where the compliance tensor $\bar{\mathbf{N}}$ has a form dependent on six material parameters, five of which are independent

$$\bar{\mathbf{N}}=\begin{bmatrix} \frac{1}{E_1} & -\frac{\nu_{12}}{E_1} & -\frac{\nu_{31}}{E_3} & 0 & 0 & 0 \\ -\frac{\nu_{12}}{E_1} & \frac{1}{E_1} & -\frac{\nu_{31}}{E_3} & 0 & 0 & 0 \\ -\frac{\nu_{13}}{E_1} & -\frac{\nu_{13}}{E_1} & \frac{1}{E_3} & 0 & 0 & 0 \\ 0 & 0 & 0 & \frac{1}{G_{31}} & 0 & 0 \\ 0 & 0 & 0 & 0 & \frac{1}{G_{31}} & 0 \\ 0 & 0 & 0 & 0 & 0 & \frac{1}{G_{12}} \end{bmatrix}. \tag{8.5.1}$$

In the investigation, we use RVE II in such a way that the geometrical parameters in the longitudinal direction are changed, different from the parameters in the transversal and the sagittal directions. In other words, the side length of the base facet is kept constant at $a = 1$ mm and the high c of the RVE took values from the interval 1–2.25 mm (Fig. 8.5.1a). The column thickness is constant and it amounts to $b_2 = 0.2$ mm. The material parameters calculated for different RVE geometries are shown in Figs. 8.5.1b-d. Bearing in mind that the larger value of c corresponds to the greater porosity, the diagrams show that increasing side length c causes decreasing Young's and shear modulus and increasing Poisson's ratio. In contrast to the isotropic case, different material parameters for different directions are obtained. The values are compared with the bounds explained in the previous section Eqs. (8.4.6)–(8.4.11) and this comparison yields similar observations: Young's modulus and the Poisson ratio (except ν_{12}) lie within the permitted domain, and the model underestimates the shear modulus.

The comparison of the values for the isotropic and anisotropic cases is shown in Fig. 8.5.2. To this end, both RVEs are considered and like for RVE II (Fig. 8.5.1), c_1 is used to denote the height of RVE I in the following discussion. Two first diagrams (8.5.2a and 8.5.2b) are related to RVE I, where the length $a = 1$ mm and the wall thickness $d = 0.1$ mm are constant. Three groups of tests are performed. In the first group of tests the wall width amounts to $b = 1/4$ mm, in the second group of tests the width is $b = 1/6$ mm and in the last series, b is 1/8 mm. These groups of tests are denoted by (a), (b) and (c) respectively. In each group of the tests, for the fixed width b, the height c of the RVE changes from 1 to 2.25 mm with a step of 0.25 mm. The values obtained for the isotropic case are represented by the red line. Figures 8.5.2c and 8.5.2d show the values obtained for RVE II. Here again, three groups of tests are performed. In each of the groups, the side length a of the basic square is fixed. The chosen values are: $a = 1$ mm for the group a, $a = 1.5$ mm for the group b and $a = 2$ mm for the group c. For the constant a, the height c of the RVE increases from 1 to 2.25 mm with a step of 0.25 mm. The rod thickness of 0.2 mm is constant. The red line again represents the values corresponding to an isotropic case.

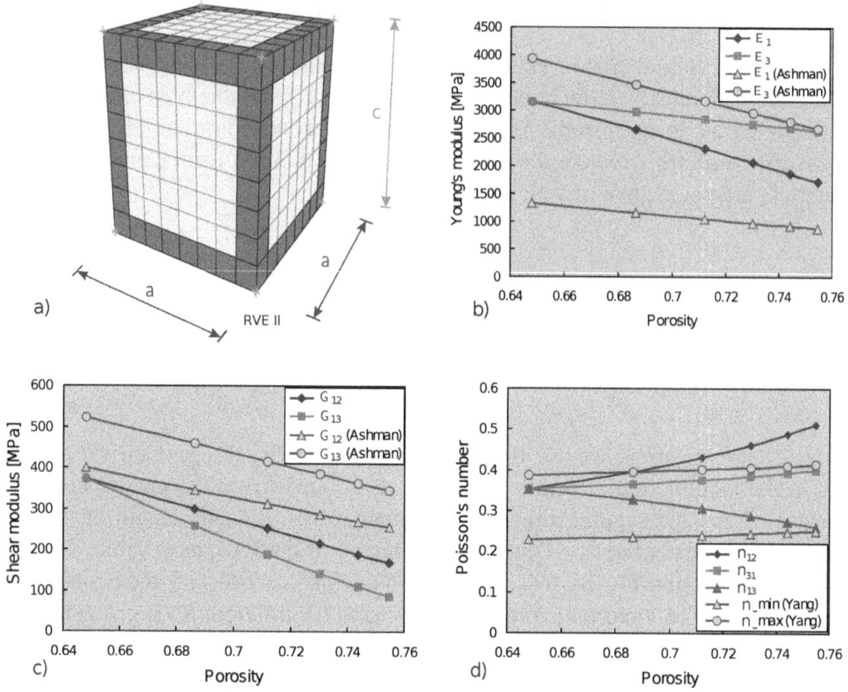

FIGURE 8.5.1 a) RVE II is used for the investigation of the transversal anisotropy. b)-d) Change of the effective material parameters with an increasing side length c of RVE II. Subscript 3 indicates the vertical direction, while 1 and 2 denote two orthogonal directions in the horizontal plane.

Another contribution to the anisotropy of cancellous bone certainly is the anisotropy of the frame material itself. This type of anisotropy is rarely taken into consideration for two reasons. Firstly, because the anisotropy due to the different geometrical parameters in different directions is predominant, secondly because the material parameters of the singular trabeculae are extremely difficult to identify. For the last reason, the material parameters characteristic for cortical-compact bone are used to simulate trabeculae behavior, although many authors suggest that trabecular material has a considerably lower Young's modulus [351].

Up to now, we also used material parameters corresponding to the cortical bone, in addition assuming the material isotropy. Conversely, in our subsequent simulations, we treat the trabecular material as the transversally anisotropic one with the properties: $E_1 = 12.8$ GPa, $G_{13} = G_{23} = 3.3$ GPa, $\nu_{12} = \nu_{21} = 0.53$ and $\nu_{31} = \nu_{32} = 0.41$. The remaining parameters are calculated according to the standard relationships

$$\frac{\nu_{13}}{E_1} = \frac{\nu_{31}}{E_3}, \qquad \frac{\nu_{23}}{E_1} = \frac{\nu_{32}}{E_3}, \qquad G_{12} = \frac{E_1}{2(1+\nu_{12})}. \tag{8.5.2}$$

The mentioned values are taken from the work of Black and Hastings [33, 348]. In the tests, the Young modulus in the third direction E_3 is taken as a variable.

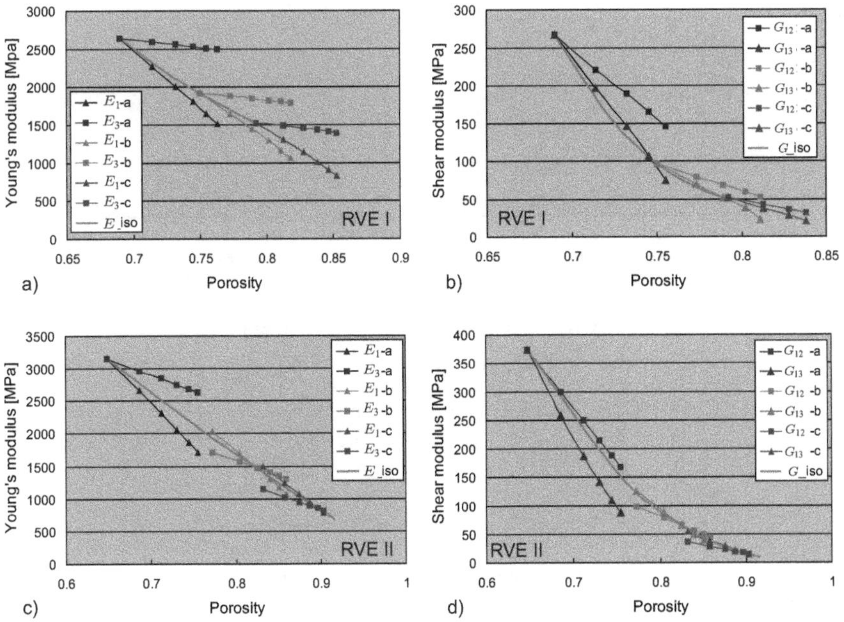

FIGURE 8.5.2 The relationships between the effective values obtained for the isotropic and anisotropic cases. Diagrams a) and b) show the results obtained by using RVE I, diagrams c) and d) show the results of the analysis of RVE II. Three groups of anisotropic tests are performed: a, b and c. The results for the isotropic case are represented by the red line.

In the investigation of the influence of Young's modulus corresponding to the sagittal and transversal direction on the effective material parameters, we carried out a group of tests where E_1 changes in the range 12–22 GPa with a step of 2 GPa. The results presented in Fig. 8.5.3 are obtained for a fixed geometry of the RVE chosen such that the side length of the representative cube amounts to 1 mm, solid rods have the width $b = 0.125$ mm and the porosity amounts to 0.844. These values can now be compared with the values corresponding to the isotropic case being: $E = 2077.734$ N/mm^2, $G = 86.243$ N/mm^2 and $\nu = 0.378$ for the mentioned porosity and column model (Fig. 8.4.3). The comparison shows that, when increasing E_1, the effective values get closer to those corresponding to material isotropy, but there is still a significant distinction as the assumed material parameters for sagittal and transversal directions are different. The values for Poisson's ratio are in the allowed range typical for transversal anisotropy: $-1.0 < \nu < 1.0$. Of course, now the same procedure can be used in order to capture the influence of orthogonal anisotropy. To this end another set of material parameters should be used, for which the values obtained by ultrasonic tests [13, 33] are often recommended.

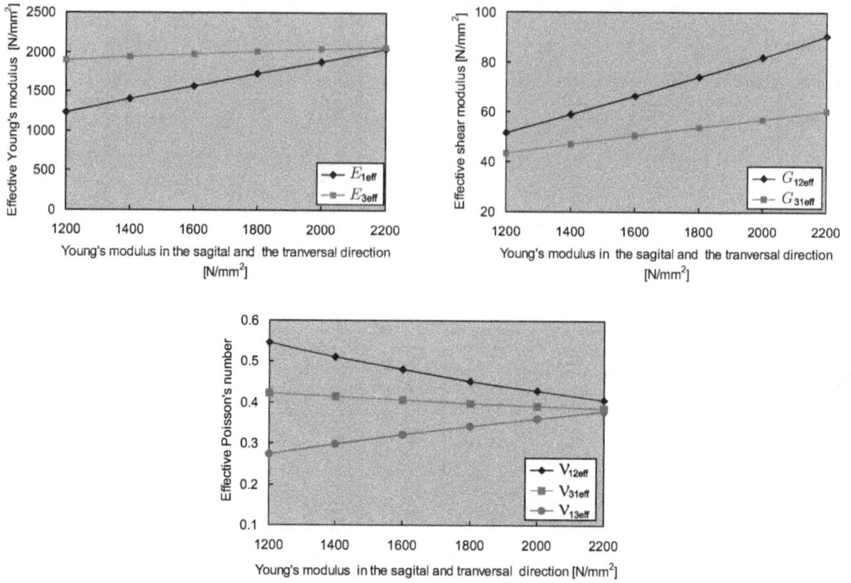

FIGURE 8.5.3 Diagrams of effective material parameters for the case of transversal anisotropy induced by the material anisotropy of the solid phase. Subscript 3 indicates the vertical direction, while 1 and 2 denote two orthogonal directions in the horizontal plane.

8.6 THE INFLUENCE OF REFLECTION ON THE ATTENUATION

8.6.1 PRINCIPLES OF THE REFLECTION PHENOMENON

By the propagation of a wave through a multiphase material, the following phenomena can be observed: each contact surface represents a barrier, whose overcoming leads to the weakening of the original wave and to the change of the original direction of the wave propagation. While the first phenomenon goes back to the reflection, the second one is known as refraction. A typical scenario of the wave propagation through two different media is shown in Fig. 8.6.1a, where the incident wave encounters the contact surface, which is partially reflected and partially transmitted. The incident angle α^{in} is equal to the reflection one α^{r}, while the refracted angle α^{tr} is calculated by using the expression

$$\frac{\sin \alpha^{\mathrm{in}}}{\sin \alpha^{\mathrm{tr}}} = \frac{c_1}{c_2}, \tag{8.6.1}$$

where c_1 and c_2 are the velocities of the wave through the first and second phase. A special case of the wave propagation through different media is shown in Fig. 8.6.1b. Here, the incident wave is perpendicular to the contact surface and the three aforementioned angles are equal to each other. It should be noted that this case is relevant for latter derivations where the sound wave, propagating perpendicularly to the sample, is simulated (Fig. 8.3.2).

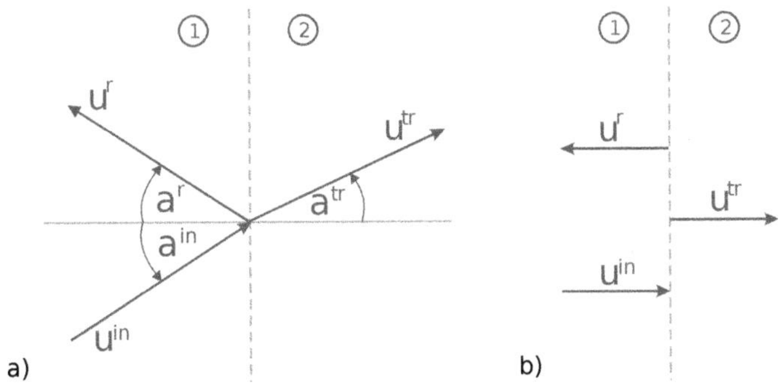

FIGURE 8.6.1 Reflection and transmission of a wave. a) An arbitrary incident wave. b) The incident wave is perpendicular to the contact surface.

Fig. 8.6.1 also indicates that the material continuity must stay preserved in spite of the branching of the wave. This is achieved by the equality of the resultant amplitude of the incoming and reflected wave with the amplitude of the transmitted wave

$$u^{\text{in}} + u^{\text{r}} = u^{\text{tr}}. \tag{8.6.2}$$

In this expression, u denotes the amplitude of the oscillating particles in the direction of the wave propagation and the superscripts "in," "r" and "tr" are used to denote the quantities related to the incoming, reflected and transmitted waves. Some more information on the reflected and transmitted waves is obtained on the basis of the acoustic impedances of the involved materials. The reflection coefficient R and the transmission coefficient T, for example, define the relationships between the amplitude of the reflected/transmitted wave and the amplitude of the incident wave

$$R = \frac{u^{\text{r}}}{u^{\text{in}}} = \frac{z_1 - z_2}{z_1 + z_2}, \qquad T = \frac{u^{\text{tr}}}{u^{\text{in}}} = \frac{2z_1}{z_1 + z_2}. \tag{8.6.3}$$

Here, $z_1 = \rho_1 c_1$ and $z_2 = \rho_1 c_2$ denote the acoustic impedances of both media. Expressions analog to Eq. (8.6.3) can also be defined for the wave pressure and wave energy [34]. However, since the displacement variational formulation is considered, the amplitude of the oscillations is of greater significance.

It is still worth mentioning that the value of the reflection coefficient has great influence on the way of wave propagation. According to (8.6.3), this coefficient can have positive as well as negative values depending on the relation of the impedances of the involved media. If $z_1 > z_2$, the reflection coefficient is positive, the reflected wave is in the phase with the incident wave, and the amplitudes of the transmitted wave are higher than the amplitudes of the incident wave. The situation is opposite if $z_2 > z_1$. In this case, the reflection coefficient is negative, the reflected wave and the incident wave are out of phase and the amplitudes of the transmitted wave are lower than those of the incident wave. The transmission scenario for different relations between acoustic impedances is presented in Fig. 8.6.2.

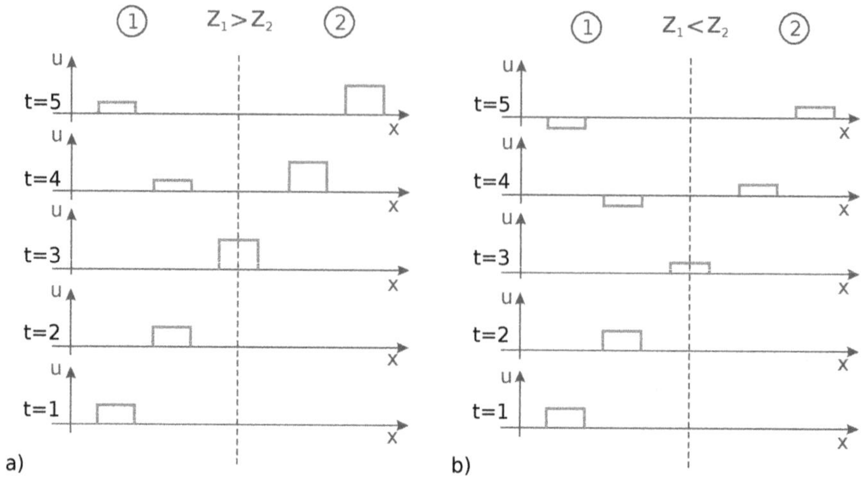

FIGURE 8.6.2 Amplitudes of the reflected and transmitted wave, dependent on the relationship between the acoustic impedances of the involved media.

8.6.2 VARIATIONAL FORMULATION FOR THE WAVE PROPAGATION TAKING THE EFFECTS OF THE REFLECTION INTO ACCOUNT

Although the amplitudes of the transmitted wave can be lower or higher than those of the original wave, the transmitted wave always has lower energy than the original one. The loss of energy of a wave upon the contact layer between different media is explained by the reflection phenomena. This aspect is used in the current contribution for the extension and the improvement of previously proposed models [220, 222].

For the purpose of a variational formulation for the reflection effects, the standard relation for the wave pressure p needs to be recalled

$$p = z\dot{u}, \qquad z = \rho c. \tag{8.6.4}$$

Here, the pressure p is linearly dependent on the velocity of the oscillating particles $\dot{u}$ and the acoustic impedance z is the proportionality constant. On the basis of this relation, two conjugate formulations for the power of a wave are defined

$$\psi(\dot{u}) = \frac{1}{2}z\dot{u}^2 \quad \text{and} \quad \bar{\psi}(p) = \frac{1}{2}\frac{p^2}{z}. \tag{8.6.5}$$

Furthermore, these equations can easily be adapted for the simulation of the reflected wave. To this end, the amplitudes of the reflected wave are expressed on the basis of the continuity equation (8.6.3)

$$u^{\text{in}} + u^{\text{r}} = u^{\text{tr}} \quad \Rightarrow \quad u^{\text{r}} = u^{\text{tr}} - u^{\text{in}}. \tag{8.6.6}$$

The obtained relation is inserted in (8.6.5)

$$\psi = \frac{1}{2}z(\dot{u}^{\text{r}})^2 \quad \Rightarrow \quad \psi = \frac{1}{2}z(\dot{u}^{\text{tr}} - \dot{u}^{\text{in}})^2. \tag{8.6.7}$$

Here, only expression (8.6.5) is transformed, as the concept based on the displacement variational formulation is used in this chapter.

In the next step, the derivative of the wave power and the action integral are considered [241]. Bearing in mind that the derivative of the wave power represents pressure

$$p = \frac{\partial \psi(\dot{u})}{\partial \dot{u}} \quad \Rightarrow \quad p^{\mathrm{r}} = \frac{\partial \psi(\dot{u}^{\mathrm{r}})}{\partial \dot{u}^{\mathrm{r}}}, \tag{8.6.8}$$

the action integral (8.2.17) is extended as follows

$$\delta I = -\delta \int_0^t L(\mathbf{u}, \dot{\mathbf{u}})\,\mathrm{d}t + \int_0^t \int_{\mathrm{A}} \delta u^{\mathrm{r}} \frac{\partial \psi(\dot{u}^{\mathrm{r}})}{\partial \dot{u}^{\mathrm{r}}}\,\mathrm{d}A\,\mathrm{d}t + \delta \int_0^t \Pi^{\mathrm{ext}}(\mathbf{u})\,\mathrm{d}t. \tag{8.6.9}$$

The new term on the right-hand side represents the effects of reflection and is based on the fact that δu^{r} is the work conjugate of the pressure (8.6.8). Here, the scenario shown in Fig. 8.5.3b is assumed such that u^{r} is the amplitude of the reflected wave in the direction perpendicular to the contact surface. For this reason, u^{r} is introduced as a scalar and not a vector quantity. Functions $\mathbf{u}$ and u^{r} depend on two arguments: $\mathbf{u} = \mathbf{u}(\mathbf{x}, t)$, $u^{\mathrm{r}} = u^{\mathrm{r}}(\mathbf{x}, t)$. The symbol A denotes the contact surface between the phases.

For the periodic excitation, integral (8.6.9) becomes a form dependent only on the amplitudes of oscillations $\mathbf{u}(\mathbf{x})$ and $u^{\mathrm{r}}(\mathbf{x})$

$$\delta I = -\delta L(\mathbf{u}, \ddot{\mathbf{u}}) + \int_{\mathrm{A}} \delta u^{\mathrm{r}} \frac{\partial \psi(\dot{u}^{\mathrm{r}})}{\partial \dot{u}^{\mathrm{r}}}\,\mathrm{d}A + \delta \Pi^{\mathrm{ext}}(\mathbf{u}), \tag{8.6.10}$$

which is also suitable for defining the boundary value problem at the microscale. In this case, the influence of external forces $\delta \Pi^{\mathrm{ext}}(\mathbf{u})$ is replaced by the virtual work due to the macroscopic deformations $\delta L(\bar{\varepsilon})$:

$$\delta I = -\delta L(\mathbf{u}, \ddot{\mathbf{u}}) + \int_{\mathrm{A}} \delta u^{\mathrm{r}} \frac{\partial \psi(\dot{u}^{\mathrm{r}})}{\partial \dot{u}^{\mathrm{r}}}\,\mathrm{d}A - \delta L(\bar{\varepsilon}). \tag{8.6.11}$$

In the final step, the emphasis is on the transformation of the reflection term, which is denoted by Ψ hereafter:

$$\Psi(\dot{u}^{\mathrm{r}}) = \int_{\mathrm{A}} \delta u^{\mathrm{r}} \frac{\partial \psi(\dot{u}^{\mathrm{r}})}{\partial \dot{u}^{\mathrm{r}}}\,\mathrm{d}A. \tag{8.6.12}$$

Here, the expressions

$$\delta u^{\mathrm{r}} = \delta u^{\mathrm{tr}} - \delta u^{\mathrm{in}}, \qquad \frac{\partial \psi(\dot{u}^{\mathrm{r}})}{\partial \dot{u}^{\mathrm{r}}} = z(\dot{u}^{\mathrm{tr}} - \dot{u}^{\mathrm{in}}), \tag{8.6.13}$$

$$u = u(x)e^{i\omega t}, \qquad \dot{u} = i\omega\, u(x)e^{i\omega t} = i\omega u, \tag{8.6.14}$$

are derived on the basis of the periodicity of the deformations as well as relations (8.6.6) and (8.6.7). Subsequently, these expressions are inserted in (8.6.12) in order to obtain the full form of the reflection term:

$$\Psi(u) = i \int_{\mathrm{A}} \delta u^{\mathrm{in}} z\omega u^{\mathrm{in}}\,\mathrm{d}A - i \int_{\mathrm{A}} \delta u^{\mathrm{in}} z\omega u^{\mathrm{tr}}\,\mathrm{d}A$$

$$-i\int_{A}\delta u^{tr}z\omega u^{in}dA+i\int_{A}\delta u^{tr}z\omega u^{tr}dA. \tag{8.6.15}$$

It should be noted that an extension analogous to that one in Eq. (8.2.19) can be performed to simulate the effects of friction forces.

8.6.3 NUMERICAL IMPLEMENTATION

The numerical implementation is explained on the basis of the example shown in Fig. 8.6.3. Here, an acoustic wave propagates through two elements with different material properties. The contact surface is perpendicular to the direction of the wave propagation, hence, there is no change in direction due to refraction (Fig. 8.6.1b).

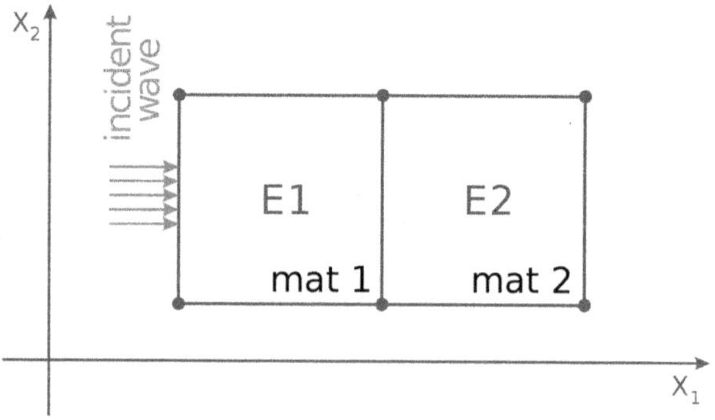

FIGURE 8.6.3 The wave propagation through two quadrilateral elements with different material properties.

According to expressions (8.6.11) and (8.6.15), the action integrals corresponding to these two elements have the form

$$\delta I^{E1}=-\delta L(\hat{\mathbf{u}}^{E1})+\frac{1}{2}\Psi(\hat{\mathbf{u}}^{in},\hat{\mathbf{u}}^{tr})-\delta L^{E1}(\bar{\varepsilon})=0, \tag{8.6.16}$$

$$\delta I^{E2}=-\delta L(\hat{\mathbf{u}}^{E2})+\frac{1}{2}\Psi(\hat{\mathbf{u}}^{in},\hat{\mathbf{u}}^{tr})-\delta L^{E2}(\bar{\varepsilon})=0, \tag{8.6.17}$$

where the symbols with the hat indicate the nodal values. The coefficient $1/2$ in front of the reflection term is introduced because this influence is split between two elements.

The expressions (8.6.16)–(8.6.17) require additional comments. According to Fig. 8.6.3, the wave passes through elements E1 and E2 so that the transformation of the wave amplitudes occurs upon contact. This implies that the DOFs of the nodes of element E1, lying on the contact surface, coincide with the DOFs corresponding to the incident wave $\hat{\mathbf{u}}^{in}$. On the other hand, the DOFs belonging to the contact nodes of element E2 coincide with the DOFs related to the transferred wave $\hat{\mathbf{u}}^{tr}$. Bearing this in mind, expressions (8.6.16) and (8.6.17) can be written as

$$\delta I^{E1}=-\delta L(\hat{\mathbf{u}}^{E1})+\frac{1}{2}\Psi(\hat{\mathbf{u}}^{E1},\hat{\mathbf{u}}^{E2})-\delta L^{E1}(\bar{\varepsilon})=0, \tag{8.6.18}$$

$$\delta I^{\mathrm{E2}} = -\delta L(\hat{\mathbf{u}}^{\mathrm{E2}}) + \frac{1}{2}\Psi(\hat{\mathbf{u}}^{\mathrm{E1}}, \hat{\mathbf{u}}^{\mathrm{E2}}) - \delta L^{\mathrm{E2}}(\bar{\varepsilon}) = 0, \tag{8.6.19}$$

which shows that each of the action integrals depends on the DOFs of two elements and hence is inconvenient for a numerical implementation. In order to overcome this problem, a transient element, suitable for the modeling of reflection effects, is introduced between elements E1 and E2 (Fig. 8.6.4).

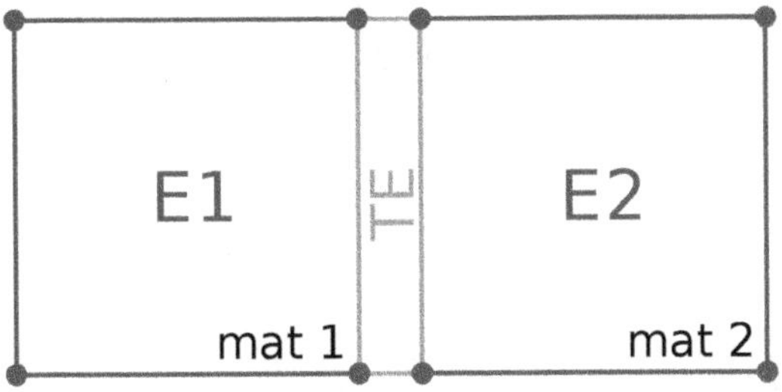

FIGURE 8.6.4 Two quadrilateral elements with different material properties E1 and E2 are separated by a transient element TE.

According to the situation presented in Fig. 8.6.4, the variational formulation related to the elements E1 and E2 remains as already defined in Section 8.4.1 and can be briefly described in the following manner:

$$\delta I^{\mathrm{E1}} = -\delta L(\hat{\mathbf{u}}^{\mathrm{E1}}) - \delta L^{\mathrm{E1}}(\bar{\varepsilon}) = 0, \tag{8.6.20}$$

$$\delta I^{\mathrm{E2}} = -\delta L(\hat{\mathbf{u}}^{\mathrm{E2}}) - \delta L^{\mathrm{E2}}(\bar{\varepsilon}) = 0. \tag{8.6.21}$$

In addition, the newly proposed element is used to implement reflection effects

$$\delta I^{\mathrm{TE}} = \Psi(\hat{\mathbf{u}}^{\mathrm{in}}, \hat{\mathbf{u}}^{\mathrm{tr}}) = 0. \tag{8.6.22}$$

This model is further simplified by defining the transient element as a double plane element (Fig. 8.6.5), which is substantiated by the fact that the transformation of the wave amplitudes occurs at the contact of phases and that the thickness of the transient layer is infinitesimally small in reality.

As shown in Fig. 8.6.5, the new element type is composed of two parts. The area determined by the nodes 1–4 represents a 4-node element with bilinear shape functions. The mentioned nodes are linked to the nodes of the element E1 so that they copy information about the amplitudes of the incident wave. The FE approximation for this part of the element has the form

$$\mathbf{u}^{\mathrm{in}} = \mathbf{N}\hat{\mathbf{u}}^{\mathrm{in}}, \tag{8.6.23}$$

where $\mathbf{u}^{\mathrm{in}}$ and $\hat{\mathbf{u}}^{\mathrm{in}}$ are the displacement function and the vector of DOFs related to an element

$$\mathbf{u}^{\mathrm{in}} = \begin{bmatrix} u^{\mathrm{inR}} & 0 & 0 & iu^{\mathrm{inI}} & 0 & 0 \end{bmatrix}^{\mathrm{T}}, \tag{8.6.24}$$

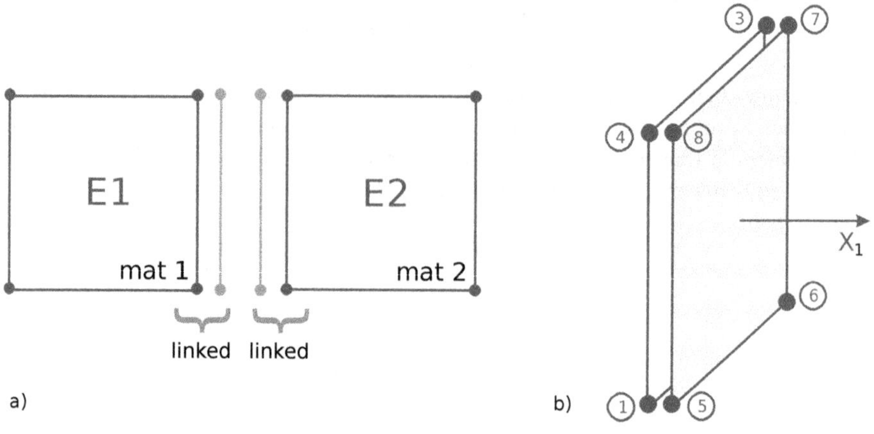

FIGURE 8.6.5 a) 2D presentation of the transient element with the infinitely small thickness and its neighboring elements. b) 3D presentation of the transient element.

$$\hat{\mathbf{u}}^{\text{in}} = \left[(\hat{\mathbf{u}}_1^{\text{in}})^{\text{T}} \quad (\hat{\mathbf{u}}_2^{\text{in}})^{\text{T}} \quad (\hat{\mathbf{u}}_3^{\text{in}})^{\text{T}} \quad (\hat{\mathbf{u}}_4^{\text{in}})^{\text{T}} \right]^{\text{T}}. \tag{8.6.25}$$

The nodal DOFs, related to the node k, are defined as follows

$$\hat{\mathbf{u}}_k^{\text{in}} = \left[\hat{u}_k^{\text{in R}} \quad 0 \quad 0 \quad i\hat{u}^{\text{in I}} \quad 0 \quad 0 \right]^{\text{T}}, \tag{8.6.26}$$

while the matrix of the shape functions $\mathbf{N}$ has the block from

$$\mathbf{N} = [\mathbf{N}_1 \quad \mathbf{N}_2 \quad \mathbf{N}_3 \quad \mathbf{N}_4], \tag{8.6.27}$$

$$\mathbf{N}_i = \begin{bmatrix} N_i & & & & & \\ & 0 & & & & \\ & & 0 & & & \\ & & & N_i & & \\ & & & & 0 & \\ & & & & & 0 \end{bmatrix}, \tag{8.6.28}$$

where all out-of-diagonal elements are equal to zero.

The surface, determined by nodes 5–8, has similar properties as the surface 1–4, but the nodes 5–8 are linked with the element E2. Thus, they import data on the transmitted wave into the transient element. The FE approximation for this part of the transient element is defined in the same way as it is done for the first part. The difference is that the notation corresponding to the incident wave is replaced by the notation related to the transmitted wave. For example, the amplitudes of the transmitted wave are analogous to (8.6.23)

$$\mathbf{u}^{\text{tr}} = \mathbf{N}\hat{\mathbf{u}}^{\text{tr}}. \tag{8.6.29}$$

The block-matrix $\mathbf{N}$ remains unchanged in both cases.

Further derivations are straightforward. After the implementation of the chosen FE approximation in (8.6.15) and (8.6.22), the action integral of the transient element turns into

$$\Psi(u) = i(\delta\hat{\mathbf{u}}^{\text{in}})^{\text{T}}\mathbf{K}\cdot\hat{\mathbf{u}}^{\text{in}} - i(\delta\hat{\mathbf{u}}^{\text{in}})^{\text{T}}\mathbf{K}\cdot\hat{\mathbf{u}}^{\text{tr}}$$

$$-i(\delta\hat{\mathbf{u}}^{\text{tr}})^{\text{T}}\mathbf{K}\cdot\hat{\mathbf{u}}^{\text{in}} + i(\delta\hat{\mathbf{u}}^{\text{tr}})^{\text{T}}\mathbf{K}\cdot\hat{\mathbf{u}}^{\text{tr}} = 0, \tag{8.6.30}$$

where the matrix $\mathbf{K}$ is defined by the expression

$$\mathbf{K} = z\omega \int_{\text{A}} \mathbf{N}^{\text{T}}\mathbf{N}\text{d}A. \tag{8.6.31}$$

Finally, the short formulation

$$\Psi(u) = \delta\hat{\mathbf{u}}\,\mathbf{K}^{\text{e}}\,\hat{\mathbf{u}} = 0, \qquad \mathbf{K}^{\text{e}}\,\hat{\mathbf{u}} = \mathbf{0} \tag{8.6.32}$$

is obtained by introducing the block matrices and vectors of the form

$$\delta\hat{\mathbf{u}} = \begin{bmatrix} \delta\hat{\mathbf{u}}^{\text{in}} \\ \delta\hat{\mathbf{u}}^{\text{tr}} \end{bmatrix}, \quad \mathbf{K}^{\text{e}} = i\begin{bmatrix} \mathbf{K} & -\mathbf{K} \\ -\mathbf{K} & \mathbf{K} \end{bmatrix}, \quad \hat{\mathbf{u}} = \begin{bmatrix} \hat{\mathbf{u}}^{\text{in}} \\ \hat{\mathbf{u}}^{\text{tr}} \end{bmatrix}. \tag{8.6.33}$$

Here, an additional remark related to the numerical implementation is necessary. The coordinate system shown in Figs. 8.6.3 and 8.6.5 is the local one and is chosen in such a way that the axis X_1 coincides with the direction of wave propagation. However, the element implemented in the FE program is adapted to the transformation from the global to the local coordinate system so that the direction of wave propagation can be chosen arbitrarily but perpendicular to the exposed contact surface.

8.6.4 NUMERICAL RESULTS

The selected numerical examples deal with the simulation of the wave propagation through cancellous bone. For this purpose, an RVE consisting of the solid frame and the fluid marrow together with the transit layer on the contact area between the phases is assumed (Fig. 8.6.6). Obviously, the configuration of the RVE coincides with the one shown in Fig. 8.4.1, but the transient layer is introduced as a new component.

In order to get an impression of the effective bone behavior, the experiment shown in Fig. 8.3.2 is simulated at the macrolevel and the obtained numerical values are compared with the experimental results. Similar to Section 8.3, the emphasis is placed on the investigation of the change of the effective attenuation with respect to the increasing excitation frequency of the acoustic wave.

The geometry of the RVE is determined by two parameters: the length of the external side of the RVE a and the width of the solid framework column b (Fig. 8.6.6). The length a amounts to 1 mm while the column width b is $b = 0.125$ mm. The unit length $a = 1$ mm is assumed as any RVE can be scaled up to the unit cube. The effective density corresponding to this geometry is 1108 kg/m^3, and it belongs to the range that is typical of the cancellous bone (1050–1250 kg/m^3). The frequency of the excitation wave is varied. The applied values are in the range of 400–1000 kHz.

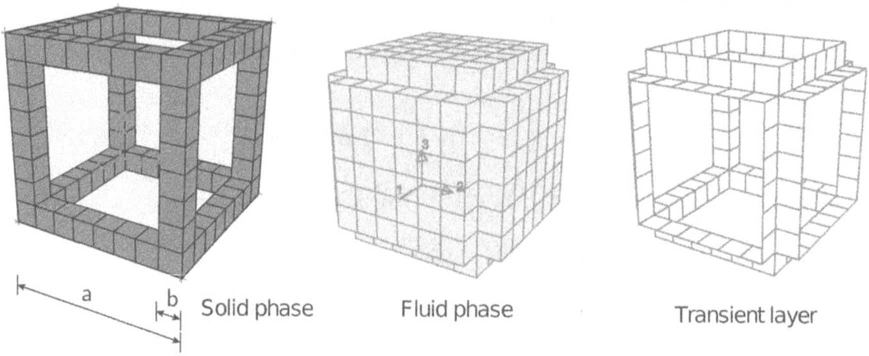

FIGURE 8.6.6 Geometry of the proposed RVE for cancellous bone.

For the modeling of both phases, the solid elements described in Section 8.4.1 and the material parameters listed in Table 8.1 are used. The simulations also require data on the wave velocities and the corresponding acoustic impedances:

$$c_s = \sqrt{E_s/\rho_s} = 3350\,\frac{\text{m}}{\text{s}}, \quad z_s = \rho_s c_s = 6.566\,\frac{\text{kg}}{\text{mm}^2\text{s}}, \tag{8.6.34}$$

$$c_f = \sqrt{K_f/\rho_f} = 1450\,\frac{\text{m}}{\text{s}}, \quad z_f = \rho_f c_f = 1.3775\,\frac{\text{kg}}{\text{mm}^2\text{s}}. \tag{8.6.35}$$

Please recall that the indices s and f are introduced to denote the solid and the fluid phases. As presented in Fig. 8.6.2, two idealized scenarios are possible. In the first case (Fig. 8.6.2a), the wave propagates through the solid bone and then passes into the fluid marrow. Since the reflected wave returns through the solid medium, the acoustic impedance z in expression (8.6.31) takes the value characteristic of bone, namely $z = z_s$. In the second case (Fig. 8.6.2b), the wave passes from the marrow into the bone and the reflected wave propagates through the fluid, which is described by the condition $z = z_f$.

At the macroscopic level, the simulations deal with the model presented in Fig. 8.6.7a. Some illustrative examples of simulations are shown in Figs. 8.6.7b-d. The focus is placed on the difference between the amplitudes of oscillations for different excitation frequencies. The reflection coefficient applied in simulations has the value typical of solid bone ($z = z_s$). Note that for the purpose of a simplified numerical implementation, the reflection coefficient at all contact surfaces takes the same value.

Fig. 8.6.7e shows some more details on the oscillation for a cut through the middle of the sample. Data presented here enable the evaluation of the attenuation coefficient by using Eq. (8.3.9). The obtained values (Fig. 8.6.8) clearly show some important advantages of the proposed model. First, it yields the attenuation in the range 5–25 dB/cm which agrees very well with the experimentally obtained values [177, 205]. The achieved improvement is more obvious if the current values are compared with the results obtained on the basis of the model without taking the reflection effects

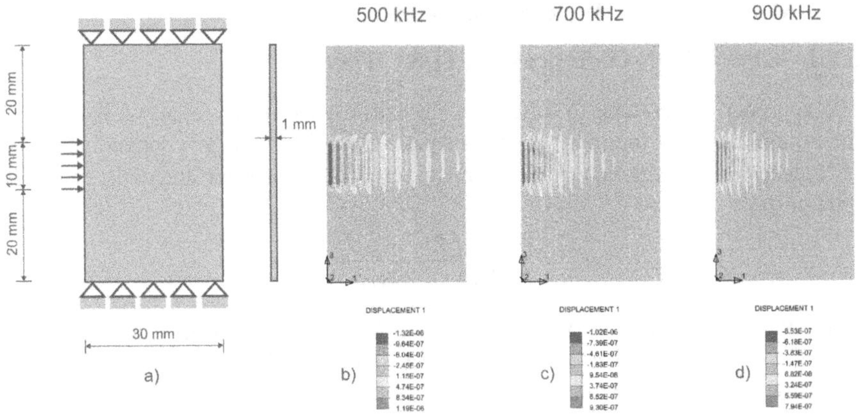

FIGURE 8.6.7 a) Geometry of the model simulated at the macrolevel. b)-d) Wave propagation through the sample.

into consideration (Section 8.3.4). The results obtained in this way are in the range 0.8–3.2 dB/cm and thus are much smaller than the real ones. Another interesting observation considering Fig. 8.6.8a is that the diagrams obtained for different values of the reflection coefficient are clearly separated. The values obtained by assuming $z = z_s$ and $z = z_f$ also represent the upper and the lower bound for the effective attenuation coefficient. According to the results for set (a) in Fig. 8.6.8, it can be concluded that numerical values might be well approximated by a linear function as predicted by some experimental investigations [75, 397]. On the other hand, the results for set (b) indicate that a nonlinear attenuation occurs as suggested in [177].

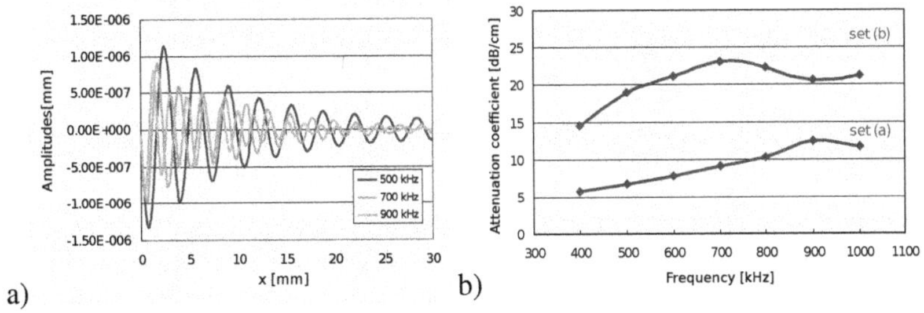

FIGURE 8.6.8 a) Amplitudes of the oscillations for a cut through the middle of the sample. b) The attenuation coefficient. Set (a) is obtained for the reflection coefficient $z = z_s$. Set (b) corresponds to the reflection coefficient $z = z_f$.

8.7 MULTISCALE INVERSE ANALYSIS

8.7.1 DEFINITION OF THE MERIT FUNCTION

The approach we assume to follow for the purpose of inverse analysis is based on the minimization of the merit or mean square error function (χ^2) depending on the difference between the experimental and numerical results according to the following relationship

$$\chi^2 = \sum_{j=1}^{N_{ts}} \sum_{i=1}^{N} \left(\frac{\tilde{g}_{j,i} - g_i(\mathbf{i}_j, \mathbf{a})}{s_{j,i}} \right)^2. \tag{8.7.1}$$

Here, N_{ts} is the number of different tests and N the number of measured values for each test, $\tilde{g}_{j,i}$ are experimental values, and $g_i(\mathbf{i}_j, \mathbf{a})$ are corresponding values obtained by using a mechanical model. Symbol $\mathbf{i}_j$ denotes the j-th set of input parameters, $s_{j,i}$ is the standard deviation which, in this case, is equal to the measurement error, and $\mathbf{a}$ is the set of material parameters, which have to be evaluated. In the particular case considered here, the mechanical model is related to a boundary value problem for simulation of a body with a heterogeneous microstructure. More precisely, $g_i(\mathbf{i}_j, \mathbf{a})$ are the real and imaginary parts of effective moduli, which are obtained as results of calculations at the microlevel. Finally, $\mathbf{i}_j$ are the macroscopic perturbations and the vector $\mathbf{a}$ contains geometrical parameters related to the bone microarchitecture. Subsequently, the numerically obtained displacements are written in a shorter form

$$g_{j,i} = g_i(\mathbf{i}_j, \mathbf{a}). \tag{8.7.2}$$

Note that there are also alternative proposals for the function χ^2, however, expression (8.7.1) has been established as the standard one, which is often explained with the following reasons: First, assumption (8.7.1) allows a simple numerical implementation. Second, it can be shown that expression (8.7.1) is not just an assumption, but also a result of the maximum likelihood estimation (MLE), i.e. in the case of the Gaussian distribution, the likelihood function has the following form

$$P = \prod_{j=1}^{N_{ts}} \prod_{i=1}^{N} \left\{ \exp\left[-\frac{1}{2} \left(\frac{\tilde{g}_{j,i} - g_i(\mathbf{i}_j, \mathbf{a})}{s_{j,i}} \right)^2 \right] \Delta g \right\}. \tag{8.7.3}$$

This expression represents the probability that the numerical values $g_i(\mathbf{i}_j, \mathbf{a})$ belong to the ranges $\tilde{g}_{j,i} + \Delta g$ around the experimental values $\tilde{g}_{j,i}$, where Δg is a constant. Since monotone transformations like the logarithm preserve the maximizer and can simplify calculations, the maximizer of (8.7.3) is simultaneously the maximizer of the logarithm of this expression and the minimizer of its negative logarithm:

$$\left[\sum_{j=1}^{N_{ts}} \sum_{i=1}^{N} \left(\frac{\tilde{g}_{j,i} - g_i(\mathbf{i}_j, \mathbf{a})}{2 s_{j,i}} \right)^2 \right] - (N_{ts} + N) \log \Delta g. \tag{8.7.4}$$

Now the minimizer of the previous expression is equal to the minimizer of the starting condition (8.7.1) as N_{ts}, N, $s_{j,i}$ and Δg are constants [344, 392].

8.7.2 THE LEVENBERG-MARQUARDT METHOD

Among different approaches, the Levenberg-Marquardt method is chosen for the minimization of the merit function (8.7.1) [240]. This method combines the steepest descent method with the concept of minimizing the Taylor approximation of the function. According to the first one, the minimization of function $\chi^2(\mathbf{a})$ is achieved by assuming an increment in the argument $\Delta\mathbf{a}$ in the direction of the negative function gradient $\mathbf{b}$:

$$\mathbf{b} = k\Delta\mathbf{a}, \qquad \Delta\mathbf{a} = \mathbf{a}_{i+1} - \mathbf{a}_i, \quad \mathbf{b} = -\nabla\chi^2|_{\mathbf{a}_i}. \tag{8.7.5}$$

Here k is a positive proportionality constant. Although suitable for the numerical implementation, this simple concept has one great disadvantage: it converges very slowly in the vicinity of the solution. Because of this, one makes use of the fact that the Taylor expansion is a good approximation of a function in the vicinity of its extremum

$$\chi^2(\mathbf{a}) \approx \chi^2(\mathbf{a}_i) + (\mathbf{a}-\mathbf{a}_i)\cdot\nabla\chi^2|_{\mathbf{a}_i} + \frac{1}{2}(\mathbf{a}-\mathbf{a}_i)\cdot\nabla\nabla\chi^2|_{\mathbf{a}_i}\cdot(\mathbf{a}-\mathbf{a}_i). \tag{8.7.6}$$

Now, by minimizing the approximation (8.7.6), an alternative expression for the increment in the argument is obtained

$$\mathbf{A}\cdot\Delta\mathbf{a} = \mathbf{b}, \quad \mathbf{A} = \nabla\nabla\chi^2|_{\mathbf{a}_i}. \tag{8.7.7}$$

Here, $\mathbf{A}$ represents the matrix of second derivatives, known as the Hessian matrix. Within the Levenberg-Marquardt method, the two previously explained concepts are combined by introducing a modified Hessian $\bar{\mathbf{A}}$ so that Eq. (8.7.7) becomes

$$\bar{\mathbf{A}}\cdot\Delta\mathbf{a} = \mathbf{b}, \qquad \bar{\mathbf{A}} = \mathbf{A} + \lambda\mathbf{I}. \tag{8.7.8}$$

This new formulation depends on the parameter λ determining which of the approaches is currently active. In the vicinity of the solution, λ has a small value leading to the minimization of the Taylor approximation. Far away from the solution, it has large values, $\mathbf{A}$ assumes a diagonal form and the method gains the properties of the steepest descent approach.

The evaluation of the parameter λ is dependent on the behavior of the function which has to be minimized. In the first iteration, an arbitrary value is assigned to this parameter. In each further iteration step, λ is increased if the function is nondecreasing and decreased if the function is also decreasing. In the latter case, an update of the arguments is performed as follows:

$$\begin{cases} \chi^2(\mathbf{a}_{i+1}) \geq \chi^2(\mathbf{a}_i), & \lambda = \lambda * c, \quad \text{no update of } \mathbf{a}, \\ \chi^2(\mathbf{a}_{i+1}) < \chi^2(\mathbf{a}_i), & \lambda = \frac{\lambda}{c}, \quad \mathbf{a} \text{ is updated: } \mathbf{a} = \mathbf{a}_{i+1}. \end{cases}$$

In both cases, c is a constant so that $c > 1$.

Some alternative interpretations of the Marquardt-Levenberg method are also possible. In the sense of the optimization, this approach represents the first trust-region method [304]. Here, the derivation of the method relies on the minimization of the approximation (8.7.6) accompanied by a subsidiary condition determining the trust region

$$\chi^2(\mathbf{a}) \approx \chi^2(\mathbf{a}_i) + (\mathbf{a}-\mathbf{a}_i)\cdot\nabla\chi^2|_{\mathbf{a}_i} + \frac{1}{2}(\mathbf{a}-\mathbf{a}_i)\cdot\nabla\nabla\chi^2|_{\mathbf{a}_i}\cdot(\mathbf{a}-\mathbf{a}_i), \tag{8.7.9}$$

$$||\Delta\mathbf{a}|| \leq \lambda. \tag{8.7.10}$$

In the last condition, λ denotes the trust region radius.

From the point of view of the inverse analysis, the Marquardt-Levenberg method can be treated as a kind of Tikhonov regularization method resulting from the minimization of the following problem

$$T_\lambda(\mathbf{a}) = ||\mathbf{g}-\tilde{\mathbf{g}}||^2 + \lambda||\mathbf{a}||^2 \tag{8.7.11}$$

where $\mathbf{g}$ and $\tilde{\mathbf{g}}$ are vectors containing the numerical and experimental values $g_{i,j}(\mathbf{i}_j,\mathbf{a})/s_{i,j}$ and $\tilde{g}_{i,j}/s_{i,j}$ respectively. In the last expression, λ represents the regularization parameter [109].

The numerical implementation of the Marquardt-Levenberg method also deserves special attention since a great number of different versions and interpretations of it exist. For example, the Hessian matrix $\mathbf{A}$ is often a discussion point. Its evaluation is based on the following derivatives

$$\frac{\partial\chi^2}{\partial a_k} = -2\sum_{j=1}^{N_{\text{ts}}}\sum_{i=1}^{N}\frac{\tilde{g}_{j,i}-g_{j,i}}{s_{j,i}^2}\frac{\partial g_{j,i}}{\partial a_k}, \tag{8.7.12}$$

$$\frac{\partial^2\chi^2}{\partial a_k\partial a_l} = 2\sum_{j=1}^{N_{\text{ts}}}\sum_{i=1}^{N}\frac{1}{s_{j,i}^2}\left[\frac{\partial g_{j,i}}{\partial a_k}\frac{\partial g_{j,i}}{\partial a_l} - (\tilde{g}_{j,i}-g_{j,i})\frac{\partial^2 g_{j,i}}{\partial a_l\partial a_k}\right]. \tag{8.7.13}$$

However, the calculation of second derivatives is always accompanied by a large computational effort so that many different approximations for the Hessian are proposed. In the case of the Gauss-Newton approximation, the last term in (8.7.13) is for example completely neglected [392]. Alternatively, the Broyden-Fletcher-Goldfarb-Shanno (BFGS) approach, the Davidon-Fletcher-Powel (DFP) formula, or the Symmetric-Rank-1 (SR1) method can be used for an approximative evaluation of (8.7.13). For an overview and a comprehensive derivation of these approaches, the book of Nocedal [322] might be recommended.

The evaluation of derivatives (8.7.12) and (8.7.13) is performed by using the forward finite difference method

$$\frac{\partial g_{j,i}}{\partial a_k} = \frac{g_{j,i}(\mathbf{a}_k)-g_{j,i}}{\Delta a_k}, \tag{8.7.14}$$

$$g_{j,i}(\mathbf{a}_k) = g_i(\mathbf{i}_j,\mathbf{a}_k),$$

$$\mathbf{a}_k = [a_1, a_2, \ldots, a_k+\Delta a_k, \ldots, a_n]^{\mathrm{T}}, \qquad k = 1,\ldots n,$$

where Δa_k is the arbitrarily chosen increment in the argument and n the number of unknown material parameters. An often used alternative to (8.7.14) is the central difference method, even though this approach goes along with an additional computer effort.

During the iteration procedure, attention is paid to the subsidiary conditions, requiring that evaluated material parameters as well as the parameter λ remain within corresponding admissible domains

$$a_{i,\min} \leq a_i \leq a_{i,\max}, \quad i = 1,...,n. \tag{8.7.15}$$

$$\lambda_{\min} \leq \lambda \leq \lambda_{\max}. \tag{8.7.16}$$

Obviously, the condition (8.7.15) prevents the material parameters from attaining physically unrealistic values, and the condition (8.7.16) prevents the iteration procedure from becoming an endless loop as well as parameter λ from exceeding the computer accuracy.

Another special feature of the method is certainly the exit criterion, requiring that the function is decreasing and that the absolute value of the function increment is below the prescribed tolerance

$$\chi^2(\mathbf{a}_{i+1}) < \chi^2(\mathbf{a}_i) \quad \text{and} \quad |\chi^2(\mathbf{a}_{i+1}) - \chi^2(\mathbf{a}_i)| < \text{tol}. \tag{8.7.17}$$

In an ideal case, such as the minimization of a convex function, the Marquardt-Levenberg method can find a solution without exceeding limits (8.7.15)–(8.7.16). However, in a case of more complex, nonconvex functions, it often happens that after finding a local minimum such that (8.7.17) is not fulfilled, the solver permanently looks for an optimum in the vicinity of this solution: it is not able to jump over the barrier separating it from another possibly global minimum. In order to avoid this disadvantage, the condition (8.7.16) allows only the final number of iteration steps in the vicinity of a minimum. If the limit is exceeded but the condition (8.7.17) is still not fulfilled, the new starting iteration vector is randomly generated according to the expression

$$a_i = a_{i,\min} + (a_{i,\max} - a_{i,\min})r_i, \qquad i = 1,...,n \tag{8.7.18}$$

where $0 \leq r_i \leq 1$ is a randomly generated value. Now, the iteration procedure goes on until (8.7.17) is fulfilled and if it is necessary, browses the admissible domain. Alternatively to the simple expression (8.7.18), some more complex relationships can be used for generating new values.

At this point, a remark concerning the well-posedness of the problem is still necessary: This property is certainly dependent on the characteristics of the operator corresponding to the forward problem. As an illustration, it can be mentioned that the compact operators related to the infinite-dimensional spaces are ill-posed [392]. The tendency observed in the numerical examples [244, 331] is that the lack of reliable experimental results leads to a non-unique solution if a greater number of parameters is determined. However, examples performed in the next section (Section 8.7.3) deals with problems with a small number of parameters to be detected.

8.7.3 NUMERICAL EXAMPLES

The application of the previously explained concept will be demonstrated on the basis of academic examples in which data on the effective shear modulus are used to evaluate missing geometrical properties of the RVE. Concerning Eq. (8.7.1), this means that $g_i(\mathbf{i}_j, \mathbf{a})$ are the real $(i = 1)$ and the imaginary $(i = 2)$ parts of the effective shear modulus obtained as final results of calculations at the microlevel. Furthermore, $\mathbf{i}_j$ is the macroscopic perturbation related to the shear strains $\bar{\varepsilon}_{12}$ $(j = 1)$ and $\bar{\varepsilon}_{23}$ $(j = 2)$. While effective values are generally determined experimentally, they are here obtained as a result of a forward analysis.

The first example is concerned with the evaluation of the average thickness of trabeculae for the RVE shown in Fig. 8.4.1. This RVE consists of the solid frame, where all trabeculae have the unit length ($a = 1$ mm), while their thickness (b) should be determined. Material parameters of both phases are listed in Table 8.1. The evaluations are performed, if the effective shear modulus for the excitation frequency of 600 kHz is known. Due to the isotropy, only one shear test is possible $(j = 1)$. The iteration path for the sought geometrical parameter is presented in Fig. 8.7.1. The following three starting values are chosen: 0.10, 0.20 and 0.25 mm and for all of them, the exact solution amounting to 0.15 mm is achieved. The procedure converges very fast and the unique solution is achieved in only 4–8 steps.

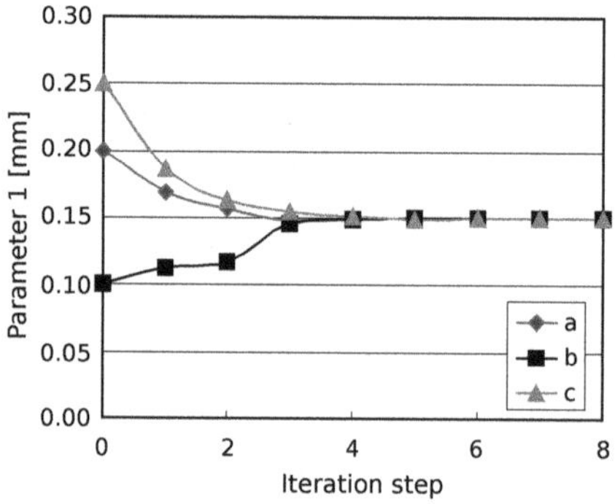

FIGURE 8.7.1 Iteration path by the determination of the characteristic thickness of the trabeculae b if the effective shear modulus is known. Applied frequency amounts to 600 kHz.

The same procedure has been applied for the parameter identification if a higher excitation frequency is applied (900 kHz). However, this time, a higher number of iteration steps was necessary because the program reached several local minima before it located the right solution. Each time, a new starting iteration vector has been generated and the value of the merit functions was controlled. The iteration was stopped when the merit function reached a value below the prescribed tolerance. The iteration path and corresponding values for the merit function are shown in Fig. 8.7.2. Obviously, the new iteration vector has been generated five times, but just in the last sequence of the iterations, the merit function is approaching zero, thus showing that the exact solution is achieved. A comparison of Figs. 8.7.1 and 8.7.2a indicates that the lower frequency is more suitable for an inverse analysis.

The following example is concerned with a simple RVE convenient for modeling cancellous bone as a transversally anisotropic medium. In general, the transversal

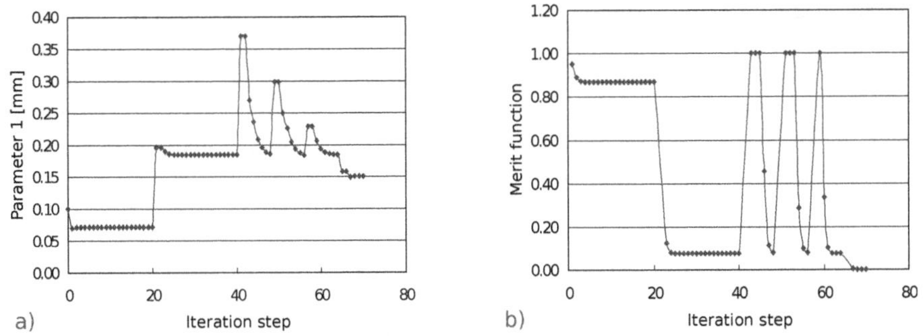

FIGURE 8.7.2 Determination of the thickness of trabeculae b if the applied excitation frequency amounts to 900 kHz. a) Iteration path. b) Merit function.

anisotropy of the cancellous bone is caused by two things: the geometrical properties and the transversal anisotropy of the solid phase, however the influence of the latter is mostly neglected (Section 8.5). The chosen RVE is presented in Fig. 8.5.1a, where two dimensions of the RVE, lying in the horizontal plane, are scaled to one ($a = 1$ mm), while the height of the RVE (c) as well as the thickness of the trabeculae (b) should be determined. The iteration path for this case has been shown in Figs. 8.7.3a,b. The applied frequency is 500 kHz and the sought values are $c = 0.9$ mm and $d = 0.15$ mm for the height of the RVE and for the thickness of trabeculae, respectively. Three groups of the starting iteration vectors $(0.7; 0.2)$, $(1.2; 0.2)$ and $(1.5; 0.2)$ have been chosen and for all of them a unique solution has been obtained. However, the number of necessary iteration steps varies and amounts to 6, 12 and 51 depending on the chosen starting iteration vector. For these calculations, the values of the effective shear modulus for two directions have been used ($j = 1, 2$).

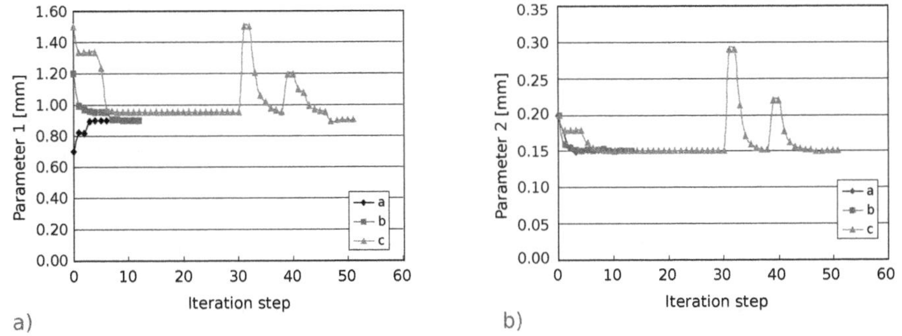

FIGURE 8.7.3 a) The iteration path for parameter c which is the height of the RVE. b) Iteration path for thickness of trabeculae b. Applied frequency amounts to 500 kHz.

An alternative RVE for modeling cancellous bone as a transversally anisotropic material is presented in Fig. 8.7.4a. Here, the transversal anisotropy is achieved by a shifted disposition of single cells, whose geometry is defined in Fig. 8.4.1. The emphasis is again placed on the determination of the thickness of the trabeculae since all of them have the same length scaled to one. Fig. 8.7.4b shows the iteration path for the excitation frequency of 600 kHz if three starting values (0.2 mm, 0.17 mm and 0.34 mm) are assumed. Here, the unique solution, amounting to 0.15 mm, has been found within a maximum of 12 steps.

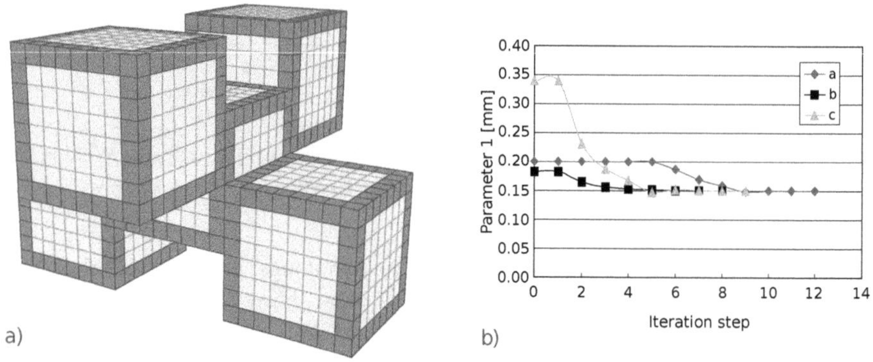

FIGURE 8.7.4 a) The RVE, consisting of several simpler cells, is suitable for an analysis of the transversal anisotropy of the bone. b) Iteration path for the average thickness of trabeculae if the applied frequency amounts to 600 kHz.

A completely different situation occurs if the applied frequency amounts to 800 kHz. In this case, a much higher number of iteration steps is necessary which is shown in Fig. 8.7.5. Similar to the example presented in Fig. 8.7.2a, the starting iteration value is regenerated 5 times before the global minimum was found. This endorses the previous conclusion that the lower frequencies yield more reliable data for an inverse analysis. In both cases (Fig. 8.7.4 and Fig. 8.7.5), the shear modulus for one direction has been used ($j = 1$) and an increase of the experimental data ($j = 1,2$) did not significantly contribute to an acceleration of the iteration.

The examples presented indicate that the shown concept for the inverse analysis can be used successfully for the identification of microscopic geometrical parameters characteristic of cancellous bone. However, it should be emphasized that here only simple academic examples are performed. Neither the influence of the measurement error nor of the large set of unknown parameters is considered. Obviously, many issues are still open. First, the method can be used in the reconstruction of the RVEs with alternative more complex geometries. Furthermore, it is suitable for the evaluation of material properties of trabeculae under the assumption that the geometry is known. The inverse homogenization theory is in this context especially advantageous as it enables parameter identification on the basis of reliable macroscopic measurements and not on the basis of direct testing of trabeculae, whose small dimensions are regularly an obstacle when speaking of experiments. A final aspect, which should be commented on, is an alternative use of data on the effective Young modulus instead of the effective shear modulus. The advantage of the use of the shear

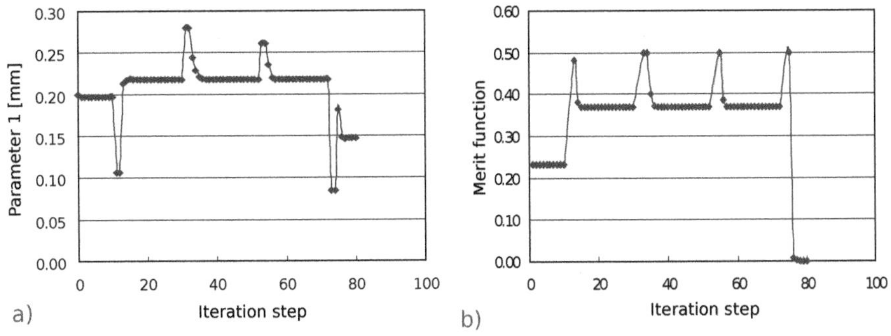

FIGURE 8.7.5 The average thickness of trabeculae is investigated if the effective shear modulus for the frequency of 800 kHz is known. a) Iteration path. b) Merit function.

modulus lies in the fact that this parameter is a direct output of the calculations at the microlevel, while Young's modulus must be evaluated from the effective elasticity tensor where also the influence of other material parameters, for example Poisson's number, must be taken into consideration. On the other hand, measurements yielding Young's modulus are related to the longitudinal waves. They are simpler to perform and thus provide more reliable data than the measurements of the complex shear modulus.

9 G-convergence and Homogenization of Viscoelastic Flows

9.1 INTRODUCTION

Formulation of constitutive equations of multiphase materials under flow is a fundamental problem of continuum mechanics. Using mathematical homogenization theory [24, 229, 363] to solve this problem is intuitively appealing, but not easy. The reasons for this can be summarized as follows: it is difficult to homogenize evolution equations, non-linear equations, and equations involving general [1] geometric distribution of the constituents. In particular, addressing the last difficulty is the necessary first step in developing a complete homogenization theory for moving interface problems.

Consider a composite material with two constituents which we call phases. During flow, the interface between the phases is advected by the velocity. Therefore, the interface motion is coupled to the flow dynamics, and one cannot expect the geometry of phases to preserve its properties such as periodicity, or, more generally, random homogeneity combined with scale separation.

Therefore, homogenization techniques that require a specific type of geometry, e.g. two-scale convergence ([7, 8, 315]) and ergodic theorems, cannot be used in general. Consequently, the only suitable tool, at least for problems that lack variational formulation, seems to be G-convergence [313, 341, 374, 375, 414, 413].

In this chapter, we first consider the definition of G-convergence and recall some basic facts about it. Then we introduce a corrector operator approach which is related to the method of oscillating test functions [113, 114, 313, 341] and also to the "condition N" [414, 413]. Then we use the corrector approach to derive effective equations of viscoelastic mixtures with time-varying interface. The fine-scale model accounts for small (in the sense to be described below) deformations of the interface. However, the deformations inside the constituent phases may be arbitrarily large. Thus our fine-scale equations are more general than linearized models used in acoustics of composites. We show that effective constitutive equations contain long-memory terms, and that the effective material parameters satisfy certain auxiliary problems which resemble cell problems of periodic homogenization.

[1] Not necessarily random homogeneous.

9.2 MAIN DEFINITIONS. CORRECTOR OPERATORS FOR G-CONVERGENCE

Let $A^\varepsilon : X \to X^\star$ denote a sequence of operators from a reflexive Banach space X into its topological dual space $X^\star$. The sequence A^ε is parametrized by a positive sequence $\varepsilon > 0$ such that $\varepsilon \to 0$. Consider the sequence of problems

$$A^\varepsilon \mathbf{u}^\varepsilon = \mathbf{f}^\varepsilon. \tag{9.2.1}$$

Definition 9.1. *The sequence A^ε G-converges to an operator A as $\varepsilon \to 0$ if both of the following conditions hold:*
(i) if $\mathbf{f}^\varepsilon$ converges to $\mathbf{f}$ strongly in $X^\star$, $\mathbf{u}^\varepsilon$ converges weakly to $\mathbf{u}$ in X and $A^\varepsilon \mathbf{u}^\varepsilon = \mathbf{f}^\varepsilon$ for infinitely many ε, then

$$A\mathbf{u} = \mathbf{f}; \tag{9.2.2}$$

(ii) if $\mathbf{f} \in X^\star$ and $\mathbf{u} \in X$ is a solution of (9.2.2), *then there exist sequences $\mathbf{f}^\varepsilon$ converging to $\mathbf{f}$ strongly in $X^\star$ and $\mathbf{u}^\varepsilon$ converging weakly to $\mathbf{u}$ in X such that $A^\varepsilon \mathbf{u}^\varepsilon = \mathbf{f}^\varepsilon$.*

Remark. In the special case when $\mathbf{f}^\varepsilon = \mathbf{f}$ in (9.2.1), and $\mathbf{f} \in X^\star$ is arbitrary, the above definition reduces to requiring that

(i) $\mathbf{u}^\varepsilon$ converges weakly to $\mathbf{u}$ in X, and
(ii) $\mathbf{u}$ is a solution of (9.2.2).

This is equivalent to pointwise convergence of the sequence of the inverse operators $(A^\varepsilon)^{-1}$ to A^{-1} in X. Note that G-convergence of an operator sequence A^ε can be identified with the Painleve-Kuratowski set convergence [250, 355] of the corresponding operator graphs Γ^ε defined as a set of pairs $(\mathbf{u}, A^\varepsilon \mathbf{u})$ (see e.g. the definition of G-convergence in [341]). Set convergence is sequentially compact provided the topology of $X \times X^\star$ has a countable base. Therefore, existence of an abstract G-limit operator A is easily obtained for A^ε that may be non-linear, non-local and multivalued. Once existence of A is established, one needs to describe the structure of A, which is a problem of *characterization*. A useful tool for solving this problem is the method of oscillating test functions [113, 114, 313, 341]. To find a suitable sequence $\mathbf{w}^\varepsilon$ of such functions, we first prescribe *corrector operators* $N^\varepsilon : X \to X$ and then define

$$\mathbf{w}^\varepsilon = N^\varepsilon \mathbf{w},$$

where $\mathbf{w}$ is an arbitrary fixed test function from a dense subset of X.

For linear operators A^ε, the characterization problem can be solved if N^ε satisfy

$$N^\varepsilon \mathbf{w} \to \mathbf{w} \quad \text{weakly in } X, \tag{9.2.3}$$

$$(A^\varepsilon)^{\mathrm{T}} N^\varepsilon \mathbf{w} \to \overline{A}^{\mathrm{T}} \mathbf{w} \quad \text{strongly in } X^\star, \tag{9.2.4}$$

possibly along a subsequence. Here, $(A^\varepsilon)^{\mathrm{T}}$ is the formal adjoint of A^ε and $\overline{A}^{\mathrm{T}}$ is a limiting operator. If $\overline{A}^{\mathrm{T}}$ can be described, then A can be characterized as follows.

If $A^\varepsilon \mathbf{u}^\varepsilon = \mathbf{f}$ and (9.2.3) holds, then

$$\langle A^\varepsilon \mathbf{u}^\varepsilon, N^\varepsilon \mathbf{w}\rangle = \langle \mathbf{f}, N^\varepsilon \mathbf{w}\rangle \to \langle \mathbf{f}, \mathbf{w}\rangle = \langle A\overline{\mathbf{u}}, \mathbf{w}\rangle, \tag{9.2.5}$$

and if (9.2.4) is true, then

$$\langle A^\varepsilon \mathbf{u}^\varepsilon, N^\varepsilon \mathbf{w}\rangle = \langle \mathbf{u}^\varepsilon, (A^\varepsilon)^{\mathsf{T}} N^\varepsilon \mathbf{w}\rangle \to \langle \overline{\mathbf{u}}, \overline{A}^{\mathsf{T}} \mathbf{w}\rangle = \langle \overline{A}\overline{\mathbf{u}}, \mathbf{w}\rangle. \tag{9.2.6}$$

Here $\langle \cdot, \cdot \rangle$ denotes the duality pairing between X and $X^\star$.

Since $\mathbf{w}$ is any function in a dense subset of X, (9.2.5), (9.2.6) imply $A\overline{\mathbf{u}} = \overline{A}\overline{\mathbf{u}}$. Thus the structure of A can be inferred by taking the pointwise limit of the product $(A^\varepsilon)^{\mathsf{T}} N^\varepsilon$.

At present, there are no general methods for constructing correctors. Known methods of G-limit characterization rely on specific features of certain classes of problems: elliptic equations in divergence form [313], elliptic and parabolic problems with coercivity [414, 413], linear elasticity [330], operator equations with monotone [81], and pseudo-monotone [341] operators.

The version of the corrector operator technique described in this chapter is inspired by "condition N" in the papers [414, 413] of Zhikov, Kozlov and Oleinik (see also the book [330] for an application to linear elasticity).

9.3 A SCALAR ELLIPTIC EQUATION IN DIVERGENCE FORM

As an example of the application of the corrector operator technique, let A^ε be elliptic differential operators in divergence form. In a bounded Lipschitz domain $\Omega \subset \mathbf{R}^n$, consider a sequence of problems

$$A^\varepsilon u^\varepsilon \equiv \operatorname{div}(a^\varepsilon \nabla u) = f, \tag{9.3.1}$$

where $f \in H^{-1}(\Omega)$.

The matrices $a^\varepsilon(x)$ are *symmetric* and satisfy standard strong ellipticity and boundedness conditions with upper and lower bounds independent of ε.

Then, for each $\varepsilon > 0$, there exists a unique weak solution $u^\varepsilon \in H_0^1(\Omega)$ of (9.3.1), and the sequence of solution operators $(A^\varepsilon)^{-1} : H^{-1}(\Omega) \to H_0^1(\Omega)$ satisfies

$$\| (A^\varepsilon)^{-1} f \|_{H_0^1(\Omega)} \leq C \| f \|_{H^{-1}(\Omega)}, \tag{9.3.2}$$

with C independent of ε.

It is then easy to see that there exists a subsequence of $\{A^\varepsilon\}$ (not relabeled) and a bounded, invertible, linear operator $\overline{A} : H_0^1(\Omega) \to H^{-1}(\Omega)$, called the *G-limit* of the sequence $\{A^\varepsilon\}$, such that $(A^\varepsilon)^{-1} f \to \left(\overline{A}\right)^{-1} f$ weakly in $H_0^1(\Omega)$, for each $f \in H^{-1}(\Omega)$, along this subsequence.

Our goal is to find an explicit representation for $\overline{A}$. Specifically, we want to prove that for each $u \in H_0^1(\Omega)$, $\overline{A}u = \operatorname{div}(\overline{a}\nabla u)$, with a certain *effective matrix* $\overline{a}$.

First note that $\| u^\varepsilon \|_{H_0^1(\Omega)}$ is bounded independent of ε, and thus there exists a subsequence $\{u^\varepsilon\}$, not relabeled, such that $u^\varepsilon \to \overline{u}$ weakly in $H_0^1(\Omega)$.

Theorem 9.1

Let $\{u^\varepsilon\}$ be such that u^ε solve (9.3.1) and $u^\varepsilon \to \overline{u}$ weakly in $H_0^1(\Omega)$. Then $\overline{u} \in H_0^1(\Omega)$ is a weak solution of

$$A\overline{u} = f, \tag{9.3.3}$$

where

$$Aw = \text{div}\,(\overline{a}\nabla w) \tag{9.3.4}$$

for each $w \in H_0^1(\Omega)$, and $\overline{a}$ is an effective matrix. ■

Proof. For $\varepsilon > 0$, consider the weak formulation of (9.3.1):

$$\langle A^\varepsilon u^\varepsilon, w\rangle = \langle f, w\rangle, \tag{9.3.5}$$

for each $w \in H_0^1(\Omega)$. Since a^ε is symmetric, (9.3.5) is equivalent to

$$\langle u^\varepsilon, A^\varepsilon w\rangle = \langle f, w\rangle. \tag{9.3.6}$$

The idea is to insert into (9.3.6) oscillating test functions

$$w^\varepsilon \equiv w + \mathbf{v}^\varepsilon \cdot \nabla w \equiv N^\varepsilon w, \tag{9.3.7}$$

where $N^\varepsilon = I + \mathbf{v}^\varepsilon \cdot \nabla$ is the corrector operator, and $w \in C_0^\infty(\Omega)$ is arbitrary, The vector $\mathbf{v}^\varepsilon$ will be chosen to satisfy an auxiliary equation that will be specified below.

Next,

$$A^\varepsilon w^\varepsilon = \text{div}\,[a^\varepsilon(I + \mathbf{v}^\varepsilon)]\,\nabla w + [a^\varepsilon(I + \mathbf{v}^\varepsilon)]\,H(w) + \text{div}g^\varepsilon, \tag{9.3.8}$$

where $H(w)$ is the Hessian matrix of w: $H(w)_{ij} = \partial_i\partial_j w$, and

$$g_i^\varepsilon = a_{ij}^\varepsilon v_k^\varepsilon \partial_k \partial_j w. \tag{9.3.9}$$

Let $\mathbf{v}^\varepsilon \in (H_0^1(\Omega))^n$ be the weak solution of the following auxiliary problem

$$\text{div}\,[a^\varepsilon(I + \nabla\mathbf{v}^\varepsilon)] = A\mathbf{h}, \tag{9.3.10}$$

where a vector $\mathbf{h} \in (H_0^1(\Omega))^n$ is chosen as follows. Consider the sequence of vectors $(A^\varepsilon)^{-1}\,[\text{div}a^\varepsilon]$. Since a^ε are bounded pointwise independent of ε, $\text{div}a^\varepsilon$ is bounded in $H^{-1}(\Omega)$ independent of ε. Further, the norms of the operators $(A^\varepsilon)^{-1}$ are bounded independent of ε. Therefore, the sequence $(A^\varepsilon)^{-1}\,[\text{div}a^\varepsilon]$ is bounded in $(H_0^1(\Omega))^n$ independent of ε. Thus, there exists a subsequence, not relabeled, and a vector $\mathbf{b} \in (H_0^1(\Omega))^n$ such that

$$(A^\varepsilon)^{-1}\,[\text{div}a^\varepsilon] \to \mathbf{b},$$

weakly in $(H_0^1(\Omega))^n$. Now choose $\mathbf{h} = \mathbf{b}$. Then we claim that $\mathbf{v}^\varepsilon \to 0$ weakly in $(H_0^1(\Omega))^n$, along this subsequence. Indeed, rewrite (9.3.10) with $\mathbf{h} = \mathbf{b}$ as

$$A^\varepsilon\mathbf{v}^\varepsilon = -\text{div}(a^\varepsilon) + A\mathbf{b}, \tag{9.3.11}$$

so that

$$\mathbf{v}^{\varepsilon} = -(A^{\varepsilon})^{-1}[\mathrm{div}a^{\varepsilon}] + (A^{\varepsilon})^{-1}A\mathbf{b}. \tag{9.3.12}$$

The first term on the right converges to $-\mathbf{b}$ weakly in $(H_0^1(\Omega))^n$. The second term on the right converges to $\mathbf{b}$ weakly in $(H_0^1(\Omega))^n$ because by definition of A,

$$(A^{\varepsilon})^{-1}A\mathbf{b} \to A^{-1}A\mathbf{b} = \mathbf{b}.$$

Next, note that from (9.3.10) it follows that $a^{\varepsilon}(I + \nabla\mathbf{v}^{\varepsilon})$ is bounded in $L^2(\Omega)$ independent of ε. Thus, there exists a matrix $\overline{a} \in \left(L^2(\Omega)\right)^{n^2}$ such that $a^{\varepsilon}(I + \nabla\mathbf{v}^{\varepsilon}) \to \overline{a}$ weakly in $\left(L^2(\Omega)\right)^{n^2}$, possibly along a subsequence. In addition, passing to the limit $\varepsilon \to 0$ in (9.3.10) we get

$$\mathrm{div}\overline{a} = A\mathbf{b}, \tag{9.3.13}$$

and thus

$$\mathrm{div}\left[a^{\varepsilon}(I + \mathbf{v}^{\varepsilon})\right] = \mathrm{div}\overline{a}. \tag{9.3.14}$$

Now return to (9.3.8). Using (9.3.14), rewrite (9.3.8) as

$$A^{\varepsilon}w^{\varepsilon} = \mathrm{div}\,(\overline{a})\,\nabla w + \overline{a}H(w) + (-\overline{a} + [a^{\varepsilon}(I + \mathbf{v}^{\varepsilon})])\,H(w) + \mathrm{div}g^{\varepsilon}. \tag{9.3.15}$$

The sum of the first two terms on the right equals $\mathrm{div}(\overline{a}\nabla w)$. The third term on the right converges to zero weakly in $L^2(\Omega)$ and thus strongly in $H^{-1}(\Omega)$. The fourth term converges to zero strongly in $H^{-1}(\Omega)$ because $g^{\varepsilon} \to 0$ in $L^2(\Omega)$ (this follows from (9.3.9), taking into account that $\mathbf{v}^{\varepsilon} \to 0$ weakly in $(H_0^1(\Omega))^n$ and a^{ε} is bounded pointwise independent of ε). Therefore, $A^{\varepsilon}w^{\varepsilon} \to \mathrm{div}\,(\overline{a}\nabla w)$ strongly in $H^{-1}(\Omega)$.

Now we can pass to the limit in (9.3.6). Since $u^{\varepsilon} \to \overline{u}$ weakly in $H_0^1(\Omega)$, and $A^{\varepsilon}w^{\varepsilon} \to \mathrm{div}\,(\overline{a}\nabla w)$ strongly in $H^{-1}(\Omega)$,

$$\langle u^{\varepsilon}, A^{\varepsilon}w\rangle \to \langle \overline{u}, \mathrm{div}\,(\overline{a}\nabla w)\rangle = \langle \mathrm{div}\,(\overline{a}\nabla\overline{u})\,, w\rangle.$$

Finally, $\langle f, w^{\varepsilon}\rangle \to \langle f, w\rangle$ because $\mathbf{v}^{\varepsilon}\cdot\nabla w \to 0$ weakly in $H_0^1(\Omega)$, and thus $w^{\varepsilon} \to w$ weakly in $H_0^1(\Omega)$.

Therefore, $\overline{u}$ is a weak solution of (9.3.3), with A defined by (9.3.4). □

Remark. The auxiliary problem (9.3.10) is similar to the well known cell problem of periodic homogenization. However, the problem is posed not in a periodicity cell, but in the whole domain Ω. The choice of the right hand side is made using an abstract G-limit operator A, and this operator is then characterized using the solution of the auxiliary problem. Therefore, the auxiliary problem is not suitable for computing correctors $\mathbf{v}^{\varepsilon}$. The useful feature here is that the corrector operator for G-limit characterization has exactly the same form as the corresponding operator in the periodic case, and a suitable choice of the right hand side in the modified cell problem allows us to extend periodic homogenization techniques to the case of arbitrary non-periodic geometries.

9.4 TWO-PHASE VISCO-ELASTIC FLOWS WITH TIME-VARYING INTERFACE

9.4.1 INTRODUCTION

In this section we derive homogenized models for flows of incompressible viscoelastic mixtures with time-varying interface. Both phases are modeled as Kelvin-Voight materials: the elastic stresses satisfy Hook's law written in the spatial (Eulerian) formulation, and the viscous stresses obey Newton's law. In particular, such models describe mixtures of fluid-like (small elasticity), and a solid-like (small viscosity) materials. We also consider the fluid-structure interaction problem, where one phase is a solid-like viscoelastic material, and the other is a viscous Newtonian fluid.

9.4.2 EQUATIONS OF BALANCE AND CONSTITUTIVE EQUATIONS

9.4.2.1 Choice of a model

We consider two-phase materials in which at least one of the phases resists shearing, and the material stress tensor can be written as a sum of an elastic (conservative) and dissipative stresses. To avoid mathematical difficulties in dealing with nonlinear elasticity we limit our investigation to flows for which Hook's law of linear elasticity is an appropriate model of the elastic stress. We further suppose that deformation of the interface is small. In this case, the equations of motion are often formally linearized, assuming that the density is constant, and spatial and referential description are identified. As a consequence, the interface (always fixed in the referential description) is fixed in the spatial description. This is unsatisfactory for mixtures in which one phase is solid-like and the other one is fluid-like. In this case, large deformations of the fluid-like phase may occur even when the deformations in the solid-like phase are small. Consequently, the contribution of the inertial terms to the overall momentum balance cannot be neglected. In addition, a correct model should respect physically realistic interface conditions (continuity of velocity and equality of tractions) as well as physically correct (formal) energy balance. These requirements make it necessary to account for inertia in both phases, with the corresponding terms that include a moving interface and densities governed by the mass conservation equations. By contrast, using Hook's law for the elastic stress may lead to non-physical energy fluctuations unless the interface in the constitutive equations is fixed. To reconcile the above requirements, the micro-scale equations contain a moving interface in the inertial terms and a frozen interface in the constitutive equations. The resulting model is more broadly applicable than a linear model with constant densities and static interface, but not as general as a fully nonlinear model.

Let $\rho^\varepsilon, \mathbf{v}^\varepsilon$ denote, respectively, the density and velocity of a particular ε-mixture in the family of such mixtures parameterized by a sequence of $\varepsilon \to 0$. We also define the displacement

$$\mathbf{u}^\varepsilon(t,\mathbf{x}) = \int_0^t \mathbf{v}^\varepsilon(\tau,\mathbf{x})d\tau. \tag{9.4.1}$$

Suppose that each mixture occupies the same fixed domain U at all times. The evolution of the above state variables is governed by the following equations.
Interface evolution equation. Evolution of the domains occupied by the phases can be described by the interface evolution equation. Let $\theta^\varepsilon(t,\mathbf{x})$ denote the characteristic function of the domain occupied by one of the phases at the time t. This function satisfies the advection equation

$$\partial_t\theta^\varepsilon + \mathbf{v}^\varepsilon \cdot \nabla\theta^\varepsilon = 0 \quad \text{in } U \tag{9.4.2}$$

which states that material derivative of θ^ε is zero, and it is therefore fixed in the referential description. Equation (9.4.2) is supplemented with the initial condition

$$\theta^\varepsilon(0,\mathbf{x}) = \theta_0^\varepsilon(\mathbf{x}) \quad \text{in } U. \tag{9.4.3}$$

Mass conservation equation.
The composite density satisfies

$$\partial_t\rho^\varepsilon + \operatorname{div}(\rho^\varepsilon\mathbf{v}^\varepsilon) = 0, \quad \text{in } U \tag{9.4.4}$$

with the initial condition

$$\rho^\varepsilon(0,\mathbf{x}) = \rho_1\theta_0^\varepsilon(\mathbf{x}) + \rho_2(1-\theta_0^\varepsilon(\mathbf{x})), \quad \mathbf{x} \in U, \tag{9.4.5}$$

where ρ_1, ρ_2 are the densities of the respective phases. These densities are assumed to be constant and bounded below by a positive constant.
Incompressibility.

$$\operatorname{div}\,\mathbf{v}^\varepsilon = 0. \tag{9.4.6}$$

Momentum balance for the composite.

$$\partial_t\mathbf{v}^\varepsilon + \operatorname{div}(\rho^\varepsilon\mathbf{v}^\varepsilon \otimes \mathbf{v}^\varepsilon) - \operatorname{div}(\mathbf{T}_1^\varepsilon - P_1^\varepsilon\mathbf{I} + \mathbf{T}_2^\varepsilon - P_2^\varepsilon\mathbf{I}) = 0. \tag{9.4.7}$$

Here $\mathbf{T}_s^\varepsilon - P_s^\varepsilon\mathbf{I}$ is the stress tensor in the phase s, P_s^ε is the pressure, and $\mathbf{I}$ denotes the unit tensor. The initial conditions for (9.4.7) are

$$\mathbf{v}^\varepsilon(0,\mathbf{x}) = \mathbf{v}_0(\mathbf{x}), \tag{9.4.8}$$

where $\mathbf{v}_0$ does not depend on ε. In addition, (9.4.1) implies that $\mathbf{u}^\varepsilon(0,\mathbf{x}) = 0$.
On the boundary ∂U, the condition

$$\mathbf{v}^\varepsilon(t,\mathbf{x}) = 0 \tag{9.4.9}$$

is imposed.
Constitutive equations. As explained above, a static interface seems to be a natural choice that is relatively easy to handle (unlike combining referential formulation for the elastic stress with the spatial formulation for the viscous stress), and compatible with the spatial form of Hook's law for the elastic part of the stress. We therefore define

$$\mathbf{T}^\varepsilon = \mathbf{T}_1^\varepsilon + \mathbf{T}_2^\varepsilon - P^\varepsilon\mathbf{I}, \text{ where } P^\varepsilon = P_1^\varepsilon + P_2^\varepsilon, \tag{9.4.10}$$

$$\mathbf{T}_1^\varepsilon(t,\mathbf{x}) = \theta_0^\varepsilon(\mathbf{x})\left(\mathbf{A}^1 e(\mathbf{u}^\varepsilon) + B^1 e(\mathbf{v}^\varepsilon)\right),$$
$$\mathbf{T}_2^\varepsilon(t,\mathbf{x}) = (1-\theta_0^\varepsilon(\mathbf{x}))\left(\mathbf{A}^2 e(\mathbf{u}^\varepsilon) + B^2 e(\mathbf{v}^\varepsilon)\right),$$

and $\mathbf{A}^s, B^s$, $s = 1,2$ are constant material tensors. In (9.4.10), $e = \frac{1}{2}(\nabla + \nabla^T)$ is the symmetric part of the gradient. We assume that both phases are isotropic. In that case

$$A_{ijkl}^s = \mu^s \delta_{ik}\delta_{jl}, \quad B_{ijkl}^s = \nu^s \delta_{ik}\delta_{jl}, \quad s = 1,2, \tag{9.4.11}$$

where μ^s are the elastic moduli and ν^s are the viscosities of the phases. All these constants are supposed to be positive.

We also assume that the tensors $\mathbf{A}^\varepsilon, \mathbf{B}^\varepsilon$ satisfy

$$\alpha_1 \xi\cdot\xi \le \mathbf{A}^\varepsilon \xi\cdot\xi \le \alpha_2 \xi\cdot\xi, \qquad \beta_1 \xi\cdot\xi \le \mathbf{B}^\varepsilon \xi\cdot\xi \le \beta_2 \xi\cdot\xi. \tag{9.4.12}$$

for each $\xi \in \mathbf{R}^{3\times 3}$, with $\alpha_i > 0, \beta_i > 0$ independent of ε.

Remark. Together, (9.4.10), (9.4.4), (9.4.7) form a closed system, and thus the interface evolution equation (9.4.2) can be dropped. This fact is important for compressible flows for which the mass conservation equation is stable with respect to weak convergence, while equation (9.4.2) is not. For incompressible flows considered in this chapter, the interface evolution equation has the same structure $\partial_t + \mathrm{div}(\mathbf{v}^\varepsilon \cdot)$ as the mass conservation. Moreover, if the initial densities ρ_1, ρ_2 are constant, then the densities of the phases remain constant during the motion. In this case, the interface evolution equation and the mass conservation equation are essentially equivalent.

Interface conditions. There are two interface conditions: the first is continuity of $\mathbf{v}$ across the interface (which is the actual moving interface governed by (9.4.2)), and the second is the equality of tractions $(\mathbf{T}_s^\varepsilon - P_s^\varepsilon \mathbf{I})\,\nu_s$ on the frozen interface. Here ν_s denotes the exterior (to the phase s) unit normal to the frozen interface.

9.4.2.2 Weak formulation of the micro-scale problem

In this section we provide the weak formulation of the problem to be homogenized. It consists of the mass conservation and momentum balance equations.

Mass conservation.

$$\int_U \rho^\varepsilon(0,\mathbf{x})\phi(0,\mathbf{x})d\mathbf{x} - \int_{I_T}\int_U \rho^\varepsilon \partial_t \phi d\mathbf{x}dt - \int_{I_T}\int_U \rho^\varepsilon \mathbf{v}^\varepsilon \cdot \nabla\phi d\mathbf{x}dt = 0, \tag{9.4.13}$$

where $\rho^\varepsilon(0,\mathbf{x})$ is given by (9.4.5), and

$$I_T = [0,T).$$

Equation (9.4.13) is supposed to hold for each smooth test function ϕ, equal to zero on ∂U and vanishing for $t \ge T$.

Momentum balance.

$$-\int_U \rho^\varepsilon(0,\mathbf{x})\mathbf{v}_0 \cdot \psi d\mathbf{x} - \int_{I_T}\int_U \rho^\varepsilon \mathbf{v}^\varepsilon \cdot \partial_t \psi d\mathbf{x}dt - \int_{I_T}\int_U \rho^\varepsilon \mathbf{v}^\varepsilon \otimes \mathbf{v}^\varepsilon \cdot \nabla\psi d\mathbf{x}dt \tag{9.4.14}$$

$$+\int_{I_T}\int_U (\mathbf{T}_1^\varepsilon+\mathbf{T}_2^\varepsilon)\cdot e(\psi)d\mathbf{x}dt=0.$$

Equation (9.4.14) holds for all smooth test functions ψ, such that div $\psi=0$, ψ is equal to zero on ∂U, and $\psi(t,\mathbf{x})=0$ when $t\geq T$. The dependence of $\mathbf{T}_i^\varepsilon$ on $\mathbf{v}_i^\varepsilon$ and θ_0^ε is given by (9.4.10).

Remark. It is important to note that, because of the condition $\mathbf{v}^\varepsilon=0$ on ∂U, the identity (9.4.14) also holds for test functions ψ with the condition $\psi=0$ replaced by a less restrictive $\psi\cdot\nu=0$ on ∂U, where ν is the exterior unit normal to ∂U. This fact will be used in Section 9.6 to construct oscillating test functions.

9.4.3 FINITE ENERGY WEAK SOLUTIONS AND BOUNDS

We suppose that the initial conditions satisfy

$$0<C_1\leq\rho^\varepsilon(0,\mathbf{x})\leq C_2, \tag{9.4.15}$$

$$\mathbf{v}_0\in H_0^1(U) \tag{9.4.16}$$

with C_1,C_2 independent of ε. The system (9.4.13), (9.4.14) closely resembles the system of density-dependent Navier-Stokes equations with density-dependent viscosity studied in [273], Chapter 2. The only difference is the presence of the strain-dependent terms of the type $\mathbf{A}e(\mathbf{u}^\varepsilon)$ in the constitutive equations. When the viscosity does not vanish, as is the case here, the existence and overall properties of the weak solutions are determined by the viscosity, and not elasticity, of the medium. The proof of existence for each fixed $\varepsilon>0$ is outlined in Appendix A.2. It yields, for each $\varepsilon\in\{\varepsilon_k\}_{k=1}^\infty$, existence of the finite energy weak solutions of (9.4.13), (9.4.14) with the same properties as in [273], Theorem 2.1, namely

$$\rho^\varepsilon\in L^\infty(I_T\times U)\cap C(I_T,L^p(U)),\quad \text{for all } 1\leq p<\infty, \tag{9.4.17}$$

$$\mathbf{v}^\varepsilon\in L^2(I_T,H_0^1(U)), \tag{9.4.18}$$

satisfying the energy inequalities

$$\begin{aligned}\frac{1}{2}\int_U\rho^\varepsilon|\mathbf{v}^\varepsilon|^2d\mathbf{x}(t)&+\int_U\left[\mathbf{A}^1\theta_0^\varepsilon(\mathbf{x})+\mathbf{A}^2(1-\theta_0^\varepsilon(\mathbf{x}))\right]|e(\mathbf{u}^\varepsilon)|^2d\mathbf{x}(t)\\ &+\int_{I_T}\int_U\left[B^1\theta_0^\varepsilon(\mathbf{x})+B^2(1-\theta_0^\varepsilon(\mathbf{x}))\right]|e(\mathbf{v}^\varepsilon)|^2d\mathbf{x}dt\leq\\ \frac{1}{2}\int_U\rho^\varepsilon|\mathbf{v}^\varepsilon|^2d\mathbf{x}(0)&+\int_U\left[\mathbf{A}^1\theta_0^\varepsilon(\mathbf{x})+\mathbf{A}^2(1-\theta_0^\varepsilon(\mathbf{x}))\right]|e(\mathbf{u}^\varepsilon)|^2d\mathbf{x}(0).\end{aligned} \tag{9.4.19}$$

Let us list the implications of the above estimates. First, renormalizing solutions of the mass conservation equations, as in [273], Section 2.3, we obtain

$$\|\rho^\varepsilon\|_{L^\infty(I_T\times U)}\leq C_2. \tag{9.4.20}$$

Next we note that (9.4.1) implies $\mathbf{u}^\varepsilon(0,\mathbf{x})=0$, and by (9.4.15), (9.4.16), the other initial conditions are bounded independent of ε. This implies that the left hand side of (9.4.19) is bounded independent of ε, so that

$$\| \mathbf{T}^{\varepsilon} \|_{L^2(I_T \times U)} \leq C, \tag{9.4.21}$$

with C independent of ε. Combining (9.4.21) with the first Korn inequality for functions with zero trace on the boundary (see, e.g. [330], Theorem 2.1), and then with Poincaré inequality, we deduce

$$\| \mathbf{v}^{\varepsilon} \|_{L^2(I_T, H_0^1(U))} \leq C, \tag{9.4.22}$$

$$\| \mathbf{u}^{\varepsilon} \|_{L^\infty(I_T, H_0^1(U))} \leq C. \tag{9.4.23}$$

with C independent of ε. Then it follows that

$$\| \rho^{\varepsilon} |\mathbf{v}^{\varepsilon}|^2 \|_{L^\infty(I_T, L^1(U))} \leq C, \tag{9.4.24}$$

which, together with (9.4.20), yields

$$\| \rho^{\varepsilon} \mathbf{v}^{\varepsilon} \|_{L^2(I_T, L^2(U))} \leq C, \tag{9.4.25}$$

with C independent of ε.

The above uniform bounds allow us to extract a subsequence, still denoted by $\{\varepsilon_k\}_{k=1}^{\infty}$ such that

$$\rho^{\varepsilon}(0,\mathbf{x}) \to \overline{\rho}_0, \quad \text{weakly} - * \text{ in } L^\infty(U), \tag{9.4.26}$$
$$\rho^{\varepsilon} \to \overline{\rho}, \quad \text{weakly} - * \text{ in } L^\infty(I_T \times U), \tag{9.4.27}$$
$$\mathbf{v}^{\varepsilon} \to \overline{\mathbf{v}}, \quad \text{weakly in } L^2(I_T, H_0^1(U)), \tag{9.4.28}$$
$$\mathbf{T}^{\varepsilon} \to \overline{\mathbf{T}}, \quad \text{weakly in } L^2(I_T, L^2(U)), \tag{9.4.29}$$
$$\mathbf{u}^{\varepsilon} \to \overline{\mathbf{u}}, \quad \text{weakly in } L^2(I_T, H_0^1(U)), \tag{9.4.30}$$
$$\rho^{\varepsilon}\mathbf{v}^{\varepsilon} \to \overline{\mathbf{z}}, \quad \text{weakly in } L^2(I_T, L^2(U)), \tag{9.4.31}$$
$$P^{\varepsilon} \to \overline{P}, \quad \text{weakly in } L^2(I_T, L^2(U)). \tag{9.4.32}$$

The last comment concerns $\partial_t \rho^{\varepsilon}$ and $\partial_t(\rho^{\varepsilon}\mathbf{v}^{\varepsilon})$. For technical reasons we need bounds on these sequences in the space $L^1(I_T, W^{-m,1}(U))$, where $m > 0$ may be large. Such bounds can be also deduced from (9.4.19) and (9.4.13), (9.4.14) (see [274]).

9.5 MAIN THEOREM AND OUTLINE OF THE PROOF

The main theorem is as follows.

Theorem 9.2

The limits $\overline{\rho}, \overline{\mathbf{v}}, \overline{\mathbf{u}}$ satisfy

$$\operatorname{div} \overline{\mathbf{v}} = 0, \tag{9.5.1}$$

and the integral identities

$$\int_U \overline{\rho}_0 \phi(0,\mathbf{x})d\mathbf{x} - \int_{I_T}\int_U \overline{\rho}\partial_t\phi d\mathbf{x}dt - \int_{I_T}\int_U \overline{\rho}\,\overline{\mathbf{v}}\cdot\nabla\phi d\mathbf{x}dt = 0, \tag{9.5.2}$$

$$-\int_U \overline{\rho}_0 \mathbf{v}_0\cdot\psi d\mathbf{x} - \int_{I_T}\int_U \overline{\rho}\,\overline{\mathbf{v}}\cdot\partial_t\psi d\mathbf{x}dt - \int_{I_T}\int_U \overline{\rho}\,\overline{\mathbf{v}}\otimes\overline{\mathbf{v}}\cdot\nabla\psi d\mathbf{x}dt \tag{9.5.3}$$
$$+\int_{I_T}\int_U \overline{\mathbf{T}}\cdot e(\psi)d\mathbf{x}dt = 0$$

for all smooth test functions ϕ, ψ, such that div $\psi = 0$, and ϕ, ψ are equal to zero on ∂U and vanish for $t \geq T$.

Moreover, there exist the effective tensors $\overline{\mathbf{A}} \in L^2(U), \overline{\mathbf{B}} \in L^2(U)$ and $\overline{\mathbf{C}} \in L^2(I_T \times U)$ such that the effective deviatoric stress $\overline{\mathbf{T}}$ satisfies

$$\overline{\mathbf{T}} = \overline{\mathbf{A}}e(\overline{\mathbf{u}}) + \overline{\mathbf{B}}e(\overline{\mathbf{v}}) + \int_0^t \overline{\mathbf{C}}(t-\tau)e(\overline{\mathbf{v}})(\tau)d\tau. \tag{9.5.4}$$

■

Remark. Equations (9.9.11)–(9.9.14) essentially mean that the effective equations for $\overline{\rho}, \overline{\mathbf{v}}, \overline{\mathbf{u}}, \overline{P}$ are

$$\partial_t\overline{\rho} + \operatorname{div}(\overline{\rho}\,\overline{\mathbf{v}}) = 0,$$
$$\operatorname{div}\overline{\mathbf{v}} = 0,$$
$$\partial_t(\overline{\rho}\,\overline{\mathbf{v}}) + \operatorname{div}(\overline{\rho}\,\overline{\mathbf{v}}\otimes\overline{\mathbf{v}}) - \operatorname{div}\overline{\mathbf{T}} + \nabla\overline{P} = 0, \tag{9.5.5}$$

with initial and boundary conditions

$$\overline{\mathbf{v}}(0,\mathbf{x}) = \mathbf{v}_0, \quad \overline{\rho}(0,\mathbf{x}) = \overline{\rho}_0, \quad \overline{\mathbf{v}} = 0 \text{ on } \partial U,$$

and $\overline{\mathbf{T}}$ given by (9.9.14).

The result follows from a number of propositions and theorems. Here, the outline of the proof is presented for the reader's convenience.

1. *Construct the corrector operators* $N^\varepsilon\mathbf{w} = \mathbf{w}^\varepsilon$.

N^ε are defined by (9.6.1)–(9.6.3). The proposed construction is non-local in t and satisfies the divergence-free constraint.

2. *Auxiliary problems*. Equation (9.6.1) includes two types of auxiliary functions: $\mathbf{n}^{pq,\varepsilon}$ and $\mathbf{m}^{pq,\varepsilon}$. An expression of the time derivative of $\mathbf{w}^\varepsilon$ additionally contains the final value $\mathbf{m}_T^{pq,\varepsilon} = \mathbf{m}^{pq,\varepsilon}(T)$. These three families of functions satisfy auxiliary problems. The right hand sides in the auxiliary problems are chosen so that the sequences of solutions converge to zero weakly in the appropriate Sobolev-type spaces. The choice of the right hand sides involves abstract G-limit operators corresponding to each of the auxiliary problems. An analysis of auxiliary problems is presented in Sections 9.6.1, 9.6.2. Propositions 9.3, 9.5 are used repeatedly in the remainder of the proof.

3. *Convergence of* $\mathbf{w}^\varepsilon$. Using estimates from Step 2, show that

$$\mathbf{w}^\varepsilon \to \mathbf{w}, \quad \partial_t \mathbf{w}^\varepsilon \to \partial_t \mathbf{w} \text{ weakly in } L^2(I_T, H_0^1(U)).$$

This is done in Proposition 9.6. This convergence corresponds to the condition (9.2.3) on N^ε.

4. *Convergence of the inertial terms.* Proposition 9.6 combined with Lemma 9.2 (Lemma 5.1 from [273]) implies that

$$\int_U \rho^\varepsilon(0,\mathbf{x})\mathbf{v}_0 \cdot \mathbf{w}^\varepsilon + \int_0^T \int_U \rho^\varepsilon \mathbf{v}^\varepsilon \cdot \mathbf{w}_t^\varepsilon + \int_0^T \int_U \rho^\varepsilon \mathbf{v}^\varepsilon \otimes \mathbf{v}^\varepsilon \cdot \nabla \mathbf{w}^\varepsilon$$

converges to

$$\int_U \overline{\rho}(0,\mathbf{x})\mathbf{v}_0 \cdot \mathbf{w} + \int_0^T \int_U \overline{\rho}\, \overline{\mathbf{v}} \cdot \mathbf{w}_t + \int_0^T \int_U \overline{\rho}\, \overline{\mathbf{v}} \otimes \overline{\mathbf{v}} \cdot \nabla \mathbf{w}.$$

This is done in Proposition 9.7.

5. *Compensated compactness of the stress.* Convergence of the inertial terms implies

$$\int_0^T \int_U T^\varepsilon \cdot e(\mathbf{w}^\varepsilon) \to \int_0^T \int_U \overline{T} \cdot e(\mathbf{w}).$$

This is shown in Proposition 9.8.

6. *Effective stress.* Characterization of $\overline{\mathbf{T}}$ is obtained in Theorem 9.3. This step corresponds to establishing condition (9.2.4) on N^ε. The main idea is to write

$$\int_0^T \int_U T^\varepsilon \cdot e(\mathbf{w}^\varepsilon) = -\langle \mathbf{u}^\varepsilon, \operatorname{div}(\mathbf{A}^\varepsilon - \mathbf{B}^\varepsilon \partial_t)\mathbf{w}^\varepsilon \rangle. \tag{9.5.6}$$

The expression for $\langle \mathbf{u}^\varepsilon, \operatorname{div}(\mathbf{A}^\varepsilon - \mathbf{B}^\varepsilon \partial_t)\mathbf{w}^\varepsilon \rangle \equiv \langle \mathbf{u}^\varepsilon, \operatorname{div}(\mathbf{A}^\varepsilon - \mathbf{B}^\varepsilon \partial_t)N^\varepsilon \mathbf{w} \rangle$ contains a number of terms depending on $\mathbf{n}^{pq,\varepsilon}, \mathbf{m}_T^{pq,\varepsilon}$, $\mathbf{m}^{pq,\varepsilon}$ and the corresponding pressures. In some of these terms we can pass to the limit using Lemma 5.1 from [273], since $\mathbf{u}^\varepsilon$ has Sobolev regularity. In other terms this is not possible, but these terms vanish by design of the auxiliary problems. The effective tensors $\overline{\mathbf{A}}, \overline{\mathbf{B}}$ and $\overline{\mathbf{C}}$ are obtained as weak limits of the three fluxes that appear in the auxiliary problems for, respectively, $\mathbf{n}^{pq,\varepsilon}, \mathbf{m}_T^{pq,\varepsilon}$ and $\mathbf{m}^{pq,\varepsilon}$.

7. *Mass conservation.* The mass conservation equation is weakly stable. This is a well-known fact (see [273]).

9.6 CORRECTOR OPERATORS AND OSCILLATING TEST FUNCTIONS

We look for corrector operators of the form

$$\begin{aligned} N^\varepsilon \mathbf{w} \equiv \mathbf{w}^\varepsilon(t,\mathbf{x}) &= \mathbf{w}(t,\mathbf{x}) + \mathbf{n}^{pq,\varepsilon}(\mathbf{x})e(\mathbf{w})_{pq} \\ &+ \int_t^T \mathbf{m}^{pq,\varepsilon}(t-\tau+T,\mathbf{x})e(\mathbf{w}_t)_{pq}(\tau,\mathbf{x})d\tau + \nabla\phi^\varepsilon, \end{aligned} \tag{9.6.1}$$

(Summation over $p,q \in \{1,2,3\}$ is assumed). Here, $\mathbf{w} \in C_0^\infty(I_T \times U), \text{div}\,\mathbf{w} = 0$ is an arbitrary test function, and $\mathbf{n}^{pq,\varepsilon} \in H_0^1(U), \mathbf{m}^{pq,\varepsilon} \in L^2(I_T, H_0^1(U))$ are to be specified. So far we require that

$$\text{div}\,\mathbf{n}^{pq,\varepsilon} = 0, \qquad \text{div}\,\mathbf{m}^{pq,\varepsilon} = 0. \tag{9.6.2}$$

More conditions will be imposed below.

The function $\phi^\varepsilon \in L^2(I_T, H^1(U))$ satisfies

$$\Delta\phi^\varepsilon = -\mathbf{n}^{pq,\varepsilon} \cdot \nabla e(\mathbf{w})_{pq} - \int_t^T \mathbf{m}^{pq,\varepsilon}(t-\tau+T,\mathbf{x}) \cdot \nabla e(\mathbf{w}_t)_{pq}(\tau,\mathbf{x})d\tau, \tag{9.6.3}$$
$$\nabla\phi^\varepsilon \cdot \nu = 0 \text{ on } \partial U.$$

The choice of the first three terms in (9.6.1) is motivated by similar expressions used in periodic [135] and scale-separated [138] homogenization. The last term is needed to enforce the divergence-free constraint. This is necessary in order to avoid dealing with pressure in (9.4.14) which is not $L^1_{loc}(U)$ in general.

Note also that $\nabla\phi^\varepsilon \cdot \nu$ is zero on ∂U for almost all t. This makes $\mathbf{w}^\varepsilon$ correctly defined test functions for (9.4.14) (see the Remark following that equation). Moreover, we have the following:

Lemma 9.1

The function $\mathbf{w}^\varepsilon$ defined by (9.6.1)–(9.6.3) satisfies

$$\text{div}\,\mathbf{w}^\varepsilon = 0.$$

■

Proof. Taking divergence of (9.6.1) and using (9.6.2) we find

$$\text{div}\left(\mathbf{w}(t,\mathbf{x}) + \mathbf{n}^{pq,\varepsilon} e(\mathbf{w})_{pq} + \int_t^T \mathbf{m}^{pq,\varepsilon}(t-\tau+T,\mathbf{x}) e(\mathbf{w}_t)_{pq}(\tau,\mathbf{x})d\tau\right)$$
$$= \mathbf{n}^{pq,\varepsilon} \cdot \nabla e(\mathbf{w})_{pq} + \int_t^T \mathbf{m}^{pq,\varepsilon}(t-\tau+T,\mathbf{x}) \cdot \nabla e(\mathbf{w}_t)_{pq}(\tau,\mathbf{x})d\tau,$$

and the claim follows from the condition (9.6.3). ■

9.6.1 AUXILIARY PROBLEM FOR $\mathbf{m}^{PQ,\varepsilon}$

In this subsection, p,q are fixed, so we drop them to simplify notations, and write $\mathbf{m}^\varepsilon$ instead of $\mathbf{m}^{pq,\varepsilon}$ and so on. We look for $\mathbf{m}^\varepsilon$ that solve the auxiliary problem

$$-\text{div}\,(\mathbf{A}^\varepsilon e(\mathbf{m}^\varepsilon) - \mathbf{B}^\varepsilon e(\mathbf{m}_t^\varepsilon)) - \nabla P_3^\varepsilon = \mathbf{f}, \qquad \text{div}\,\mathbf{m}^\varepsilon = 0, \tag{9.6.4}$$

satisfying the condition

$$\mathbf{m}^\varepsilon(T,\mathbf{x}) = \mathbf{m}_T^\varepsilon. \tag{9.6.5}$$

The objective of this section is to show that the right hand side $\mathbf{f}$ can be chosen so that $\mathbf{m}^\varepsilon$ converges weakly to zero in an appropriate space.

Let $\psi(t) \in C^\infty(I_T)$ satisfy $\psi(0) = 0$ and $\psi(T) = 1$. We use this function to reduce (9.6.4), (9.6.5) to a problem with a different right hand side and zero condition at $t = T$. Writing

$$\mathbf{m}^\varepsilon = \hat{\mathbf{m}}^\varepsilon + \psi \mathbf{m}_T^\varepsilon$$

we deduce that $\hat{\mathbf{m}}^\varepsilon$ solves

$$-\mathrm{div}\left(\mathbf{A}^\varepsilon e(\hat{\mathbf{m}}^\varepsilon) - \mathbf{B}^\varepsilon e(\hat{\mathbf{m}}_t^\varepsilon)\right) - \nabla P^\varepsilon = \mathbf{f} + \mathrm{div}\left(\mathbf{A}^\varepsilon e(\psi \mathbf{m}_T^\varepsilon) - \mathbf{B}^\varepsilon e(\psi_t \mathbf{m}_T^\varepsilon)\right), \qquad \mathrm{div}\, \hat{\mathbf{m}}^\varepsilon = 0, \tag{9.6.6}$$

with the initial condition

$$\hat{\mathbf{m}}(T, \mathbf{x}) = 0. \tag{9.6.7}$$

Define the spaces

$$\mathscr{V} = \{\mathbf{v} \in L^2(I_T, H_0^1(U)),\ \mathrm{div}\, \mathbf{v} = 0\}, \tag{9.6.8}$$

$$\mathscr{W} = \{\mathbf{v} \in \mathscr{V} : \mathbf{v}_t \in \mathscr{V}\}, \tag{9.6.9}$$

$$\mathscr{W}_T = \{\mathbf{v} \in \mathscr{W} : \mathbf{v}(T) = 0\}. \tag{9.6.10}$$

The space $\mathscr{V}$ is equipped with a norm

$$\| \mathbf{v} \|_{\mathscr{V}} = \left(\int_0^T \int_U e(\mathbf{v}) \cdot e(\mathbf{v}) d\mathbf{x} dt \right)^{1/2}. \tag{9.6.11}$$

This norm is induced by the norm

$$\| \mathbf{v} \|_{H_0^1(U)} \equiv \left(\int_U e(\mathbf{v}) \cdot e(\mathbf{v}) d\mathbf{x} \right)^{1/2}. \tag{9.6.12}$$

By Korn inequality ([330]), (9.6.12) is a norm equivalent to the standard one. Also, $\mathscr{W}_T$ is dense in $\mathscr{V}$. This can be proved in the same way as e.g. Thm. 2.1 in [272].

A weak solution $\hat{\mathbf{m}}^\varepsilon$ of (9.6.13), (9.6.14) is an element of $\mathscr{W}_T$ satisfying

$$\int_0^T \int_U (\mathbf{A}^\varepsilon e(\hat{\mathbf{m}}^\varepsilon) - \mathbf{B}^\varepsilon e(\hat{\mathbf{m}}_t^\varepsilon)) \cdot e(\phi) d\mathbf{x} dt = \langle \mathbf{f} - \mathbf{g}^\varepsilon, \phi \rangle_{\mathscr{V}, \mathscr{V}^\star} \tag{9.6.13}$$

for all $\phi \in \mathscr{V}$. Here,

$$\mathbf{g}^\varepsilon = -\mathrm{div}\,(\mathbf{A}^\varepsilon e(\psi \mathbf{m}_T^\varepsilon) - \mathbf{B}^\varepsilon e(\psi_t \mathbf{m}_T^\varepsilon)).$$

Equation (9.6.13) can be stated as

$$G^\varepsilon \hat{\mathbf{m}}^\varepsilon = \mathbf{f} - \mathbf{g}^\varepsilon, \tag{9.6.14}$$

with the operator $G^\varepsilon : \mathscr{W}_T \to \mathscr{V}^\star$. We consider it as an unbounded operator on $\mathscr{V}$ with the domain $\mathscr{W}_T$. The corresponding bilinear form is defined as

$$\langle G^\varepsilon \mathbf{u}, \mathbf{v} \rangle = \int_0^T \int_U (\mathbf{A}^\varepsilon e(\mathbf{u}) - \mathbf{B}^\varepsilon e(\mathbf{u}_t)) \cdot e(\mathbf{v}) d\mathbf{x} dt \tag{9.6.15}$$

for each $\mathbf{u} \in \mathscr{W}_T$, $\mathbf{v} \in \mathscr{V}$. Finally, we note that the adjoint operator $G^{\varepsilon,\star}$ with the domain

$$\mathscr{W}_0 = \{\mathbf{v} \in \mathscr{W} : \mathbf{v}(0) = 0\}$$

is defined by

$$\langle G^{\varepsilon,\star}\mathbf{u}, \mathbf{v}\rangle = \int_0^T \int_U (\mathbf{A}^\varepsilon e(\mathbf{u}) + \mathbf{B}^\varepsilon e(\mathbf{u}_t)) \cdot e(\mathbf{v}) d\mathbf{x} dt \tag{9.6.16}$$

Proposition 9.1. *(i) G^ε is strongly coercive:*

$$\langle G^\varepsilon \mathbf{u}, \mathbf{u}\rangle \geq \alpha_1 \parallel \mathbf{u} \parallel_{\mathscr{V}}^2 \tag{9.6.17}$$

where α_1 is a constant from (9.4.12) (and thus independent of ε);
(ii) G^ε has a bounded inverse satisfying

$$\parallel (G^\varepsilon)^{-1}\mathbf{f} \parallel_{\mathscr{V}} \leq \frac{1}{\alpha_1} \parallel \mathbf{f} \parallel_{\mathscr{V}^\star} \tag{9.6.18}$$

for each $\mathbf{f} \in \mathscr{V}^\star$.

Proof. If $\mathbf{u}$ is sufficiently smooth, then after integrating by parts in (9.6.16) we would have

$$\langle G^\varepsilon \mathbf{u}, \mathbf{u}\rangle = \int_0^T \int_U \mathbf{A}^\varepsilon e(\mathbf{u}) \cdot e(\mathbf{u}) d\mathbf{x} dt + \frac{1}{2}\left(\int_U \mathbf{B}^\varepsilon e(\mathbf{u}) \cdot e(\mathbf{u}) d\mathbf{x}\right)(0),$$

where we took into account that $\mathbf{u}(T) = 0$. However, for an arbitrary $\mathbf{u} \in \mathscr{W}_T$ the second term in the right hand side may not be well defined. To bypass this difficulty, observe that for almost all $t \in I_T$,

$$\frac{1}{2}\left(\int_U \mathbf{B}^\varepsilon e(\mathbf{u}) \cdot e(\mathbf{u}) d\mathbf{x}\right)(t)$$

is finite. For such t we have

$$\begin{aligned}
&\int_t^T \int_U (\mathbf{A}^\varepsilon e(\mathbf{u}) \cdot e(\mathbf{u}) - \mathbf{B}^\varepsilon e(\mathbf{u}_t) \cdot e(\mathbf{u})\, d\mathbf{x} dt \\
= &\int_t^T \int_U \mathbf{A}^\varepsilon e(\mathbf{u}) \cdot e(\mathbf{u}) d\mathbf{x} dt + \frac{1}{2}\left(\int_U \mathbf{B}^\varepsilon e(\mathbf{u}) \cdot e(\mathbf{u}) d\mathbf{x}\right)(t) \\
\geq &\int_t^T \int_U \mathbf{A}^\varepsilon e(\mathbf{u}) \cdot e(\mathbf{u}) d\mathbf{x} dt \geq \alpha_1 \int_t^T \int_U e(\mathbf{u}) \cdot e(\mathbf{u}) d\mathbf{x} dt
\end{aligned}$$

The last inequality follows from (9.4.12). Using absolute continuity in t of the first and last terms in the above inequality, we can pass to the limit $t \to 0^+$ and obtain

$$\langle G^\varepsilon \mathbf{u}, \mathbf{u}\rangle = \lim_{t \to 0+} \int_t^T \int_U (\mathbf{A}^\varepsilon e(\mathbf{u}) \cdot e(\mathbf{u}) - \mathbf{B}^\varepsilon e(\mathbf{u}_t) \cdot e(\mathbf{u})\, d\mathbf{x} dt$$

$$\geq \lim_{t \to 0+} \alpha_1 \int_t^T \int_U e(\mathbf{u}) \cdot e(\mathbf{u}) d\mathbf{x} dt = \alpha_1 \parallel \mathbf{u} \parallel_{\mathscr{V}}^2,$$

which proves (i).

(ii) follows from (i). This is known (see, e.g. [413], Lemma 1). We only sketch the proof for completeness. Since G^{ε} is closed, passing to the limit in (9.6.17) we obtain that the image of G^{ε} is closed in $\mathscr{V}^{\star}$. If this image does not contain all $\mathscr{V}^{\star}$, then, because of the density of $\mathscr{W}_T$ in $\mathscr{V}$, there is $\mathbf{g} \in \mathscr{V}$ such that $\langle G^{\varepsilon}\mathbf{u}, \mathbf{g} \rangle = 0$ for all $\mathbf{u} \in \mathscr{W}_T$. This yields $\mathbf{g} \in \mathscr{W}_0$ (domain of $G^{\varepsilon,\star}$) and $G^{\varepsilon,\star}\mathbf{g} = 0$. Next we observe that $G^{\varepsilon,\star}$ satisfies (9.6.17) which yields $\mathbf{g} = 0$ and gives a contradiction. Thus G^{ε} is onto. The estimate (9.6.18) follows from (9.6.17). ■

Remark. (ii) implies existence of the pressure $P_3^{\varepsilon} \in L^2(I_T, L^2(U))$. This follows using standard arguments from [383] combined with the inclusion $\mathbf{A}^{\varepsilon} e(\mathbf{m}^{pq}) + \mathbf{B}^{\varepsilon} e(\mathbf{m}_t^{pq}) \in L^2(I_T, L^2(U))$. Moreover, P_3^{ε} is bounded in $L^2(I_T, L^2(U))$ independent of ε. Therefore, extracting a subsequence if necessary, we can assume that $P_3^{\varepsilon} \to \overline{P}_3$ weakly in $L^2(I_T, L^2(U))$.

Definition 9.2. *We say that the sequence G^{ε} G-converges to an operator $G : D(G) \subset \mathscr{W}_T \to \mathscr{V}^{\star}$, if for each $\mathbf{f} \in \mathscr{V}^{\star}$ the sequence $\mathbf{u}^{\varepsilon} = (G^{\varepsilon})^{-1}\mathbf{f}$ converges to some $\mathbf{u} \in D(G)$ weakly in $\mathscr{W}_T$. In this case we define $G\mathbf{u} = \mathbf{f}$.*

Proposition 9.2. *The sequence G^{ε} contains a G-convergent subsequence. The limiting operator G has the following properties:*
(i)

$$\langle G\mathbf{u}, \mathbf{u} \rangle \geq \alpha_1 \parallel \mathbf{u} \parallel_{\mathscr{V}}^2 \tag{9.6.19}$$

for each $\mathbf{u} \in D(G)$;
(ii) $D(G) = \mathscr{W}_T$.

Proof. Let us write $G^{\varepsilon} = A^{\varepsilon} - B^{\varepsilon} \partial_t$ where $A^{\varepsilon}, B^{\varepsilon} : \mathscr{V} \to \mathscr{V}^{\star}$ are operators induced by the bilinear forms

$$a^{\varepsilon}(\mathbf{u}, \mathbf{v}) \equiv \int_0^T \int_U \mathbf{A}^{\varepsilon} e(\mathbf{u}) \cdot e(\mathbf{v}) d\mathbf{x} dt,$$

$$b^{\varepsilon}(\mathbf{u}, \mathbf{v}) \equiv \int_0^T \int_U \mathbf{B}^{\varepsilon} e(\mathbf{u}) \cdot e(\mathbf{v}) d\mathbf{x} dt,$$

respectively. Ellipticity assumptions (9.4.12) imply that $A^{\varepsilon}, B^{\varepsilon}$ are coercive and bounded with coercivity constants and bounds independent of ε. In particular coercivity of B^{ε} implies that there exists a bounded inverse $(B^{\varepsilon})^{-1}$ defined on $\mathscr{V}^{\star}$, satisfying $\parallel (B^{\varepsilon})^{-1} \parallel \leq \frac{1}{\beta_1}$, where β_1 is the lower bound from (9.4.12). Therefore, if $(A^{\varepsilon} - B^{\varepsilon}\partial_t)\mathbf{u}^{\varepsilon} = \mathbf{f}$ then $\partial_t \mathbf{u}^{\varepsilon} = (B^{\varepsilon})^{-1}(A^{\varepsilon}\mathbf{u} - \mathbf{f})$. This implies

$$\parallel \partial_t \mathbf{u}^{\varepsilon} \parallel_{\mathscr{V}} \leq \frac{\alpha_2}{\beta_1} \parallel (G^{\varepsilon})^{-1}\mathbf{f} \parallel_{\mathscr{V}} + \parallel \mathbf{f} \parallel_{\mathscr{V}^{\star}} \leq (\frac{\alpha_2}{\alpha_1 \beta_1} + 1) \parallel \mathbf{f} \parallel_{\mathscr{V}^{\star}}.$$

Thus, if $\mathbf{u}^\varepsilon \to \mathbf{u}$ weakly in $\mathscr{W}_T$ then

$$\| \mathbf{u} \|_{\mathscr{W}} \leq C(\alpha_1, \alpha_2, \beta_1) \| \mathbf{f} \|_{\mathscr{V}^\star} . \tag{9.6.20}$$

Since $\mathscr{V}^\star$ is separable, we can use diagonal procedure to find a subsequence, non relabeled, such that $\mathbf{u}^\varepsilon = (G^\varepsilon)^{-1}\mathbf{f}$ converges weakly in $\mathscr{W}_T$ to $\mathbf{u} = G^{-1}\mathbf{f}$ for all $\mathbf{f}$ in a dense subset of $\mathscr{V}^\star$. Inequality (9.6.20) implies that convergence also holds for all $\mathbf{f} \in \mathscr{V}^\star$, and the operator G^{-1} is bounded.

Next, consider a sequence $\mathbf{u}^\varepsilon$ such that $G^\varepsilon \mathbf{u}^\varepsilon = \mathbf{f}$. Then, by the preceding, $\mathbf{u}^\varepsilon$ converges weakly to $\mathbf{u}$, and $G\mathbf{u} = \mathbf{f}$ by definition of the G-limit. Since $\langle G^\varepsilon \mathbf{u}^\varepsilon, \mathbf{u}^\varepsilon \rangle = \langle \mathbf{f}, \mathbf{u}^\varepsilon \rangle = \langle G\mathbf{u}, \mathbf{u}^\varepsilon \rangle$, we can pass to the limit and obtain

$$\lim_{\varepsilon \to 0} \langle G^\varepsilon \mathbf{u}^\varepsilon, \mathbf{u}^\varepsilon \rangle = \langle G\mathbf{u}, \mathbf{u} \rangle .$$

This, together with lower semicontinuity of the norm with respect to weak convergence, allows passage to the limit in (9.6.17) which yields (9.6.19).

To prove (ii), observe first that by (9.6.20), $D(G) \subset \mathscr{W}_T$. To prove equality, we first show that G^{-1} is injective: $G^{-1}\mathbf{f} = 0$ implies $\mathbf{f} = 0$. Arguing by contradictions, suppose that there is $\mathbf{g} \in \mathscr{V}^\star, \mathbf{g} \neq 0$ such that $G^{-1}\mathbf{g} = 0$. Consider the sequence $\mathbf{u}_g^\varepsilon = (G^\varepsilon)^{-1}\mathbf{g}$. By definition of G, $\mathbf{u}_g^\varepsilon$ converges to zero weakly in $\mathscr{W}_T$. Then by (9.6.17),

$$\langle \mathbf{g}, \mathbf{u}_g^\varepsilon \rangle = \langle G^\varepsilon \mathbf{u}_g^\varepsilon, \mathbf{u}_g^\varepsilon \rangle \geq \alpha_1 \| \mathbf{u}_g^\varepsilon \|_{\mathscr{V}}^2 .$$

Passing to the limit $\varepsilon \to 0$ we obtain that $\| \mathbf{u}_g^\varepsilon \|_{\mathscr{V}} \to 0$ and thus $\mathbf{u}_g^\varepsilon$ converges to zero strongly in $\mathscr{V}$. Next, we use $\mathbf{u}_{g,t}^\varepsilon$ as the test function in the weak formulation of $G^\varepsilon \mathbf{u}_g^\varepsilon = \mathbf{g}$. Integrating by parts (which can be justified as in the proof of Proposition 9.1) and using coercivity of B^ε we obtain

$$\left| \langle \mathbf{g}, \mathbf{u}_{g,t}^\varepsilon \rangle \right| = \left| \langle G^\varepsilon \mathbf{u}_g^\varepsilon, \mathbf{u}_{g,t}^\varepsilon \rangle \right| \geq \beta_1 \| \mathbf{u}_{g,t}^\varepsilon \|_{\mathscr{V}}^2 ,$$

and thus $\mathbf{u}_{g,t}^\varepsilon$ converges to zero strongly in $\mathscr{V}$. Next, using uniform boundedness of $A^\varepsilon, B^\varepsilon$, we write

$$\| \mathbf{g} \|_{\mathscr{V}^\star} = \| A^\varepsilon \mathbf{u}_g^\varepsilon - B^\varepsilon \mathbf{u}_{g,t}^\varepsilon \|_{\mathscr{V}^\star} \leq \alpha_2 \| \mathbf{u}_g^\varepsilon \|_{\mathscr{V}} + \beta_2 \| \mathbf{u}_{g,t}^\varepsilon \|_{\mathscr{V}}$$

where α_2, β_2 are constants from (9.4.12). Passing to the limit in the above we deduce $\mathbf{g} = 0$, which contradicts assumption $\mathbf{g} \neq 0$. Thus G^{-1} is injective.

Next we show that $D(G)$ (equivalently, the range of G^{-1}) is dense in $\mathscr{V}$. If this were false, there would be $\mathbf{h} \neq 0, \mathbf{h} \in \mathscr{V}^\star$ such that $\langle \mathbf{h}, G^{-1}\mathbf{f} \rangle = 0$ for all $\mathbf{f} \in \mathscr{V}^\star$. Let $\mathbf{u}_h = G^{-1}\mathbf{h}$. Choosing $\mathbf{f} = \mathbf{h}$, we obtain using (9.6.19):

$$0 = \langle \mathbf{h}, G^{-1}\mathbf{h} \rangle = \langle G\mathbf{u}_h, \mathbf{u}_h \rangle \geq \alpha_1 \| \mathbf{u}_h \|_{\mathscr{V}} .$$

Therefore, $\mathbf{u}_h = 0$. Then $\mathbf{h} = 0$ by injectivity of G^{-1}. This contradicts the assumption $\mathbf{h} \neq 0$.

Finally, observe that the norm of $\mathscr{W}_T$ induces a scalar product

$$(\mathbf{u}, \mathbf{v})_{\mathscr{W}_T} = \int_0^T \int_U (e(\mathbf{u}_t) \cdot e(\mathbf{v}_t) + e(\mathbf{u}) \cdot e(\mathbf{v}))\, d\mathbf{x} dt.$$

In search of a contradiction, suppose that $D(G)$ is a proper subset of $\mathscr{W}_T$. Then there is $\overline{\mathbf{u}} \neq 0, \overline{\mathbf{u}} \in \mathscr{W}_T$ such that $(\overline{\mathbf{u}}, G^{-1}\mathbf{f})_{\mathscr{W}_T} = 0$ for all $\mathbf{f} \in \mathscr{V}^\star$. Expression $(\overline{\mathbf{u}}, \mathbf{v})_{\mathscr{W}_T}$ defines a bounded linear functional $l_{\overline{\mathbf{u}}}(\mathbf{v})$ on $D(G)$ which by the Hahn-Banach theorem can be extended to a bounded linear functional $L_{\overline{\mathbf{u}}}$ on $\mathscr{V}$ and this extension has the same norm as $l_{\overline{\mathbf{u}}}(\mathbf{v})$. Therefore, $(\overline{\mathbf{u}}, G^{-1}\mathbf{f})_{\mathscr{W}_T} = 0$ implies $\langle L_{\overline{\mathbf{u}}}, G^{-1}\mathbf{f}\rangle = 0$ for all $\mathbf{f} \in \mathscr{V}^\star$. Density of the range of G^{-1} implies $L_{\overline{\mathbf{u}}} = 0$. But then $(\overline{\mathbf{u}}, \overline{\mathbf{u}})_{\mathscr{W}_T} = 0$, which contradicts the assumption $\overline{\mathbf{u}} \neq 0$. Thus (ii) is proved. ∎

Proposition 9.3. *There exists* $\mathbf{f} \in \mathscr{V}^\star$ *such that the sequence of solutions* $\mathbf{m}^\varepsilon$ *of (9.6.4), with this choice of the right hand side, contains a subsequence (not relabeled) satisfying*
(i) $\mathbf{m}^\varepsilon \to 0$ *weakly in* $\mathscr{W}_T$,
(ii) $\mathbf{m}^\varepsilon \to 0$ *strongly in* $L^2(I_T, L^2(U))$.

Proof. By Proposition 9.5, proved below in Sect. 9.6.2, we can assume that $\mathbf{m}_T^\varepsilon$ converges to zero weakly in $H_0^1(U)$. Then $\psi \mathbf{m}_T^\varepsilon \to 0$ and $\partial_t \psi \mathbf{m}_T^\varepsilon \to 0$ weakly in $\mathscr{V}$.

Since $\mathbf{m}^\varepsilon = \hat{\mathbf{m}}^\varepsilon + \psi \mathbf{m}_T^\varepsilon$, (i) will be proved if we show that there is a choice of $\mathbf{f}$ such that $\hat{\mathbf{m}}^\varepsilon \to 0$ weakly in $\mathscr{W}_T$.

Consider (9.6.14). In view of (9.4.12) and uniform bounds on $\mathbf{m}_T^\varepsilon$, the sequence $\mathbf{g}^\varepsilon$ is bounded in $\mathscr{V}^\star$. Therefore, the sequence $(G^\varepsilon)^{-1}\mathbf{g}^\varepsilon$ is bounded in $\mathscr{W}_T$, and we can extract a subsequence that converges weakly to some $\mathbf{q} \in \mathscr{W}_T$. By Proposition 9.2, (ii), $\mathbf{q} \in D(G)$. Therefore, we can choose

$$\mathbf{f} = G\mathbf{q}. \tag{9.6.21}$$

Then

$$\hat{\mathbf{m}}^\varepsilon = (G^\varepsilon)^{-1}(\mathbf{f} - \mathbf{g}^\varepsilon) = (G^\varepsilon)^{-1} G\mathbf{q} - (G^\varepsilon)^{-1}\mathbf{g}^\varepsilon.$$

By Proposition 9.2, $(G^\varepsilon)^{-1} G\mathbf{q} \to \mathbf{q}$ weakly in $\mathscr{W}_T$, up to extraction of a subsequence. Hence, $\hat{\mathbf{m}}^\varepsilon \to 0$. Thus (i) is proved.

To prove (ii), observe first that $\mathbf{m}_T^\varepsilon$ converges to zero strongly in $L^2(U)$ and thus $\psi \mathbf{m}_T^\varepsilon \to 0$ and $\partial_t \psi \mathbf{m}_T^\varepsilon \to 0$ strongly in $L^2(I_T, L^2(U))$. Therefore, to prove strong convergence of $\mathbf{m}^\varepsilon$ it is enough to prove strong convergence of $\hat{\mathbf{m}}^\varepsilon$.

Next, note that $\hat{\mathbf{m}}_t^\varepsilon$ is bounded in $\mathscr{V}$ independent of ε. Now strong convergence of $\hat{\mathbf{m}}^\varepsilon$ is deduced from (i) and J. L. Lions' compactness theorem (see e.g. [383], Theorem 2.1, Chapter III). ∎

9.6.2 AUXILIARY PROBLEMS FOR $\mathbf{n}^{PQ,\varepsilon}, \mathbf{m}_T^{PQ,\varepsilon}$

In this section, p, q are once again fixed, so we drop them to simplify notations, and write $\mathbf{n}^\varepsilon$ instead of $\mathbf{n}^{pq,\varepsilon}$ and so on.

We seek $\mathbf{n}^\varepsilon$, $\mathbf{m}_T^\varepsilon$ satisfying, respectively, the auxiliary problems

$$-\text{div}\left(\mathbf{A}^\varepsilon\left(\mathbf{I}^{pq} + e(\mathbf{n}^\varepsilon)\right)\right) - \nabla P_1^\varepsilon = \mathbf{f}_1, \tag{9.6.22}$$

$$-\text{div}\left(\mathbf{B}^\varepsilon\left(\mathbf{I}^{pq} + e(\mathbf{n}^\varepsilon)\right) - e(\mathbf{m}_T^\varepsilon)\right) - \nabla P_2^\varepsilon = \mathbf{f}_2, \tag{9.6.23}$$

where

$$\mathbf{I}^{pq} = \frac{1}{2}\left(\mathbf{e}_p \otimes \mathbf{e}_q + \mathbf{e}_q \otimes \mathbf{e}_p\right),$$

and $\mathbf{f}_1, \mathbf{f}_2$ are to be specified. First, we find $\mathbf{n}^\varepsilon$ from (9.6.22). Then this $\mathbf{n}^\varepsilon$ should be plugged into (9.6.23), and then $\mathbf{m}_T^\varepsilon$ can be found. The goal, as before, is to choose $\mathbf{f}_1, \mathbf{f}_2$ so that $\mathbf{n}^\varepsilon, \mathbf{m}_T^\varepsilon$ would have subsequences that converge weakly to zero. Let

$$V = \{\mathbf{v} \in H_0^1 : \text{div}\,\mathbf{v} = 0\},$$

equipped with the norm (9.6.12). Given $\mathbf{f}_1 \in V^\star$, $\mathbf{n}^\varepsilon \in V$ is a weak solution of (9.6.22) provided

$$\int_U \mathbf{A}^\varepsilon \left(\mathbf{I}^{pq} + e(\mathbf{n}^\varepsilon)\right) \cdot e(\mathbf{w}) d\mathbf{x} = \langle \mathbf{f}_1, \mathbf{w} \rangle_{V,V^\star} \tag{9.6.24}$$

for all $\mathbf{w} \in V$. Weak solutions of (9.6.23) are defined similarly. This identity can be written as an operator equation

$$A^\varepsilon \mathbf{n}^\varepsilon = \mathbf{f}_1 - \mathbf{g}_1^\varepsilon, \tag{9.6.25}$$

where

$$\mathbf{g}_1^\varepsilon = -\text{div}\left(\mathbf{A}^\varepsilon \mathbf{I}^{pq}\right) \in V^\star,$$

and $A^\varepsilon : V \to V^\star$ is the operator induced by the bilinear form

$$a^\varepsilon(\mathbf{u}, \mathbf{v}) = \int_U \mathbf{A}^\varepsilon e(\mathbf{u}) \cdot e(\mathbf{v}) d\mathbf{x}.$$

Similarly, (9.6.23) can be written as

$$B^\varepsilon \mathbf{m}_T^\varepsilon = \mathbf{f}_2 - \mathbf{g}_2^\varepsilon, \tag{9.6.26}$$

where the operator B^ε is induced by the form

$$b^\varepsilon(\mathbf{u}, \mathbf{v}) = \int_U \mathbf{B}^\varepsilon e(\mathbf{u}) \cdot e(\mathbf{v}) d\mathbf{x},$$

and

$$\mathbf{g}_2^\varepsilon = -\text{div}\left(\mathbf{B}^\varepsilon \left(\mathbf{I}^{pq} + e(\mathbf{n}^\varepsilon)\right)\right).$$

By (9.4.12), operators $A^\varepsilon, B^\varepsilon$ satisfy

$$\langle A^\varepsilon \mathbf{u}, \mathbf{u} \rangle_{V,V^\star} \geq \alpha_1 \| \mathbf{u} \|_V^2 \tag{9.6.27}$$

$$\| A^\varepsilon \mathbf{u} \|_{V^\star} \leq \alpha_2 \| \mathbf{u} \|_V \tag{9.6.28}$$

$$\langle B^\varepsilon \mathbf{u}, \mathbf{u} \rangle_{V,V^\star} \geq \beta_1 \| \mathbf{u} \|_V^2 \tag{9.6.29}$$

$$\| B^\varepsilon \mathbf{u} \|_{V^\star} \leq \beta_2 \| \mathbf{u} \|_V \tag{9.6.30}$$

The Lax-Milgram lemma implies the existence of unique solutions of (9.6.25), (9.6.26). These solutions satisfy

$$\mathbf{n}^\varepsilon = \left(A^\varepsilon\right)^{-1}\left(\mathbf{f}_1 - \mathbf{g}_1^\varepsilon\right), \quad \| \mathbf{n}^\varepsilon \|_V \leq \frac{1}{\alpha_1} \| \mathbf{f}_1 - \mathbf{g}_1^\varepsilon \|_{V^\star},$$

$$\mathbf{m}_T^\varepsilon = (B^\varepsilon)^{-1}(\mathbf{f}_2 - \mathbf{g}_2^\varepsilon), \quad \| \mathbf{m}_T^\varepsilon \|_V \leq \frac{1}{\beta_1} \| \mathbf{f}_2 - \mathbf{g}_2^\varepsilon \|_{V^\star}.$$

Remark. Existence of the pressures $P_1^\varepsilon \in L^2(U), P_2^\varepsilon \in L^2(U)$ follows using standard arguments from [383]. Moreover, these pressures are bounded in $L^2(U)$ independent of ε. Therefore, extracting a subsequence if necessary, we can assume that $P_j^\varepsilon \to \overline{P}_j, j = 1,2$ weakly in $L^2(U)$.

Definition 9.3. *The sequence of operators $A^\varepsilon : V \to V^\star$ G-converges to an operator A if $(A^\varepsilon)^{-1}\mathbf{f}$ converges to some $\mathbf{u} \in V$ weakly in V, for each $\mathbf{f} \in V^\star$. We also define* $\mathbf{f} = A\mathbf{u}$.

Proposition 9.4. *The sequences $A^\varepsilon, B^\varepsilon$ contain G-convergent subsequences.*

Proof. This is known [313], Theorem 2.

Proposition 9.5. *There exists $\mathbf{f}_1 \in V^\star$ (respectively $\mathbf{f}_2$) such that, up to extraction of a subsequence, $\mathbf{n}^\varepsilon$ (respectively $\mathbf{m}_T^\varepsilon$) converge to zero weakly in V.*

Proof. Consider (9.6.25). Since $\mathbf{g}_2^\varepsilon$ is bounded in $V^\star$, and $(A^\varepsilon)^{-1}$ is bounded independent of ε, the sequence $(A^\varepsilon)^{-1}\mathbf{g}_1^\varepsilon$ is bounded in V. Thus we can extract a subsequence that converges weakly in V to some $\overline{\mathbf{u}}_1 \in V$. Choose

$$\mathbf{f}_1 = A\overline{\mathbf{u}}_1. \tag{9.6.31}$$

Then

$$\mathbf{n}^\varepsilon = (A^\varepsilon)^{-1}A\overline{\mathbf{u}}_1 - (A^\varepsilon)^{-1}\mathbf{g}_1.$$

By definition of A, the first term in the right hand side converges to $\overline{\mathbf{u}}_1$ weakly in V, and so does the second. Hence $\mathbf{n}^\varepsilon \to 0$ weakly in V. For (9.6.26) the procedure is the same. Up to extraction of a subsequence, $(B^\varepsilon)^{-1}\mathbf{g}_2^\varepsilon \to \overline{\mathbf{u}}_2$ weakly in V, and we choose

$$\mathbf{f}_2 = B\overline{\mathbf{u}}_2. \tag{9.6.32}$$

■

9.7 INERTIAL TERMS IN THE MOMENTUM BALANCE EQUATION

In the remainder of the chapter, we assume that the oscillating test functions $\mathbf{w}^\varepsilon$ are defined as follows.

Definition 9.4. *Let $\mathbf{w} \in C^\infty(I_T \times U), \mathrm{div}\mathbf{w} = 0$ be arbitrary, and define $\mathbf{w}^\varepsilon$ by (9.6.1). In (9.6.1), choose $\mathbf{n}^{pq,\varepsilon}, \mathbf{m}_T^{pq,\varepsilon}$ that solve, respectively (9.6.22), (9.6.23) with the right hand sides chosen according to (9.6.31), (9.6.32). Also, let $\mathbf{m}^{pq,\varepsilon}$ satisfy (9.6.4) with the right hand side chosen according to (9.6.21).*

Proposition 9.6. *The sequence* $\mathbf{w}^\varepsilon$ *defined as above satisfies*

$$\mathbf{w}^\varepsilon \to \mathbf{w} \ \text{ in } L^2(I_T, L^2(U)), \quad \mathbf{w}_t^\varepsilon \to \mathbf{w}_t \ \text{ in } L^2(I_T, L^2(U)). \tag{9.7.1}$$

Also, $\mathbf{w}^\varepsilon \in L^\infty(I_T, H_0^1(U))$, *and*

$$\| \mathbf{w}^\varepsilon \|_{L^\infty(I_T, H_0^1(U))} \leq T^{1/2} \| \mathbf{w}_t^\varepsilon \|_{L^2(I_T, H_0^1(U))} \leq CT^{1/2} \tag{9.7.2}$$

with C independent of ε.

Proof. First we need a formula for the time derivative of $\mathbf{w}^\varepsilon$. After taking time derivative of (9.6.1), integrating by parts in the time convolution (which involves putting time differentiation on $\mathbf{w}_t$ instead of $\mathbf{m}^{pq,\varepsilon}$) and using $\mathbf{w}(T) = 0$ we obtain

$$\begin{aligned} \mathbf{w}_t^\varepsilon &= \mathbf{w}_t + \mathbf{n}^{pq,\varepsilon}(\mathbf{x}) e(\mathbf{w}_t)_{pq} + \\ &\int_t^T \mathbf{m}^{pq,\varepsilon}(t - \tau + T) e(\mathbf{w}_{tt})_{pq}(\tau) d\tau + \nabla \phi_t^\varepsilon. \end{aligned} \tag{9.7.3}$$

To prove strong convergence of $\mathbf{w}_t^\varepsilon$ we need to prove that all terms in the right hand side of (9.7.3) converge to zero strongly in $L^2(I_T, L^2(U))$.
Step 1. Show that

$$\mathbf{n}^{pq,\varepsilon} e(\mathbf{w})_{pq} + \int_t^T \mathbf{m}^{pq,\varepsilon}(t - \tau + T) e(\mathbf{w}_t)_{pq}(\tau) d\tau \to 0,$$

and

$$\mathbf{n}^{pq,\varepsilon}(\mathbf{x}) e(\mathbf{w}_t)_{pq} + \int_t^T \mathbf{m}^{pq,\varepsilon}(t - \tau + T) e(\mathbf{w}_{tt})_{pq}(\tau) d\tau \to 0$$

strongly in $L^2(I_T \times U)$.

By Proposition 9.5, $\mathbf{n}^{pq,\varepsilon} \to 0$ weakly in $H_0^1(U))$ and thus strongly in $L^2(U)$. By Proposition 9.3, $\mathbf{m}^{pq,\varepsilon} \to 0$ strongly in $L^2(I_T, L^2(U))$, and we conclude.
Step 2. Show that $\nabla \phi^\varepsilon \to 0$ *and* $\partial_t \nabla \phi^\varepsilon \to 0$ *strongly in* $L^2(I_T \times U)$.

First we estimate $e(\nabla \phi^\varepsilon)$. Note that $e(\nabla \phi^\varepsilon)_{ij} = \partial_i \partial_j \phi^\varepsilon$ and write $\partial_i \partial_j \phi^\varepsilon = \partial_i \partial_j E \star (\Delta \phi^\varepsilon) + \partial_i \partial_j K \star (\Delta \phi^\varepsilon)$, where $E = c\frac{1}{|x|}$ is a fundamental solution of the Laplacian, and K is a harmonic function, which depends only on U. To estimate $\partial_i \partial_j E \star (\Delta \phi^\varepsilon)$ we use Calderón-Zygmund inequality (see, e.g. [133], Theorem 9.9) with $p = 2$ and obtain

$$\| \partial_i \partial_j E \star (\Delta \phi^\varepsilon) \|_{L^2(I_T \times U)} \leq \| \Delta \phi^\varepsilon \|_{L^2(I_T \times U)}$$

Since $\partial_i \partial_j K$ is a smooth function, there exists a constant $C(U)$ depending only on U such that

$$\| \partial_i \partial_j K \star (\Delta \phi^\varepsilon) \|_{L^2(I_T \times U)} \leq C(U) \| \Delta \phi^\varepsilon \|_{L^2(I_T \times U)}$$

Thus

$$\| \partial_i \partial_j \phi^\varepsilon \|_{L^2(I_T \times U)} \leq (1 + C(U)) \| \Delta \phi^\varepsilon \|_{L^2(I_T \times U)} . \tag{9.7.4}$$

Combining (9.7.4) with (9.6.3) we find

$$\int_{I_T}\int_U |e(\nabla\phi^\varepsilon)|^2 d\mathbf{x}dt \le C(\mathbf{w},U)\sum_{p,q=1}^{3}\left(\| \mathbf{n}^{pq,\varepsilon} \|^2_{L^2(U)} + \| \mathbf{m}^{pq,\varepsilon} \|^2_{L^2(I_T\times U)}\right) \quad (9.7.5)$$

so $e(\nabla\phi^\varepsilon)$ converges to zero strongly in $L^2(I_T \times U)$. This also implies that the components of the Hessian of ϕ^ε converge to zero strongly in $L^2(I_T \times U)$.

Finally, the standard a priori estimate for the Neumann problem $\Delta\phi^\varepsilon = f^\varepsilon, f \in L^2(U)$, satisfying $\nabla\phi^\varepsilon \cdot \nu = 0$ on the boundary, yields

$$\int_U \nabla\phi^\varepsilon \cdot \nabla\phi^\varepsilon d\mathbf{x} \le \| f^\varepsilon \|_{L^2(U)} \| \phi^\varepsilon \|_{L^2(U)} \le C \| f^\varepsilon \|_{L^2(U)} \left(\int_U \nabla\phi^\varepsilon \cdot \nabla\phi^\varepsilon d\mathbf{x}\right)^{1/2}. \quad (9.7.6)$$

Here, C is the constant in Poincaré inequality. Poincaré inequality applies after we impose the condition $\int_{\partial U}\phi^\varepsilon dS = 0$, standard for the Neumann problem. Since f^ε, given by the right hand side of (9.6.3), converges to zero strongly in $L^2(I_T, L^2(U))$, (9.7.6) implies that $\nabla\phi^\varepsilon$ converges to zero strongly in $L^2(I_T, L^2(U))$.

Differentiating (9.6.3) in t and integrating by parts as in (9.7.3) we find

$$\Delta\phi_t^\varepsilon = -\mathbf{n}^{pq,\varepsilon} \cdot \nabla e(\mathbf{w}_t)_{pq} - \int_t^T \mathbf{m}^{pq,\varepsilon}(t-\tau+T) \cdot \nabla e(\mathbf{w}_{tt})_{pq}(\tau)d\tau. \quad (9.7.7)$$

Therefore, arguing as above we have

$$\int_{I_T}\int_U |e(\nabla\phi_t^\varepsilon)|^2 d\mathbf{x}dt \le C(\mathbf{w},U)\sum_{p,q=1}^{3}\left(\| \mathbf{n}^{pq,\varepsilon} \|^2_{L^2(U)} + \| \mathbf{m}^{pq,\varepsilon} \|^2_{L^2(U)}\right), \quad (9.7.8)$$

which yields $e(\nabla\phi_t^\varepsilon) \to 0$, and then $\nabla\phi_t^\varepsilon \to$ strongly in $L^2(I_T \times U)$.

Step 4. Prove (9.7.2).

Since $\mathbf{w}^\varepsilon(t) = -\int_t^T \mathbf{w}_t^\varepsilon(\tau)d\tau$, we obtain for almost all $t \in I_T$

$$\| \mathbf{w}^\varepsilon \|_{H_0^1(U)}(t) \le \int_0^T \| \mathbf{w}_t^\varepsilon \|_{H_0^1(U)}(\tau)d\tau \le T^{1/2}\left(\int_0^T \| \mathbf{w}_t^\varepsilon \|^2_{H_0^1(U)}(\tau)d\tau\right)^{1/2},$$

and (9.7.2) follows. ■

Next we will need the following lemma ([273], Lemma 5.1).

Lemma 9.2

Let g^n, h^n converge weakly to g, h, respectively in $L^{p_1}(0,T;L^{p_2}(\Omega)), L^{q_1}(0,T;L^{q_2}(\Omega))$, where $1 \le p_1, p_2 \le \infty$,

$$\frac{1}{p_1} + \frac{1}{q_1} = \frac{1}{p_2} + \frac{1}{q_2} = 1.$$

We assume in addition that $\partial_t g^n$ is bounded in $L^1(0,T;W^{-m,1}(\Omega))$ for some $m \geq 0$ independent of n and

$$\| h^n - h^n(\cdot + \xi, t) \|_{L^{q_1}(0,T;L^{q_2}(\Omega))} \to 0$$

as $|\xi| \to 0$, uniformly in n.

Then $g^n h^n$ converges to gh in the sense of distributions on $\Omega \times (0,T)$. ■

This lemma can be used to obtain the effective mass conservation equation and to pass to the limit in the inertial terms in the momentum equation (9.4.14).

Proposition 9.7. *Let* $\mathbf{w}^\varepsilon$ *be functions from Definition 9.4. Then*

$$\lim_{\varepsilon \to 0} \int_U \rho^\varepsilon(0,\mathbf{x})\mathbf{v}_0 \cdot \mathbf{w}^\varepsilon \, d\mathbf{x} = \int_U \overline{\rho}(0,\mathbf{x}) \, \mathbf{v}_0 \cdot \mathbf{w} \, d\mathbf{x}, \tag{9.7.9}$$

$$\lim_{\varepsilon \to 0} \int_{I_T \times U} \rho^\varepsilon \mathbf{v}^\varepsilon \cdot \partial_t \mathbf{w}^\varepsilon \, d\mathbf{x}dt = \int_{I_T \times U} \overline{\rho} \, \overline{\mathbf{v}} \cdot \partial_t \mathbf{w} \, d\mathbf{x}dt, \tag{9.7.10}$$

and

$$\lim_{\varepsilon \to 0} \int_{I_T \times U} \rho^\varepsilon \mathbf{v}^\varepsilon \otimes \mathbf{v}^\varepsilon \cdot \nabla \mathbf{w}^\varepsilon \, d\mathbf{x}dt = \int_{I_T \times U} \overline{\rho} \, \overline{\mathbf{v}} \otimes \overline{\mathbf{v}} \cdot \nabla \mathbf{w} \, d\mathbf{x}dt. \tag{9.7.11}$$

Proof. By Lemma 9.2, $\rho^\varepsilon \mathbf{v}^\varepsilon \to \overline{\rho} \, \overline{\mathbf{v}}$ in the sense of distributions, and thus also weakly in $L^2(I_T \times U)$. By (9.7.1), $\mathbf{w}^\varepsilon, \partial_t \mathbf{w}^\varepsilon$ converge to respectively $\mathbf{w}, \partial_t \mathbf{w}$ strongly in $L^2(I_T \times U)$. This permits passage to the limit in the products and yields (9.7.10).

Since $\mathbf{w}^\varepsilon - \mathbf{w} \in C(I_T, L^2(U))$,

$$\begin{aligned} \| \mathbf{w}^\varepsilon - \mathbf{w} \|_{L^2(U)} (0) &\leq \int_0^T \| \mathbf{w}_t^\varepsilon - \mathbf{w}_t \|_{L^2(U)} (\tau) d\tau \\ &\leq T^{1/2} \left(\int_0^T \| \mathbf{w}_t^\varepsilon - \mathbf{w}_t \|_{L^2(U)} (\tau) d\tau \right)^{1/2}. \end{aligned}$$

Noting that $\mathbf{w}_t^\varepsilon \to \mathbf{w}_t$ strongly in $L^2(I_T \times U)$, we obtain that $(\mathbf{w}^\varepsilon - \mathbf{w})(0) \to 0$ strongly in $L^2(U)$. Since $\rho^\varepsilon(0,\mathbf{x})$ converges weakly-$\star$ in $L^\infty(U)$ to $\overline{\rho}(0,\mathbf{x})$, strong convergence of $\mathbf{w}^\varepsilon(0,\mathbf{x})$ permits passage to the limit in the product $\rho^\varepsilon(0,\mathbf{x})\mathbf{v}_0(\mathbf{x}) \cdot \mathbf{w}^\varepsilon(0,\mathbf{x})$ and yields (9.7.9).

Next, fix $j \in \{1,2,3\}$, pick a function $\eta \in C_0^\infty(I_T \times U)$, and insert the test function $(\mathbf{w}^\varepsilon - \mathbf{w})\eta$ into the weak formulation of the mass balance equation. This yields

$$\begin{aligned} \textstyle\int_U \rho^\varepsilon(0,\mathbf{x})(\mathbf{w}^\varepsilon - \mathbf{w})\eta(0,\mathbf{x})d\mathbf{x} &- \int_{I_T} \int_U \rho^\varepsilon \partial_t ((\mathbf{w}^\varepsilon - \mathbf{w})\eta) \, d\mathbf{x}dt \\ &- \int_{I_T} \int_U \rho^\varepsilon \mathbf{v}^\varepsilon \cdot \nabla ((\mathbf{w}^\varepsilon - \mathbf{w})\eta) \, d\mathbf{x}dt = 0. \end{aligned} \tag{9.7.12}$$

Strong convergence of $(\mathbf{w}^\varepsilon - \mathbf{w})(0)$ to zero implies

$$\lim_{\varepsilon \to 0} \int_U \rho^\varepsilon(0,\mathbf{x})(\mathbf{w}^\varepsilon - \mathbf{w})\eta(0,\mathbf{x})d\mathbf{x} = 0. \tag{9.7.13}$$

Next, note that

$$\lim_{\varepsilon\to 0}\int_{I_T}\int_U \rho^\varepsilon \partial_t\left((\mathbf{w}^\varepsilon-\mathbf{w})\eta\right)d\mathbf{x}dt = 0 \tag{9.7.14}$$

because $(\mathbf{w}_t^\varepsilon - \mathbf{w}_t) \to 0$ strongly in $L^2(I_T \times U)$, and ρ^ε is bounded in $L^\infty(I_T \times U)$ independent of ε. Now from (9.7.12), (9.7.13) and (9.7.14) we deduce

$$\lim_{\varepsilon\to 0}\int_{I_T}\int_U \rho^\varepsilon \mathbf{v}^\varepsilon\cdot\nabla\left((\mathbf{w}^\varepsilon-\mathbf{w})\eta\right)d\mathbf{x}dt = 0. \tag{9.7.15}$$

Since $\rho^\varepsilon\mathbf{v}^\varepsilon$ is bounded in $L^2(I_T\times U)$, and $\mathbf{w}^\varepsilon \to \mathbf{w}$ strongly in $L^2(I_T\times U)$, (9.7.15) implies

$$\lim_{\varepsilon\to 0}\int_{I_T}\int_U \eta\rho^\varepsilon \mathbf{v}^\varepsilon\cdot\nabla\left(\mathbf{w}^\varepsilon-\mathbf{w}\right)d\mathbf{x}dt = 0. \tag{9.7.16}$$

Since $\eta \in C_0^\infty(I_T\times U)$ is an arbitrary test function, $\rho^\varepsilon\mathbf{v}^\varepsilon\cdot\nabla(\mathbf{w}^\varepsilon-\mathbf{w}) \to 0$ in $\mathscr{D}'(I_T\times U)$.

Next we claim that $\rho^\varepsilon\mathbf{v}^\varepsilon\cdot\nabla(\mathbf{w}^\varepsilon-\mathbf{w})$ is bounded in $L^2(I_T, L^{5/6}(U))$ independent of ε. This follows from Sobolev imbedding for $\mathbf{v}^\varepsilon$ and Hlder inequality. Application of Hlder inequality yields

$$\left(\int_U |\rho^\varepsilon v_k^\varepsilon \partial_k w_j^\varepsilon|^s d\mathbf{x}\right)(t) \le \| \rho^\varepsilon \|^s_{L^\infty(U)}(t)\left(\int_U |v_k^\varepsilon|^{sq}\right)^{\frac{1}{q}}(t)\left(\int_U |\partial_k w^\varepsilon|^{sq'}\right)^{\frac{1}{q'}}(t).$$

Here $s,q \ge 1$ and $\frac{1}{q}+\frac{1}{q'} = 1$. Hence,

$$\begin{aligned}\int_{I_T}\left(\int_U \left|\rho^\varepsilon v_k^\varepsilon\partial_k w_j^\varepsilon\right|^s d\mathbf{x}\right)^{\frac{2}{s}} dt &\le \| \rho^\varepsilon \|^2_{L^\infty(I_T\times U)}\int_{I_T}\left(\int_U \left|\partial_k w_j^\varepsilon\right|^{sq'}\right)^{\frac{2}{sq'}}(t)\left(\int_U |v_k^\varepsilon|^{sq} d\mathbf{x}\right)^{\frac{2}{sq}}(t)dt \\ &\le \| \rho^\varepsilon \|^2_{L^\infty(I_T\times U)}\| \partial_k w_j^\varepsilon \|^2_{L^\infty(L^{sq'})}\int_{I_T}\left(\int_U |v_k^\varepsilon|^{sq} d\mathbf{x}\right)^{\frac{2}{sq}} dt\end{aligned}$$

Therefore

$$\| \rho^\varepsilon v_k^\varepsilon \partial_k w_j^\varepsilon \|_{L^2(I_T,L^s(U))} \le \| \rho^\varepsilon \|_{L^\infty(I_T\times U)} \| \partial_k w_j^\varepsilon \|_{L^\infty(L^{sq'})} \| v_k^\varepsilon \|_{L^2(I_T,L^{sq}(U))} \tag{9.7.17}$$

We need to choose s,q so that (i) the right hand side of (9.7.17) is finite; and (ii) $\mathbf{v}^\varepsilon \in L^2(I_T, L^{s'}(U))$, where $\frac{1}{s}+\frac{1}{s'} = 1$. By Sobolev imbedding, $s' \le 6$, and therefore

$$s \ge \frac{6}{5}. \tag{9.7.18}$$

By (9.7.2), $\| \partial_k w_j^\varepsilon \|^{\frac{2}{sq'}}_{L^\infty(L^{sq'})}$ is finite if

$$1 \le sq' \le 2 \Longleftrightarrow 1-\frac{1}{q} \le s \le 2-\frac{2}{q}, \tag{9.7.19}$$

($sq' < 2$ are allowed because U is bounded). Also, by Sobolev imbedding

$$1 \le sq \le 6 \Longleftrightarrow \frac{1}{q} \le s \le \frac{6}{q}. \tag{9.7.20}$$

The solution set of inequalities (9.7.18)–(9.7.20) is a non-empty, convex quadrilateral in the $1/q-s$-plane. For example, we can choose $s=\frac{6}{5}$ and any q satisfying $\frac{1}{5}\leq\frac{1}{q}\leq\frac{2}{5}$. If, for example, $1/q=\frac{2}{5}$, then $sq=3$, $sq'=\frac{9}{5}$ and (9.7.17) becomes

$$\| \rho^\varepsilon \mathbf{v}^\varepsilon \cdot \nabla \mathbf{w}^\varepsilon \|_{L^2(I_T,L^{\frac{6}{5}}(U))} \leq C \| \rho^\varepsilon \|_{L^\infty(I_T\times U)} \| \nabla \mathbf{w}^\varepsilon \|_{L^\infty(I_T,L^{\frac{9}{5}}(U))} \| \mathbf{v}^\varepsilon \|_{L^2(I_T,L^3(U))} \cdot$$

Since the right hand side of (9.7.17) is bounded independent of ε, the claim is proved.

Together with (9.7.16), this yields $\rho^\varepsilon\mathbf{v}^\varepsilon\cdot\nabla(\mathbf{w}^\varepsilon-\mathbf{w})\to 0$ weakly in $L^2(I_T,L^{6/5}(U))$. Therefore, the weak limit of $\rho^\varepsilon\mathbf{v}^\varepsilon\cdot\nabla\mathbf{w}^\varepsilon$ is the same as the weak limit of $\rho^\varepsilon\mathbf{v}^\varepsilon\cdot\nabla\mathbf{w}$ in $L^2(I_T,L^{6/5}(U))$. By Lemma 9.2, $\rho^\varepsilon\mathbf{v}^\varepsilon\cdot\nabla\mathbf{w}\to\overline{\rho\mathbf{v}}\cdot\nabla\mathbf{w}$. Thus

$$\rho^\varepsilon \mathbf{v}^\varepsilon \cdot \nabla \mathbf{w}^\varepsilon \to \overline{\rho}\, \overline{\mathbf{v}} \cdot \nabla \mathbf{w} \tag{9.7.21}$$

weakly in $L^2(I_T,L^{6/5}(U))$. The bound on $\partial_t[\rho^\varepsilon\mathbf{v}^\varepsilon\cdot\nabla\mathbf{w}^\varepsilon]$ in a negative Sobolev space follows from the corresponding bounds on $\rho^\varepsilon\mathbf{v}^\varepsilon$ and the fact that $\mathbf{w}_t^\varepsilon$ is bounded in $L^2(I_T,H_0^1(U))$.

Now application of Lemma 9.2 with $g^\varepsilon=\rho^\varepsilon\mathbf{v}^\varepsilon\cdot\nabla\mathbf{w}^\varepsilon$, $h^\varepsilon=\mathbf{v}^\varepsilon$ yields

$$\lim_{\varepsilon\to 0}\int_{I_T\times U} \rho^\varepsilon \mathbf{v}^\varepsilon \otimes \mathbf{v}^\varepsilon \cdot \nabla \mathbf{w}^\varepsilon \eta \, d\mathbf{x}dt = \int_{I_T\times U} \overline{\rho}\, \overline{\mathbf{v}} \otimes \overline{\mathbf{v}} \cdot \nabla \mathbf{w} \eta \, d\mathbf{x}dt. \tag{9.7.22}$$

For each $\eta\in C_0^\infty(I_T\times U)$. From Sobolev imbedding and bounds on $\nabla\mathbf{w}^\varepsilon$, in the same way as (9.7.17) was analyzed, choosing $1/q=2/5, s=6/5$, we obtain

$$\| \rho^\varepsilon \mathbf{v}^\varepsilon \otimes \mathbf{v}^\varepsilon \cdot \nabla \mathbf{w}^\varepsilon \|_{L^2(I_T,L^{\frac{6}{5}}(U))} \leq C \| \rho^\varepsilon \|_{L^\infty(I_T\times U)} \| \nabla \mathbf{w}^\varepsilon \|_{L^\infty(I_T,L^2(U))} \| \mathbf{v}^\varepsilon \otimes \mathbf{v}^\varepsilon \|_{L^2(I_T,L^3(U))} \tag{9.7.23}$$

with C independent of ε. Note that the right hand side is bounded independent of ε. Therefore, $\rho^\varepsilon\mathbf{v}^\varepsilon\otimes\mathbf{v}^\varepsilon\cdot\nabla\mathbf{w}^\varepsilon$ converges to $\overline{\rho}\,\overline{\mathbf{v}}\otimes\overline{\mathbf{v}}\cdot\nabla\mathbf{w}$ weakly in $L^2(I_T,L^{\frac{6}{5}}(U))$. This implies convergence of integrals of $\rho^\varepsilon\mathbf{v}^\varepsilon\otimes\mathbf{v}^\varepsilon\cdot\nabla\mathbf{w}^\varepsilon$ over subsets of $I_T\times U$, and in particular (9.7.11). ■

9.8 EFFECTIVE DEVIATORIC STRESS. PROOF OF THE MAIN THEOREM

Theorem 9.3

There exists a subsequence, not relabeled, and effective material tensors $\overline{\mathbf{A}}\in L^2(U), \overline{\mathbf{B}}\in L^2(U)$ and $\overline{\mathbf{C}}\in L^2(I_T\times U)$ such that for each $\mathbf{w}\in C_0^\infty(I_T\times U)$ with $\operatorname{div}\mathbf{w}=0, \mathbf{w}(T,\mathbf{x})=0$,

$$\int_{I_T}\int_U \overline{\mathbf{T}}\cdot e(\mathbf{w})d\mathbf{x}dt = \lim_{\varepsilon\to 0}\int_{I_T}\int_U \mathbf{T}^\varepsilon\cdot e(\mathbf{w}^\varepsilon)d\mathbf{x}dt$$

$$= \int_{I_T} \int_U \left(\overline{\mathbf{A}} e(\overline{\mathbf{u}}) + \overline{\mathbf{B}} e(\overline{\mathbf{v}}) + \int_0^t \overline{\mathbf{C}}(t-\tau,\cdot) e(\overline{\mathbf{v}})(\tau,\cdot) \right) \cdot e(\mathbf{w}) d\mathbf{x} d\tau$$

as $\varepsilon \to 0$ along this subsequence. ■

Proof. The theorem follows from several propositions. First, we prove that convergence of inertial terms implies compensated compactness of stress.

Proposition 9.8. *Let* $\mathbf{w}^\varepsilon$ *be test functions from Definition 9.4. Then*

$$\lim_{\varepsilon \to 0} \int_{I_T} \int_U \mathbf{T}^\varepsilon \cdot e(\mathbf{w}^\varepsilon) d\mathbf{x} dt = \int_{I_T} \int_U \overline{\mathbf{T}} \cdot e(\mathbf{w}) d\mathbf{x} dt.$$

Proof of the proposition. First, use $\mathbf{w}$ as a test function in (9.4.14) and pass to the limit $\varepsilon \to 0$. Repeated application of Lemma 9.2 in the inertial terms yields

$$\begin{aligned} &-\int_U \overline{\rho}(0,\mathbf{x})\, \mathbf{v}_0 \cdot \mathbf{w}\, d\mathbf{x} - \int_{I_T \times U} \overline{\rho}\, \overline{\mathbf{v}} \cdot \partial_t \mathbf{w}\, d\mathbf{x} dt \\ &- \int_{I_T \times U} \overline{\rho}\, \overline{\mathbf{v}} \otimes \overline{\mathbf{v}} \cdot \nabla \mathbf{w}\, d\mathbf{x} dt + \int_{I_T \times U} \overline{\mathbf{T}} \cdot e(\mathbf{w}) d\mathbf{x} dt = 0 \end{aligned} \tag{9.8.1}$$

Then insert $\mathbf{w}^\varepsilon$ into (9.4.14) and pass to the limit $\varepsilon \to 0$. By Proposition 9.7, the integrals corresponding to the inertial terms will converge to the corresponding integrals of the limiting functions $\bar{\rho}, \bar{\mathbf{v}}$. This yields

$$\begin{aligned} &-\int_U \overline{\rho}(0,\mathbf{x})\, \mathbf{v}_0 \cdot \mathbf{w}\, d\mathbf{x} - \int_{I_T \times U} \overline{\rho}\, \overline{\mathbf{v}} \cdot \partial_t \mathbf{w}\, d\mathbf{x} dt \\ &- \int_{I_T \times U} \overline{\rho}\, \overline{\mathbf{v}} \otimes \overline{\mathbf{v}} \cdot \nabla \mathbf{w}\, d\mathbf{x} dt + \lim_{\varepsilon \to 0} \int_{I_T \times U} \mathbf{T}^\varepsilon \cdot e(\mathbf{w}^\varepsilon) d\mathbf{x} dt = 0 \end{aligned} \tag{9.8.2}$$

Comparison of (9.8.1) and (9.8.2) finishes the proof. ■

Next, using symmetry of $\mathbf{A}^\varepsilon, \mathbf{B}^\varepsilon$ we have

$$\int_{I_T \times U} \mathbf{T}^\varepsilon \cdot e(\mathbf{w}^\varepsilon) d\mathbf{x} dt = \int_{I_T \times U} e(\mathbf{u}^\varepsilon) \cdot [\mathbf{A}^\varepsilon e(\mathbf{w}^\varepsilon) - \mathbf{B}^\varepsilon e(\mathbf{w}_t^\varepsilon)]\, d\mathbf{x} dt.$$

Differentiation in (9.6.1) yields

$$\begin{aligned} \mathbf{A}^\varepsilon e(\mathbf{w}^\varepsilon) - \mathbf{B}^\varepsilon e(\mathbf{w}_t^\varepsilon) \;=\; & \mathscr{F}_1^{pq,\varepsilon} e(\mathbf{w})_{pq} + \mathscr{F}_2^{pq,\varepsilon} e(\mathbf{w}_t)_{pq} \\ &+ \int_t^T \mathscr{F}_3^{pq,\varepsilon}(t-\tau+T) e(\mathbf{w}_t)_{pq}(\tau) d\tau \\ &+ \mathbf{A}^\varepsilon(\mathbf{g}_1^\varepsilon + \mathbf{g}_2^\varepsilon) - \mathbf{B}^\varepsilon(\partial_t \mathbf{g}_1^\varepsilon + \partial_t \mathbf{g}_2^\varepsilon) \\ &+ \mathbf{A}^\varepsilon e(\nabla \phi^\varepsilon) - \mathbf{B}^\varepsilon e(\nabla \phi_t^\varepsilon), \end{aligned} \tag{9.8.3}$$

where

$$\mathscr{F}_1^{pq,\varepsilon}(\mathbf{x}) = \mathbf{A}^\varepsilon \left(\mathbf{I}^{pq} + e\left(\mathbf{n}^{pq,\varepsilon}\right) \right), \quad \mathbf{I}^{pq} = \frac{1}{2} \left(\mathbf{e}_p \otimes \mathbf{e}_q + \mathbf{e}_q \otimes \mathbf{e}_p \right),$$

$$\mathscr{F}_2^{pq,\varepsilon}(\mathbf{x}) = \mathbf{B}^\varepsilon\left(\mathbf{I}^{pq} + e\left(\mathbf{n}^{pq,\varepsilon}\right) - e\left(\mathbf{m}_T^{pq,\varepsilon}\right)\right),$$

$$\mathscr{F}_3^{pq,\varepsilon}(t,\mathbf{x}) = \mathbf{A}^\varepsilon e\left(\mathbf{m}^{pq,\varepsilon}\right) - \mathbf{B}^\varepsilon e\left(\mathbf{m}_t^{pq,\varepsilon}\right),$$

$$\mathbf{g}_1^\varepsilon = \frac{1}{2}\left(\mathbf{n}^{pq,,\varepsilon} \otimes \nabla e(\mathbf{w})_{pq} + (\mathbf{n}^{pq,\varepsilon} \otimes \nabla e(\mathbf{w})_{pq})^{\mathrm{T}}\right), \tag{9.8.4}$$

$$\mathbf{g}_2^\varepsilon = \frac{1}{2}\int_t^T \left(\mathbf{m}^{pq,\varepsilon}(t-\tau+T) \otimes \nabla e(\mathbf{w})_{pq}(\tau) + (\mathbf{m}^{pq,\varepsilon}(t-\tau+T) \otimes \nabla e(\mathbf{w})_{pq}(\tau))^{\mathrm{T}}\right) d\tau. \tag{9.8.5}$$

Next we show that the only terms in (9.8.3) that contribute to the effective stress are the terms containing $\mathscr{F}_j^{pq,\varepsilon}$.

Proposition 9.9. *Let* $\mathbf{w}^\varepsilon$ *be as in Definition 9.4. Then*

$$\lim_{\varepsilon \to 0} \int_{I_T} \int_U e(\mathbf{u}^\varepsilon) \cdot [\mathbf{A}^\varepsilon(\mathbf{g}_1^\varepsilon + \mathbf{g}_2^\varepsilon) - \mathbf{B}^\varepsilon(\partial_t \mathbf{g}_1^\varepsilon + \partial_t \mathbf{g}_2^\varepsilon)]\, d\mathbf{x}dt = 0, \tag{9.8.6}$$

$$\lim_{\varepsilon \to 0} \int_{I_T} \int_U e(\mathbf{u}^\varepsilon) \cdot [\mathbf{A}^\varepsilon e(\nabla \phi^\varepsilon) - \mathbf{B}^\varepsilon e(\nabla \phi_t^\varepsilon)]\, d\mathbf{x}dt = 0. \tag{9.8.7}$$

Proof of the proposition. Since $e(\mathbf{u}^\varepsilon)$ converges to $e(\overline{\mathbf{u}})$ weakly in $L^2(I_T \times U)$, it is enough to show that all terms in brackets in (9.8.6), (9.8.7) converge to zero strongly in $L^2(I_T \times U)$.

Step 1. Prove (9.8.6).

By Proposition 9.5, $\mathbf{n}^{pq,\varepsilon}$ converges to zero strongly in $L^2(I_T \times U)$ Therefore, $\mathbf{g}_1^\varepsilon, \partial_t \mathbf{g}_1^\varepsilon$ in (9.9.3) converge to zero strongly in $L^2(I_T \times U)$. By Proposition 9.3, $\mathbf{m}^{pq,\varepsilon}$ converge to zero strongly in $L^2(I_T \times U)$. Hence, $\mathbf{g}_2^\varepsilon$ in (9.9.4) converges to zero strongly in $L^2(I_T \times U)$. Next, differentiate $\mathbf{g}_2^\varepsilon$ in t and integrate by parts in the time convolution exactly as in the proof of Proposition 9.6. Then

$$\partial_t \mathbf{g}_2^\varepsilon = \frac{1}{2}\int_t^T \left(\mathbf{m}^{pq,\varepsilon}(t-\tau+T) \otimes \nabla e(\mathbf{w}_{tt})_{pq}(\tau) + \left(\mathbf{m}^{pq,\varepsilon}(t-\tau+T) \otimes \nabla e(\mathbf{w}_{tt})_{pq}(\tau)\right)^{\mathrm{T}}\right) d\tau.$$

which converges to zero strongly in $L^2(I_T \times U)$. Next, since $\mathbf{A}^\varepsilon, \mathbf{B}^\varepsilon$ are bounded pointwise independent of ε, we deduce that

$$\mathbf{A}^\varepsilon(\mathbf{g}_1^\varepsilon + \mathbf{g}_2^\varepsilon) - \mathbf{B}^\varepsilon(\partial_t \mathbf{g}_1^\varepsilon + \partial_t \mathbf{g}_2^\varepsilon)$$

converges to zero strongly in $L^2(I_T \times U)$.

Step 2. Prove (9.8.7).

From (9.7.5) we have $e(\nabla \phi^\varepsilon) \to 0$, and by (9.7.8), $e(\nabla \phi_t^\varepsilon) \to 0$ strongly $L^2(I_T \times U)$. Hence, $\mathbf{A}^\varepsilon e(\nabla \phi^\varepsilon)$, $\mathbf{B}^\varepsilon e(\nabla \phi_t^\varepsilon)$ also converge to zero strongly in $L^2(I_T \times U)$. ■

By Proposition 9.9,

$$\int_{I_T} \int_U \overline{\mathbf{T}} \cdot e(\mathbf{w}) d\mathbf{x}dt = \lim_{\varepsilon \to 0} \int_{I_T} \int_U \mathbf{T}^\varepsilon \cdot e(\mathbf{w}^\varepsilon) d\mathbf{x}dt = \lim_{\varepsilon \to 0} I(\mathbf{u}^\varepsilon, \mathbf{w}^\varepsilon),$$

where

$$I(\mathbf{u}^\varepsilon, \mathbf{w}^\varepsilon) = \int_{I_T}\int_U e(\mathbf{u}^\varepsilon)\cdot\Big(\mathscr{F}_1^{pq,\varepsilon} e(\mathbf{w})_{pq} + \mathscr{F}_2^{pq,\varepsilon} e(\mathbf{w}_t)_{pq} + \int_t^T \mathscr{F}_3^{pq,\varepsilon}(t-\tau+T) e(\mathbf{w}_t)_{pq}(\tau)d\tau\Big)\, d\mathbf{x}dt \tag{9.8.8}$$

Proposition 9.10. *There exist the effective tensors $\overline{\mathbf{A}} \in L^2(U), \overline{\mathbf{B}} \in L^2(U)$ and $\overline{\mathbf{C}} \in L^2(I_T \times U)$ such that, up to extraction of a subsequence,*

$$\begin{aligned}\lim_{\varepsilon\to 0} I(\mathbf{u}^\varepsilon, \mathbf{w}^\varepsilon) &= -\langle \overline{\mathbf{u}}, \operatorname{div}\left(\overline{\mathscr{F}}_1^{pq} e(\mathbf{w})_{pq}\right) + \operatorname{div}\left(\overline{\mathscr{F}}_2^{pq} e(\mathbf{w}_t)_{pq}\right)\rangle \\ &\quad - \langle \mathbf{u}^\varepsilon, \operatorname{div}\left(\int_t^T \overline{\mathscr{F}}_3^{pq}(t-\tau+T)e(\mathbf{w}_t)_{pq}(\tau)d\tau d\mathbf{x}dt\right)\rangle \\ &= \int_{I_T}\int_U \left(\overline{\mathbf{A}} e(\overline{\mathbf{u}}) + \overline{\mathbf{B}} e(\overline{\mathbf{u}}_t) + \int_0^t \overline{\mathbf{C}}(t-\tau,\cdot)e(\overline{\mathbf{u}}_t)(\tau,\cdot)d\tau\right)\cdot e(\mathbf{w})d\mathbf{x}dt.\end{aligned} \tag{9.8.9}$$

Proof of the proposition. First, we note that $\mathscr{F}_j^{pq,\varepsilon}, P_j^{pq,\varepsilon}, j = 1,2$ are bounded in $L^2(U)$ independent of ε. Therefore, extracting weakly convergent subsequences, not relabeled, we have that $\mathscr{F}_j^{pq,\varepsilon} \rightharpoonup \overline{\mathscr{F}}_j^{pq}$, $P_j^{pq,\varepsilon} \rightharpoonup \overline{P}^{pq}$, and passing to the limit in (9.6.22) and (9.6.23) we obtain

$$\operatorname{div}\mathscr{F}_j^{pq,\varepsilon} - \nabla P_j^{pq,\varepsilon} = \mathbf{f}_j = \operatorname{div}\overline{\mathscr{F}}_j^{pq} - \nabla \overline{P}_j^{pq},$$

and thus

$$\operatorname{div}\mathscr{F}_j^{pq,\varepsilon} = \nabla\left(P_j^{pq,\varepsilon} - \overline{P}_j^{pq}\right) + \operatorname{div}\overline{\mathscr{F}}_j^{pq}, \quad j = 1,2. \tag{9.8.10}$$

Similarly, $\mathscr{F}_3^{pq,\varepsilon}, P^{pq,\varepsilon}$ are bounded in $L^2(I_T \times U)$ independent of ε. Therefore, extracting weakly convergent subsequences as before, we obtain

$$\operatorname{div}\mathscr{F}_3^{pq,\varepsilon} = \nabla\left(P_3^{pq,\varepsilon} - \overline{P}_3^{pq}\right) + \operatorname{div}\overline{\mathscr{F}}_3^{pq}. \tag{9.8.11}$$

From (9.8.10), (9.8.11) we deduce

$$\begin{aligned}\operatorname{div}\left(\mathscr{F}_1^{pq,\varepsilon} e(\mathbf{w})_{pq}\right) = \ &\operatorname{div}\left(\overline{\mathscr{F}}_1^{pq} e(\mathbf{w})_{pq}\right) + \nabla\left[(P_1^{pq,\varepsilon} - \overline{P}_1^{pq})e(\mathbf{w})_{pq}\right] \\ &- \left(P_1^{pq,\varepsilon} - \overline{P}_1^{pq}\right)\nabla e(\mathbf{w})_{pq} + \left(\mathscr{F}_1^{pq,\varepsilon} - \overline{\mathscr{F}}_1^{pq}\right)\cdot\nabla e(\mathbf{w})_{pq},\end{aligned} \tag{9.8.12}$$

and

$$\begin{aligned}\operatorname{div}\left(\mathscr{F}_j^{pq,\varepsilon} e(\mathbf{w}_t)_{pq}\right) &= \operatorname{div}\left(\overline{\mathscr{F}}_j^{pq} e(\mathbf{w}_t)_{pq}\right) + \nabla\left[(P_j^{pq,\varepsilon} - \overline{P}_j^{pq})e(\mathbf{w}_t)_{pq}\right] \\ &- \left(P_j^{pq,\varepsilon} - \overline{P}_j^{pq}\right)\nabla e(\mathbf{w}_t)_{pq} + \left(\mathscr{F}_j^{pq,\varepsilon} - \overline{\mathscr{F}}_j^{pq}\right)\cdot\nabla e(\mathbf{w}_t)_{pq}, \quad j = 2,3.\end{aligned} \tag{9.8.13}$$

Next, since $\operatorname{div}\mathbf{u}^\varepsilon = 0$,

$$\langle \mathbf{u}^\varepsilon, \nabla\left[(P_1^{pq,\varepsilon} - \overline{P}_1^{pq})e(\mathbf{w})_{pq}\right]\rangle = 0, \tag{9.8.14}$$

$$\langle \mathbf{u}^\varepsilon, \nabla \left[(P_2^{pq,\varepsilon} - \overline{P}_2^{pq})e(\mathbf{w}_t)_{pq}\right]\rangle = 0,$$
$$\langle \mathbf{u}^\varepsilon, \nabla \left[\int_t^T (P_3^{pq,\varepsilon} - \overline{P}_3^{pq})(t-\tau)e(\mathbf{w}_t)_{pq}(\tau)d\tau\right]\rangle = 0.$$

Integrating by parts and using (9.8.12), (9.8.13) and (9.8.14) we have

$$\begin{aligned} I(\mathbf{u}^\varepsilon, \mathbf{w}^\varepsilon) = \; & -\langle \mathbf{u}^\varepsilon, \operatorname{div}\left(\overline{\mathscr{F}}_1^{pq} e(\mathbf{w})_{pq}\right) + \operatorname{div}\left(\overline{\mathscr{F}}_2^{pq} e(\mathbf{w}_t)_{pq}\right)\rangle \\ & -\langle \mathbf{u}^\varepsilon, \operatorname{div}\left(\int_t^T \overline{\mathscr{F}}_3^{pq}(t-\tau+T)e(\mathbf{w}_t)_{pq}(\tau)d\tau d\mathbf{x}dt\right)\rangle + \mathscr{R}^\varepsilon \end{aligned}$$

where

$$\begin{aligned} \mathscr{R}^\varepsilon = \; & -\langle \mathbf{u}^\varepsilon, (\mathscr{F}_1^{pq,\varepsilon} - \overline{\mathscr{F}}_1^{pq}) \cdot \nabla e(\mathbf{w})_{pq}\rangle - \langle (\mathscr{F}_2^{pq,\varepsilon} - \overline{\mathscr{F}}_2^{pq}) \cdot \nabla e(\mathbf{w}_t)_{pq}\rangle \\ & -\langle \mathbf{u}^\varepsilon, \int_t^T \left(\mathscr{F}_3^{pq,\varepsilon} - \overline{\mathscr{F}}_3^{pq}\right)(t-\tau+T) \cdot \nabla e(\mathbf{w}_t)_{pq}(\tau)d\tau d\mathbf{x}dt\rangle \\ & -\langle \mathbf{u}^\varepsilon, (P_1^{pq,\varepsilon} - \overline{P}_1^{pq}) \cdot \nabla e(\mathbf{w})_{pq}\rangle - \langle \mathbf{u}^\varepsilon, (P_2^{pq,\varepsilon} - \overline{P}_2^{pq}) \cdot \nabla e(\mathbf{w}_t)_{pq}\rangle \\ & -\langle \mathbf{u}^\varepsilon, \int_t^T (P_3^{pq,\varepsilon} - \overline{P}_3^{pq})(t-\tau) \cdot \nabla e(\mathbf{w}_t)_{pq}(\tau)d\tau d\mathbf{x}dt\rangle. \end{aligned}$$

Since $\mathbf{u}^\varepsilon \in L^2(I_T, H_0^1(U))$, we can apply Lemma 9.2 which yields $\lim_{\varepsilon\to 0}\mathscr{R}^\varepsilon = 0$. Then we have (9.8.9), where the components of the effective tensors $\overline{\mathbf{A}}(\mathbf{x}), \overline{\mathbf{B}}(\mathbf{x}), \overline{\mathbf{C}}(t,\mathbf{x})$ are defined by

$$\overline{A}_{pqij} = \overline{\mathscr{F}}_{1,ij}^{pq}, \qquad \overline{B}_{pqij} = \overline{\mathscr{F}}_{2,ij}^{pq}, \qquad \overline{C}_{pqij} = \overline{\mathscr{F}}_{3,ij}^{pq}. \tag{9.8.15}$$

■

This completes the proof of the Theorem 9.3.

Proof of the main theorem.

To obtain (9.9.12), we pass to the limit in (9.4.13) using Lemma 9.2. Next, insert $\mathbf{w}^\varepsilon = N^\varepsilon \mathbf{w}$ into (9.4.14). The limit of the inertial terms is given in Proposition 9.7, and the limit of the term containing $\mathbf{T}^\varepsilon \cdot e(\mathbf{w}^\varepsilon)$ is provided by Theorem 9.3. Together, these results yield (9.9.13). The divergence-free constraint (9.9.11) is obtained by straightforward passing to the limit in $\operatorname{div} \mathbf{v}^\varepsilon = 0$.

■

9.9 FLUID-STRUCTURE INTERACTION

Compared to the previous sections, the main difference now is lack of ellipticity in $\mathbf{A}^\varepsilon$. In this section we assume $\mathbf{A}^\varepsilon = \mathbf{A}_1\theta_0^\varepsilon$. This means that phase one is a Kelvin-Voight viscoelastic material, and phase two is a Newtonian fluid. To deal with degeneration of $\mathbf{A}^\varepsilon$ we modify (9.6.1) as follows.

$$N^\varepsilon \mathbf{w} \equiv \mathbf{w}^\varepsilon \;=\; \mathbf{w} + \int_t^T \mathbf{n}^{pq,\varepsilon}(t-\tau+T)e(\mathbf{w})_{pq}(\tau)d\tau + \tag{9.9.1}$$

$$\int_t^T \mathbf{m}^{pq,\varepsilon}(t-\tau+T)e(\mathbf{w}_t)_{pq}(\tau)d\tau + \nabla\phi^\varepsilon.$$

Here, $\mathbf{n}^{pq,\varepsilon}, \mathbf{m}^{pq,\varepsilon}, \phi^\varepsilon$ are as in (9.6.1), (9.6.3), respectively. We note, however, that the auxiliary problems for $\mathbf{n}^{pq,\varepsilon}, \mathbf{m}^{pq,\varepsilon}$ will be different. Differentiating (9.9.1) we obtain

$$\begin{aligned} \mathbf{A}^\varepsilon e(\mathbf{w}^\varepsilon) - \mathbf{B}^\varepsilon e(\mathbf{w}_t^\varepsilon) = {} & \mathscr{F}_1^{pq,\varepsilon} e(\mathbf{w})_{pq} + \left(\mathscr{F}_2^{pq,\varepsilon}\right)(\mathbf{w}_t)_{pq} \\ & + \int_t^T \mathscr{F}_3^{pq,\varepsilon}(t-\tau+T)e(\mathbf{w})_{pq}(\tau)d\tau \\ & + \int_t^T \mathscr{F}_4^{pq,\varepsilon}(t-\tau+T)e(\mathbf{w}_t)_{pq}(\tau)d\tau \\ & + \mathbf{A}^\varepsilon(\mathbf{g}_1^\varepsilon + \mathbf{g}_2^\varepsilon) - \mathbf{B}^\varepsilon(\partial_t \mathbf{g}_1^\varepsilon + \partial_t \mathbf{g}_2^\varepsilon) \\ & + \mathbf{A}^\varepsilon e(\nabla\phi^\varepsilon) - \mathbf{B}^\varepsilon e(\nabla\phi_t^\varepsilon), \end{aligned} \tag{9.9.2}$$

where

$$\mathscr{F}_1^{pq,\varepsilon} = \mathbf{A}^\varepsilon \mathbf{I}^{pq} - \mathbf{B}^\varepsilon e\left(\mathbf{n}_T^{pq,\varepsilon}\right), \quad \mathbf{I}^{pq} = \frac{1}{2}\left(\mathbf{e}_p \otimes \mathbf{e}_q + \mathbf{e}_q \otimes \mathbf{e}_p\right),$$

$$\mathscr{F}_2^{pq,\varepsilon} = -\mathbf{B}^\varepsilon\left(\mathbf{I}^{pq} + e\left(\mathbf{m}_T^{pq,\varepsilon}\right)\right),$$

$$\mathscr{F}_3^{pq,\varepsilon}(\mathbf{x}) = \mathbf{A}^\varepsilon e\left(\mathbf{n}^{pq,\varepsilon}\right) - \mathbf{B}^\varepsilon e\left(\mathbf{n}_t^{pq,\varepsilon}\right),$$

$$\mathscr{F}_4^{pq,\varepsilon}(t,\mathbf{x}) = \mathbf{A}^\varepsilon e\left(\mathbf{m}^{pq,\varepsilon}\right) - \mathbf{B}^\varepsilon e\left(\mathbf{m}_t^{pq,\varepsilon}\right),$$

$$\mathbf{g}_1^\varepsilon = \frac{1}{2}\int_t^T \left(\mathbf{n}^{pq,\varepsilon}(t-\tau+T)\otimes\nabla e(\mathbf{w})_{pq}(\tau) + \left(\mathbf{n}^{pq,\varepsilon}(t-\tau+T)\otimes\nabla e(\mathbf{w})_{pq}(\tau)\right)^{\mathrm{T}}\right)d\tau. \tag{9.9.3}$$

$$\mathbf{g}_2^\varepsilon = \frac{1}{2}\int_t^T \left(\mathbf{m}^{pq,\varepsilon}(t-\tau+T)\otimes\nabla e(\mathbf{w}_t)_{pq}(\tau) + \left(\mathbf{m}^{pq,\varepsilon}(t-\tau+T)\otimes\nabla e(\mathbf{w}_t)_{pq}(\tau)\right)^{\mathrm{T}}\right)d\tau. \tag{9.9.4}$$

In the statement of the following auxiliary problems we drop p,q to simplify notations.

- *First auxiliary problem.* Find $\mathbf{n}_T^\varepsilon \in H_0^1(U)$ satisfying

$$\operatorname{div}\left(\mathbf{A}^\varepsilon \mathbf{I}^{pq} - \mathbf{B}^\varepsilon e(\mathbf{n}_T^\varepsilon)\right) - \nabla P_1^\varepsilon = \mathbf{f}_1, \quad \operatorname{div}\mathbf{n}_T^\varepsilon = 0. \tag{9.9.5}$$

- *Second auxiliary problem.* Find $\mathbf{m}_T^\varepsilon \in H_0^1(U)$ satisfying

$$-\operatorname{div}\left(\mathbf{B}^\varepsilon(\mathbf{I}^{pq} + e(\mathbf{m}_T^\varepsilon))\right) - \nabla P_2^\varepsilon = \mathbf{f}_2, \quad \operatorname{div}\mathbf{m}_T^\varepsilon = 0. \tag{9.9.6}$$

- *Third auxiliary problem.* Find $\mathbf{n}^\varepsilon \in \mathscr{W}$ satisfying

$$-\operatorname{div}\left((\mathbf{A}^\varepsilon - \mathbf{B}^\varepsilon\partial_t)e(\mathbf{n}^\varepsilon)\right) - \nabla P_3^\varepsilon = \mathbf{f}_3, \quad \operatorname{div}\mathbf{n}^\varepsilon = 0, \quad \mathbf{n}^\varepsilon(T) = \mathbf{n}_T^\varepsilon. \tag{9.9.7}$$

- *Fourth auxiliary problem.* Find $\mathbf{m}^\varepsilon \in \mathscr{W}$ satisfying

$$-\mathrm{div}\,((\mathbf{A}^\varepsilon - \mathbf{B}^\varepsilon \partial_t) e(\mathbf{m}^\varepsilon)) - \nabla P_4^\varepsilon = \mathbf{f}_4, \qquad \mathrm{div}\,\mathbf{m}^\varepsilon = 0, \qquad \mathbf{m}^\varepsilon(T) = \mathbf{m}_T^\varepsilon. \tag{9.9.8}$$

Since $\mathbf{B}^\varepsilon$ is still elliptic, problems (9.9.5), (9.9.6) can be analyzed exactly as problems in Section 9.6.2. All the results in that section apply without change. The problems (9.9.7), (9.9.8) were dealt with in Section 9.6.1. The most important condition is still ellipticity of $\mathbf{B}^\varepsilon$, and the only change that is needed is in the proof of (i) in Proposition 9.1 where we used ellipticity of $\mathbf{A}^\varepsilon$. To prove (i), note that

$$\int_t^T \int_U \mathbf{A}^\varepsilon e(\mathbf{u}) \cdot e(\mathbf{u}) d\mathbf{x}dt + \frac{1}{2} \int_U \mathbf{B}^\varepsilon e(\mathbf{u}) \cdot e(\mathbf{u}) d\mathbf{x}(t) = \int_t^T \langle \mathbf{f}, \mathbf{u} \rangle_{H^{-1}(U), H_0^1(U)}(\tau) d\tau \tag{9.9.9}$$

holds for almost all $t \in I_T$. Estimating the right hand side of (9.9.9) we have

$$\int_t^T \langle \mathbf{f}, \mathbf{u} \rangle_{H^{-1}(U), H_0^1(U)}(\tau) d\tau \leq \int_t^T \| \mathbf{f} \|_{H^{-1}(U)} (\tau) \| \mathbf{u} \|_{H_0^1(U)} (\tau) d\tau \tag{9.9.10}$$

$$\leq \left(\int_t^T \| \mathbf{f} \|^2_{H^{-1}(U)} (\tau) d\tau \right)^{1/2} \left(\int_t^T \| \mathbf{u} \|^2_{H_0^1(U)} (\tau) d\tau \right)^{1/2}$$

$$\leq \| \mathbf{f} \|_{L^2(I_T, H^{-1}(U))} \| \mathbf{u} \|_{L^2(I_T, H_0^1(U))}$$

Since the first term in the left hand side of (9.9.9) is non-negative, from (9.9.9), (9.9.10) we deduce, using ellipticity of $\mathbf{B}^\varepsilon$,

$$\frac{1}{2}\beta_1 \sup_{t \in I_T} \int_U e(\mathbf{u}) \cdot e(\mathbf{u}) d\mathbf{x}(t) \leq \| \mathbf{f} \|_{L^2(I_T, H^{-1}(U))} \| \mathbf{u} \|_{L^2(I_T, H_0^1(U))} \cdot$$

This yields (i) (with a different constant $\frac{1}{2}\beta_1 T^{-1}$) after observing that

$$\| \mathbf{u} \|^2_{L^2(I_T, H_0^1(U))} \equiv \int_0^T \int_U e(\mathbf{u}) \cdot e(\mathbf{u}) d\mathbf{x} \leq T \sup_{t \in I_T} \int_U e(\mathbf{u}) \cdot e(\mathbf{u}) d\mathbf{x}(t).$$

All the arguments in Sections 9.7, 9.8 apply with minor changes due to the presense of four fluxes $\mathscr{F}_j$, $j = 1, \ldots, 4$, instead of three. The result can be summarized as a theorem.

Theorem 9.4

In the case of fluid-structure interaction, the limits $\overline{\rho}, \overline{\mathbf{v}}, \overline{\mathbf{u}}$ satisfy

$$\mathrm{div}\,\overline{\mathbf{v}} = 0, \tag{9.9.11}$$

and the integral identities

$$\int_U \overline{\rho}_0 \phi(0, \mathbf{x}) d\mathbf{x} - \int_{I_T} \int_U \overline{\rho} \partial_t \phi d\mathbf{x}dt - \int_{I_T} \int_U \overline{\rho}\, \overline{\mathbf{v}} \cdot \nabla \phi d\mathbf{x}dt = 0, \tag{9.9.12}$$

$$-\int_U \overline{\rho}_0 \mathbf{v}_0 \cdot \psi d\mathbf{x} - \int_{I_T}\int_U \overline{\rho}\, \overline{\mathbf{v}} \cdot \partial_t \psi d\mathbf{x}dt - \int_{I_T}\int_U \overline{\rho}\, \overline{\mathbf{v}} \otimes \overline{\mathbf{v}} \cdot \nabla \psi d\mathbf{x}dt \tag{9.9.13}$$
$$+ \int_{I_T}\int_U \overline{\mathbf{T}} \cdot e(\psi) d\mathbf{x}dt = 0.$$

for all smooth test functions ϕ, ψ, such that div $\psi = 0$, and ϕ, ψ are equal to zero on ∂U and vanish for $t \geq T$.

Moreover, there exist the effective tensors $\overline{\mathbf{A}} \in L^2(U), \overline{\mathbf{B}} \in L^2(U)$ and $\overline{\mathbf{C}} \in L^2(I_T \times U), \overline{\mathbf{D}} \in L^2(I_T \times U)$ such that the effective deviatoric stress $\overline{\mathbf{T}}$ satisfies

$$\overline{\mathbf{T}} = \overline{\mathbf{A}} e(\overline{\mathbf{u}}) + \overline{\mathbf{B}} e(\overline{\mathbf{v}}) + \int_0^t \overline{\mathbf{C}}(t-\tau) e(\overline{\mathbf{u}})(\tau) d\tau + \int_0^t \overline{\mathbf{D}}(t-\tau) e(\overline{\mathbf{v}})(\tau) d\tau \tag{9.9.14}$$

■

Remark. The effective tensors are obtained as weak $L^2(I_T \times U)$ limits of the four fluxes:

$$\overline{A}_{pqij} = \overline{\mathscr{F}}^{pq}_{1,ij}, \quad \overline{B}_{pqij} = \overline{\mathscr{F}}^{pq}_{2,ij}, \quad \overline{C}_{pqij} = \overline{\mathscr{F}}^{pq}_{3,ij}, \quad \overline{D}_{pqij} = \overline{\mathscr{F}}^{pq}_{3,ij}. \tag{9.9.15}$$

10 Biot-Type, Models for Bone Mechanics

10.1 BONE RIGIDITY

In this chapter we discuss a particular application of the effective equations for a poroelastic material using ultrasonics for assessing bone mineral density. This method uses the travel time of sound between two transducers with a bone sample between them and the travel time in the absence of the specimen to compute the compressional wave velocity in the bone. Ultrasound (**US**) has been considered as a means to characterize the elastic properties of cortical (compact) and cancellous (trabecular) bone, Figure 10.1.1. We shall explore the use of the effective equations for poroelastic materials to explore the possibility of determining characteristic bone parameters using **US** methodology. This is an inverse problem, namely, we use the effective equations for propagating a mechanical vibration through a bone sample and then try to determine the bone parameters from the signal data.

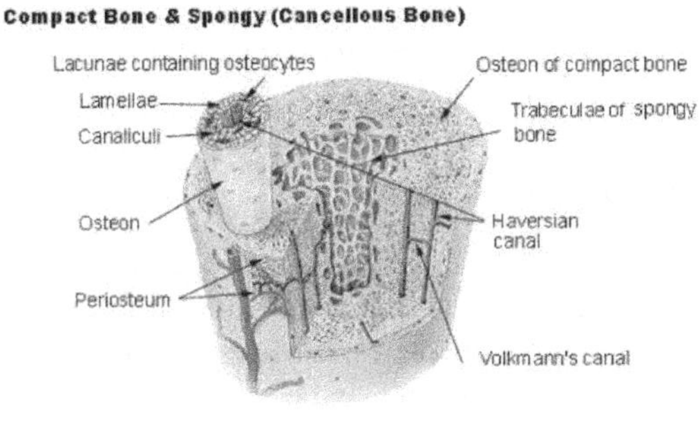

FIGURE 10.1.1 Cortical/trabecular bone. Public domain image. https://commons.wikimedia.org/wiki/File:Illu compact spongy bone.jpg

The strength of bone depends heavily on its micro-structure [187, 188, 289, 342], which is constantly changing via the bone remodeling process. Indeed, it is observed that at an early stage, osteoporosis manifests itself as an increase in anisotropy of the trabecular structure [13, 15, 16, 94, 95, 408, 373]. After a fracture, the bone is damaged, including the soft tissue and the periosternum and surrounding muscles. The cells involved in the repair of fractures are of mesenchymal origin. The reparative cells invade the hematoma and produce callus, which is made up of fibrous tissue, cartilage and young fiber bone [100]. Cortical and cancellous bone respond to fractures in different ways; cancellous bone unites very rapidly because there are many points of contact which are rich in blood and cells. In fractured, cortical bone, depending on whether contact of the severed portions is close and immobilized, healing takes place with very little external callus [79, 84]. On the other hand, if immobilization is not rigid then there is a buildup of external callus. In our models of bone, the matrix (solid part) is considered to be linear and elastic and the interstitial region saturated by a viscous fluid described by the compressible Navier-Stokes equations.

A typical scale of trabecular bone spacing is 0.5-2mm with thickness 50-150μm. In the low frequency range ($\leq$ 100 kHz), the wavelength is longer than 15mm; hence, homogenization theory can be applied there. Our use of homogenization led to a Biot-like model with several additional parameters; however, the *non-Biot* parameters were of an order 10^{-3} smaller than the Biot parameters. This implies that the Biot model is reasonable and suggests an accurate procedure for calculating these parameters other than ad hoc methods [116, 119].

We begin by mathematically emulating the experiments of Hosokowa and Otani[203, 205] and Williams et al. [399, 400, 401] for obtaining the bone parameters. We consider a long bone, such as a femur where a two-dimensional (2-D) cross-section provides a reasonable physical approximation. Hence, it is reasonable to consider the **inverse problem** for 2-dimensional samples of cancellous bone using acoustic pressure data at different locations in the water tank.[1] The inversion involves minimizing the difference, with respect to an appropriate norm, between the given acoustic pressure data and the acoustic pressure predicted by the homogenized equations over a set of effective parameters. This requires solving the effective equations and solving a nonlinear inverse problem [35, 36, 38, 76, 75, 77, 78, 107, 106, 337, 336, 338, 367, 361][2].

10.1.1 THE ISOTROPIC, BIOT MODEL

The Biot model [28, 29, 30, 31] treats a poroelastic medium as an elastic frame with interstitial pore fluid. Actually cancellous bone is anisotropic (orthotropic), however, as pointed out by Williams, if the acoustic waves passing through it travel in the trabecular direction an isotropic model may be a reasonable approximation. The motion of the frame and fluid in the pores are tracked by position vectors $\mathbf{u} = [u_x, u_y]$ and $\mathbf{U} = [U_x, U_y]$. The constitutive equations used by Biot are those of a linear elastic material with terms added to account for the interaction of the frame and interstitial fluid

$$\begin{aligned} \sigma_{xx} &= 2\mu e_{xx} + \lambda e + Q\varepsilon, \\ \sigma_{yy} &= 2\mu e_{yy} + \lambda e + Q\varepsilon, \\ \sigma_{xy} &= \mu e_{xy}, \quad \sigma_{yx} = \mu e_{yx}, \\ s &= Qe + R\varepsilon. \end{aligned} \tag{10.1.1}$$

[1] In this regard see the paper by Laugier et al. [361].

[2] Preliminary studies on determining the Biot coefficients of cancellous bone [48, 49, 51, 54, 145, 144, 56, 57, 58] by inverting the acoustic have been promising. However, the two-dimensional inverse inversion technique. [51, 52] is at this time, state of the art for **inverse problems** for cancellous bone.

TABLE 10.1
Parameters in the Biot model.

Symbol	Parameter
ρ_f	Density of the pore fluid
ρ_r	Density of frame material
K_b	Complex frame bulk modulus
μ	Complex frame shear modulus
K_f	Fluid bulk modulus
K_r	Frame material bulk modulus
β	Porosity
η	Viscosity of pore fluid
k	Permeability
α	Structure constant
a	Pore size parameter

Here R is a parameter measuring the pressure on the fluid required to force a certain volume of fluid into the sediment at constant volume, and Q measures the coupling of changes in the volume of the solid and fluid [28, 30, 31]. The dilatations are

$$e = \nabla \cdot \mathbf{u} = \frac{\partial u_x}{\partial x} + \frac{\partial u_y}{\partial y}, \quad \varepsilon = \nabla \cdot \mathbf{U} = \frac{\partial U_x}{\partial x} + \frac{\partial U_y}{\partial y}. \tag{10.1.2}$$

As usual the strains are defined as

$$e_{xx} = \frac{\partial u_x}{\partial x}, \quad e_{xy} = e_{yx} = \frac{\partial u_x}{\partial y} + \frac{\partial u_y}{\partial x}, \quad e_{yy} = \frac{\partial u_y}{\partial y}. \tag{10.1.3}$$

The parameter μ is the complex frame shear modulus and it is measured. The other parameters λ, R and Q occurring in the constitutive equations are calculated from the measured or estimated values of the parameters given in Table 10.1 using the formulas

$$\begin{aligned} \lambda &= K_b - \frac{2}{3}\mu + \frac{(K_r - K_b)^2 - 2\beta K_r (K_r - K_b) + \beta^2 K_r^2}{D - K_b} \\ R &= \frac{\beta^2 K_r^2}{D - K_b} \\ Q &= \frac{\beta K_r \left((1-\beta) K_r - K_b\right)}{D - K_b}. \end{aligned} \tag{10.1.4}$$

where

$$D = K_r (1 + \beta (K_r / K_f - 1)). \tag{10.1.5}$$

The bulk and shear moduli K_b and μ are often given imaginary parts to account for frame inelasticity. Using an argument based upon Lagrangian dynamics along with the equations (10.1.1), (10.1.2) and (10.1.3) are shown in [28] to lead to the following equations of motion for the displacements $\mathbf{u}, \mathbf{U}$ and dilatations e, ε

$$\mu\nabla^2\mathbf{u}+\nabla[(\lambda+\mu)e+Q\varepsilon] = \frac{\partial^2}{\partial t^2}(\rho_{11}\mathbf{u}+\rho_{12}\mathbf{U})+b\frac{\partial}{\partial t}(\mathbf{u}-\mathbf{U}) \tag{10.1.6}$$
$$\nabla[Qe+R\varepsilon] = \frac{\partial^2}{\partial t^2}(\rho_{12}\mathbf{u}+\rho_{22}\mathbf{U})-b\frac{\partial}{\partial t}(\mathbf{u}-\mathbf{U}).$$

Here ρ_{11} and ρ_{22} are density parameters for the solid and fluid, ρ_{12} is a density coupling parameter, and b is a dissipation parameter. These are calculated from the inputs of Table 10.1 using the formulas (10.1.4)

$$\rho_{11} = (1-\beta)\rho_r-\beta(\rho_f-m\beta)$$
$$\rho_{12} = \beta(\rho_f-m\beta)$$
$$\rho_{22} = m\beta^2$$

where

$$m = \frac{\alpha\rho_f}{\beta}.$$

As we will usually be considering the time-harmonic case, we list the Biot equations for that case too:

$$\mu\nabla^2\mathbf{u}+\nabla[(\lambda+\mu)e+Q\varepsilon]+\omega^2(\rho_{11}\mathbf{u}+\rho_{12}\mathbf{U}) = i\omega b(\mathbf{u}-\mathbf{U}), \tag{10.1.7}$$
$$\nabla[Qe+R\varepsilon]+\omega^2(\rho_{12}\mathbf{u}+\rho_{22}\mathbf{U}) = -i\omega b(\mathbf{u}-\mathbf{U}).$$

Following Fellah et al. [121, 122], a high-frequency adjustment for the Biot-Stoll model is obtained by replacing the assumption of cylindrical pores in the dissipation term by a more realistic configuration. In this regard see [227]. We rename the new coefficients, as before, as $\rho_{11}, \rho_{12}, \rho_{22}$;

$$\rho_{11} := \omega^2\Big[(1-\beta)\rho_r+\beta\rho_f(\alpha(\omega)-1)\Big], \quad \rho_{12} := -\omega^2\beta\rho_f(\alpha(\omega)-1),$$
$$\rho_{22} := \omega^2\beta\rho_f\alpha(\omega),$$

and

$$\alpha(\omega) = \alpha_\infty\left(1+\frac{i\eta\beta}{\omega\alpha_\infty\rho_f k}\sqrt{1+\frac{4\alpha_\infty^2k^2\rho_f\omega}{i\eta\Lambda^2\beta^2}}\right).$$

In [49, 54], the bone sample was situated in a water tank and we used a boundary element method to model the direct problem and the inversion procedure. The direct problem was solved using a finer mesh size than in the inversion procedure. Some noise was also added to the solution of the direct problem. In these works, simulations were performed for bone specimens of relatively high porosities, and five Biot parameters (porosity β, permeability k, pore size a and real parts of the bulk and shear moduli, Re K_b and Re μ) were determined. The algorithm was uniformly successful in finding the porosity to within 3%. Errors for the remaining parameters were often higher, but the target values of these parameters varied over at least one order of magnitude. The procedure took much CPU time because of the eigenvalues introduced by a finite water tank.[3] The papers [51, 52] were an attempt to do away with the water tank and we modeled the problem by replacing the bone sample with an infinite slab, which is a reasonable approximation for a long bone such as a femur. In this case, we were able to numerically compute, to great accuracy, Green's function for the source in the water using residue calculus. The results were, for the most part, reasonable; however, inversion times were long, ranging from six to nine hours. Buchanan et al. [58] investigated an indirect inversion approach based on the numerical solution for a set of effective velocities and transmission coefficients in order to ameliorate the difficulties posed by a direct minimization.

10.1.2 THE DIRECT PROBLEM

Let us consider the acoustic response of a bone segment situated in a water tank and illuminated by an acoustic transponder. This is referred to as the **direct problem**. Even though bone is primarily orthotropic, for didactive reasons and simplicity, let us start with an isotropic model, i.e. the two materials forming the poroelastic bone material, i.e. the mineralized collagen, and the blood-marrow fluid, are considered to be isotropic, as well as the distribution of the fluid filled pores. In the anisotropic case the materials are also considered to be isotropic and the anisotropy is due to the pore size distribution. As bone really is orthotropic, the eventual goal is to investigate parameter retrieval in this case.

The parameter μ, the complex frame shear modulus, is a measured quantity. The other parameters λ, R and Q occurring in the constitutive equations are calculated from measured or estimated values of the parameters given in Table 10.1, using the formulas given above. The bulk and shear moduli K_b and μ are often given imaginary parts to account for frame viscoelasticity. In contrast to our assumption of isotropy, we attempt to account for dissipation due to tortuosity (i.e. due to the presence of pores of arbitrary size).

Without loss of generality and for computational reasons it is assumed that the bone sample is a square. Clearly the bone sample can be cut to an arbitrary parallelpiped. An efficient numerical scheme shall be provided using a boundary integral formulation in the water tank. In this way we need not solve for the entire domain; whereas, the region occupied by the bone sample is discretized as in [49, 54, 147].

Let the bone specimen occupy the region Ω^b and the exterior region in the water tank by Ω^w. In Ω^w the equation for fluid pressure p, in the time-harmonic case, can be reduced to the two-dimensional non-homogeneous Helmholz equation, that is p

---

[3] Recently Chen, Gilbert and Guyenne [82] have used a boundary element method where the fundamental solution used in the boundary integral is Green's function for the Helmholtz equation in an empty water tank, thus reducing the problem to simpler calculations.

and $\mathbf{U}^w$ satisfying the equations

$$-(\nabla^2 p^{(w)} + k_0^2 p^{(w)}) = f \quad \text{in} \quad \Omega^w, \tag{10.1.8}$$

$$\nabla p^{(w)} - \rho^w \omega^2 \mathbf{U}^w = \mathbf{f} \quad \text{in} \quad \Omega^w, \tag{10.1.9}$$

where $\mathbf{f}$ is a given function with compact support, and $f = -div\,\mathbf{f}$; usually we take the $f(\mathbf{x}-\mathbf{x}_0) = \delta(\mathbf{x}-\mathbf{x}_0)$, $\mathbf{x}_0 \in \Omega^w$ [147]. In order to formulate a well-posed boundary value problem, one must reformulate the time-harmonic, Biot equations (10.1.7) in terms of the two unknown solid displacements u_x, u_y, and the fluid pressure s. This is necessary as there are not enough transmission conditions for the two displacements fields $\mathbf{u}$ and $\mathbf{U}$ describing the frame and fluid within the bone respectively. The main idea is to replace the fluid displacements $\mathbf{U}$ within the bone specimen Ω^b by the fluid stress s in (10.1.1) in the equations (10.1.7). To see this, we first express ε and $\mathbf{U}$ in terms of s from (10.1.1) and (10.1.7),

$$\varepsilon = \frac{1}{R}(s - Qe), \quad \mathbf{U} = -\frac{1}{p_{22}}(\nabla s + p_{12}\mathbf{u}). \tag{10.1.10}$$

By taking the divergence of the second equation of (10.1.7), we obtain

$$\nabla^2 s + p_{12}\, e + p_{22}\, \varepsilon = 0,$$

which reduces to

$$\nabla^2 s + \frac{p_{22}}{R} s + (p_{12} - \frac{p_{22}Q}{R}) e = 0, \tag{10.1.11}$$

by making use of (10.1.10). Similarly, the first equation of (10.1.7) can be written in the form:

$$\mu\nabla^2\mathbf{u} + \nabla\Big[(\lambda + \mu - \frac{Q^2}{R})e + (\frac{Q}{R} - \frac{p_{12}}{p_{22}})s\Big] + (p_{11} - \frac{p_{12}^2}{p_{22}})\mathbf{u} = \mathbf{0}. \tag{10.1.12}$$

Equations (10.1.11) and (10.1.12) then form the **modified Biot equations** for $\mathbf{u}$ and s in the bone specimen Ω^b. We are now in a position to formulate the transmission problem for the bone-fluid interaction:

Definition 10.1. *The non-homogeneous transition problem (TP_f) consists of finding the triplet ($\mathbf{u}, s, p$) such that*

$$(E_b) \quad \mu\nabla^2\mathbf{u} + \nabla\Big((\lambda + \mu - \frac{Q^2}{R})e + (\frac{Q}{R} - \frac{p_{12}}{p_{22}})s\Big) + (p_{11} - \frac{p_{12}^2}{p_{22}})\mathbf{u} = \mathbf{0} \quad in \quad \Omega^b,$$

$$(E_s) \qquad \nabla^2 s + \frac{p_{22}}{R} s + (p_{12} - \frac{p_{22}Q}{R}) e = 0 \quad in \quad \Omega^b$$

$$(E_p), \qquad -\Big(\Delta p^{(w)} + k_0^2 p^{(w)}\Big) = f \quad \text{in} \quad \Omega^w, \; f := -div\,\mathbf{f},$$

f having compact support in Ω^w, subject to $(\mathbf{u}, s, p)$ satisfying the transmission conditions B_1, B_2 and B_3 below

$$(B_1) \qquad \left(\underline{\underline{\sigma}}(\mathbf{u}) + Q \operatorname{div} \mathbf{U} + s\right)\mathbf{n} = -p^{(w)}\sigma\,\mathbf{n} \quad \text{on} \quad \Gamma = \partial\Omega^b,$$

with vanishing of the tangent frame stress $\sigma_{12} = \sigma_{21} = 0$, where $\sigma(\mathbf{u})$ and $\underline{\underline{\varepsilon}}(\mathbf{u})$ denote the stress and strain tensors, and div $\mathbf{U}$ the fluid dilatation

$$\underline{\underline{\sigma}}(\mathbf{u}) = \lambda \operatorname{div}\mathbf{u} + 2\mu\underline{\underline{\varepsilon}}(\mathbf{u}), \quad \underline{\underline{\varepsilon}}(\mathbf{u}) = \frac{1}{2}\left(\nabla\mathbf{u} + \nabla\mathbf{u}^T\right),$$

$$\operatorname{div}\mathbf{U} = \frac{1}{R}(s - Q\,e),$$

$$(B_2) \quad \rho^w\omega^2\left[1 - \beta\left(1 + \frac{p_{12}}{p_{22}}\right)\right]\mathbf{u}\cdot\mathbf{n} - \frac{\beta\rho^w\omega^2}{p_{22}}\frac{\partial s}{\partial n} = \left(\frac{\partial p^{(w)}}{\partial n} - \mathbf{n}\cdot\mathbf{f}\right) \quad \text{on} \quad \Gamma,$$

$$(B_3). \qquad s = -\beta p \quad \text{on} \quad \Gamma.$$

Moreover, we assume that the Sommerfeld radiation condition holds for $p^{(w)}$, in the case when Ω^w is unbounded.

The transmission conditions (B_1) and (B_2) represent the continuity of the flux and continuity of the aggregate pressure, respectively; whereas condition (B_3) expresses the continuity of pore pressure with external pressure. In order to pose the variational formulation, we now reduce the partial differential equation (E_p) for p,

$$-\left(p^{(w)} + k_0^2 p^{(w)}\right) = f \quad \text{in} \quad \Omega^w, \tag{10.1.13}$$

to a non-local boundary value problem for p on Γ i.e, we seek a solution p in the form of a simple-layer potential, namely

$$p^{(w)} = -\mathbf{S}\phi + p_f \quad \text{in} \quad \Omega^w, \tag{10.1.14}$$

where ϕ is a, to-be-determined, density function and $\mathbf{S}\phi$ is the simple layer potential

$$\mathbf{S}\phi(x) := \int_\Gamma \frac{i}{4} H_0^{(1)}(k_0|x-y|)\phi(y)ds_y, \quad x \in \Omega^w, \tag{10.1.15}$$

where $-\frac{i}{4}H_0^{(1)}(k_0|x-y|)$, the outgoing Hankel function of zeroth order, is the fundamental solution of the Helmholtz operator $\Delta + k_0$ in $\mathbb{R}^2$. A special solution of the non-homogeneous Helmholtz equation, namely, p_f is given in terms of the integral

$$p_f(x) := \frac{i}{4}\int_{\operatorname{supp} f} H_0^{(1)}(k_0|\mathbf{x}-\mathbf{y}|)\,f(\mathbf{y})\,d\mathbf{y}, \quad x \in \Omega^w,$$

is a particular solution of (10.2.19), which is considered to be known. Hence, if $p|_\Gamma$ is known, applying the trace operator γ_0 to (10.2.20), we then obtain a boundary integral equation for the known density ϕ

$$p^{(w)}(\mathbf{x})|_\Gamma = -\mathbf{V}\phi + \gamma_0 p_p, \tag{10.1.16}$$

where $\mathbf{V} = \gamma_0\mathbf{S}$ is the simple layer boundary integral operator. Using the transmission condition (B_3), one obtains the boundary integral equation

$$(E_{pb}) \qquad \mathbf{V}\phi - \frac{1}{\beta}s = \gamma_0 p_f.$$

A solution may be found by the above to a **nonlocal** boundary value problem. We begin with the boundary condition

$$(B_{2b}) \quad \rho^w\omega^2\left[1-\beta\left(1+\frac{p_{12}}{p_{22}}\right)\right]\mathbf{u}\cdot\mathbf{n} - \frac{\beta\rho^w\omega^2}{p_{22}}\frac{\partial s}{\partial n} = \left(\frac{\partial p^{(w)}}{\partial n} - \mathbf{n}\cdot\mathbf{f}\right) \quad \text{on} \quad \Gamma$$

and then by using

$$\frac{\partial p}{\partial n} = \frac{1}{2}\left(\phi - \mathbf{K}'\phi\right) + \frac{\partial p_f}{\partial n},$$

where the boundary integral operator $\mathbf{K}'$ appearing in (B_{2b}) is defined by

$$\mathbf{K}'\phi(x) := \frac{i}{4}\int_\Gamma \frac{\partial}{\partial n_x} H_0^{(1)}(k_0|x-y|)\phi(y)ds_y, \; x \in \Gamma.$$

We note that condition (B_{2b}) can be explicitly written in terms of ϕ

$$\frac{\partial s}{\partial n} = \frac{p_{22}}{\beta}\left\{\left[1-\beta\left(1+\frac{p_{12}}{p_{22}}\right)\right]\mathbf{u}\cdot\mathbf{n} - \frac{1}{\rho^w\omega^2}\left(\frac{1}{2}\phi - \mathbf{K}'\phi\right)\right\} + \frac{p_{22}}{\beta\rho^w\omega^2}(\mathbf{n}\cdot\mathbf{f} - \frac{\partial}{\partial n}p_f),$$

which we use for the variational formulation. It is shown in [148] that the first of the Biot equations can be put in the weak form

$$\begin{aligned} a(\mathbf{u},\mathbf{v}) + \int_{\Omega^b}\left(\frac{Q}{R} - \frac{p_{12}}{p_{22}}\right)s(\operatorname{div}\bar{\mathbf{v}})d\mathbf{x} - \int_{\Omega^b}\left(p_{11} - \frac{p_{12}^2}{p_{22}}\right)\mathbf{u}\cdot\bar{\mathbf{v}}\,d\mathbf{x} \\ -\left[1-\beta(1+\frac{p_{12}}{p_{22}})\right]\langle\mathbf{V}\phi\mathbf{n},\bar{\mathbf{v}}\rangle_\Gamma \\ = -\left[1-\beta(1+\frac{p_{12}}{p_{22}})\right]\langle\gamma_0 p_f,\bar{\mathbf{v}}\rangle_\Gamma, \quad \forall\mathbf{v}\in\left(H^1(\Omega^b)\right)^2 \end{aligned} \tag{10.1.17}$$

whereas, the second Biot equation, has the weak form

$$b(s,\tau) + p_{22}\int_{\Omega^b}\left(\frac{Q}{R} - \frac{p_{12}}{p_{22}} - \right)\nabla\cdot(\mathbf{u})\bar{\tau}d\mathbf{x} - \int_{\Omega^b}\frac{p_{22}}{R}s\bar{\tau}d\mathbf{x}$$

$$-\frac{p_{22}}{\beta}\left[1-\beta\left(1+\frac{p_{12}}{p_{22}}\right)\right]\langle \mathbf{u}\cdot\mathbf{n},\overline{\tau}\rangle_\Gamma+\frac{p_{22}}{\beta\rho^w\omega^2}\langle\left(\frac{1}{2}\phi-\mathbf{K}'\phi\right),\overline{\tau}\rangle_\Gamma \quad (10.1.18)$$
$$=\frac{p_{22}}{\beta\rho^w\omega^2}<(\mathbf{n}\cdot\mathbf{f}-\frac{\partial}{\partial n}p_f),\overline{\tau}>_\Gamma,\quad \forall\tau\in H^1(\Omega^b).$$

In [148] the nonlocal boundary value problem is formulated as

Definition 10.2 (Variational formulation). *Given* **f**, *find the triple* $(\mathbf{u},s,\phi)\in (H^1(\Omega^b))^2\times H^1(\Omega^b)\times H^{-1/2}(\Gamma)$ *such that*

$$\mathscr{A}\Big((\mathbf{u},s,\phi),(\mathbf{v},\tau,\psi)\Big)=\ell_f\,(\mathbf{v},\tau,\phi) \quad (10.1.19)$$

for all $(\mathbf{v},\tau,\psi)\in(H^1(\Omega^b))^2\times H^1(\Omega^b)\times H^{-1/2}(\Gamma)$, *where* $\mathscr{A}$ *and* ℓ_f *are respectively the sesquilinear, form and linear functional defined by*

$$\mathscr{A}\Big((\mathbf{u},s,\phi),(\mathbf{v},\tau,\psi)\Big):=a(\mathbf{u},\mathbf{v})+b(s,\tau)+\frac{p_{22}}{2\rho^w\omega^2}\langle\mathbf{V}\phi,\overline{\psi}\rangle_\Gamma$$
$$+\left(\frac{Q}{R}-\frac{p_{12}}{p_{22}}\right)\Big[\int_{\Omega^b}s(\nabla\cdot\overline{\mathbf{v}})d\mathbf{x}+p_{22}\int_{\Omega^b}\div(\mathbf{u})\overline{\tau}\,d\mathbf{x}\Big]$$
$$-\left(p_{11}-\frac{p_{12}^2}{p_{22}}\right)\int_{\Omega^b}\mathbf{u}\cdot\overline{\mathbf{v}}\,d\mathbf{x}-\frac{p_{22}}{R}\int_{\Omega^b}s\overline{\tau}\,d\mathbf{x} \quad (10.1.20)$$
$$-\Big[1-\beta(1+\frac{p_{12}}{p_{22}})\Big]\Big\{\langle\mathbf{V}\phi\mathbf{n},\overline{\mathbf{v}}\rangle_\Gamma+\frac{p_{22}}{\beta}<\mathbf{u}\cdot\mathbf{n},\overline{\tau}>_\Gamma\Big\}$$
$$+\frac{p_{22}}{\beta\rho^w\omega^2}\Big[<\langle\left(\frac{1}{2}\phi-\mathbf{K}'\phi\right),\overline{\tau}\rangle_\Gamma-\frac{1}{2}\langle s,\overline{\psi}\rangle_\Gamma\Big]$$

$$\ell_f\,(\mathbf{v},\tau,\phi):=-\Big[1-\beta(1+\frac{p_{12}}{p_{22}})\Big]\langle\gamma_0p_f,\overline{\mathbf{v}}\rangle_\Gamma+\frac{p_{22}}{\beta\rho^w\omega^2}\langle\mathbf{n}\cdot\mathbf{f}-\frac{\partial}{\partial n}p_f)\overline{\tau}\rangle_\Gamma$$
$$+\frac{p_{22}}{2\rho^w\omega^2}\langle\gamma_0p_p,\overline{\psi}\rangle \quad (10.1.21)$$

The proof of the existence and uniqueness is then shown using Garding's inequality. See [148] for further details and the proof of the theorems.

10.2 ANISOTROPIC BIOT SYSTEMS

10.2.1 CARCIONE REPRESENTATION

Bone is certainly not isotropic; hence, it behooves us to consider non-isotropic models. For the completely anisotropic case it is convenient to use the Voigt notation adopted by Carcione in his book and papers, etc. [68, 69]. We present an introduction to his formulation here before going further.[4] As in the previous section we

---

[4]Here we formulate the problem in $\mathbb{R}^3$; hence, the number of unknowns needs to be reduced from the six displacement variables to four unknowns, the three solid displacements and the fluid pressure.

model a bone sample in a water bath and our model for the bone remains the Biot poroelastic formulation.

The Carcione formulation appears different, but is equivalent to that of the previous section. First, one introduces the variation of fluid content in the bone as

$$\zeta := \nabla \cdot \left[\beta \left(\mathbf{u}^{(f)} - \mathbf{u}^{(m)}\right)\right] = -\beta \left(\theta_f - \theta_m\right), \tag{10.2.1}$$

where β is the porosity. The stress components for the fluid are given by

$$\sigma_{ij}^{(f)} = -\beta p^{(f)} \delta_{ij}, \tag{10.2.2}$$

where $p^{(f)}$ is the fluid pressure and θ_m and θ_f are the solid matrix and the fluid dilatations, respectively. It is convenient to use Voigt notation to describe the elastic frame [17, 68, 83, 262], namely

$$\mathbf{e_I} = \mathbf{e}_{i(i)}; = e_{ii}, I = 1,2,3. \quad \mathbf{e_I} = e_{ij} = \partial_j u_i + \partial_i u_j, i \neq j, I = 4,5,6,$$

where the strain and the stress are represented as 6-vectors:

$$\mathbf{e} = (e_1, e_2, e_3, e_4, e_5, e_6)^T := (e_{11}, e_{22}, e_{33}, e_{23}, e_{13}, e_{12})^T,$$

$$\sigma = (\sigma_1, \sigma_2, \sigma_3, \sigma_4, \sigma_5, \sigma_6)^T := (\sigma_{11}, \sigma_{22}, \sigma_{33}, \sigma_{23}, \sigma_{13}, \sigma_{12})^T.$$

The strain energy potential is assumed to have a quadratic form similar to that in Biot's model

$$2V = c_{\mathbf{I}}(e_{\mathbf{I}}^{(m)})^2 + 2\left[c_{\mathbf{I},\mathbf{J}} e_{\mathbf{I}}^{(m)} e_{\mathbf{J}}^{(m)}\right]_{\mathbf{I} \neq \mathbf{J}} + 2\theta_f C_{\mathbf{I}} e_{\mathbf{I}}^{(m)} + F\theta_f^2. \tag{10.2.3}$$

Here the coefficients $C_1, \ldots, C_6$ are determined by solid and fluid properties. The choice of this particular strain energy quadratic (10.2.3) forces the fluid to be viscous and hence, it can sustain shear. The fluid and solid stress tensors may be determined from the strain energy by differentiating with respect to the strains θ_f and $e_{\mathbf{I}}^{(m)}$ respectively, i.e.

$$\sigma_{\mathbf{I}}^{(m)} := \frac{\partial V}{\partial e_{\mathbf{I}}} = c_{\mathbf{I}} e_{\mathbf{I}}^{(m)} + c_{\mathbf{I},\mathbf{J}} e_{\mathbf{J}}^{(m)} + \theta_f C_{\mathbf{I}},$$

$$\sigma^{(f)} := \frac{\partial V}{\partial \theta_f} = C_{\mathbf{I}} e_{\mathbf{I}}^{(m)} + F\theta_f. \tag{10.2.4}$$

The pore pressure may be written as

$$p_f = M'\left(\zeta - \alpha_{\mathbf{I}} e_{\mathbf{I}}^{(m)}\right), \tag{10.2.5}$$

where

$$\alpha_i = \beta\left(1 + \frac{C_i}{F}\right), \; i = 1,2,3 \quad \text{and } \alpha_i = \beta\frac{C_i}{F}, \; i = 4,5,6.$$

The total stress is given by

$$\sigma_{\mathbf{I}} = c^{(m)}_{\mathbf{IJ}} - \alpha_{\mathbf{I}} p_f. \tag{10.2.6}$$

In Cartesian coordinates we have an alternate, but equivalent relation

$$\sigma_{ij} = \mathscr{C}_{ijkl} e_{kl}, \tag{10.2.7}$$

where Einstein summation notation is used. There are certain symmetries associated with the elastic stiffness tensor $\mathscr{C}$, which are obvious from its definition, namely [263]

$$\mathscr{C}_{ijkl} = \mathscr{C}_{jikl} = \mathscr{C}_{ijlk} = \mathscr{C}_{klij}.$$

The connection between the Voigt and Cartesian representation of the constitutive coefficients is given by

$$\begin{bmatrix} \mathscr{C}_{1111} & \mathscr{C}_{1122} & \mathscr{C}_{1133} & \mathscr{C}_{1123} & \mathscr{C}_{1131} & \mathscr{C}_{1112} \\ \mathscr{C}_{2211} & \mathscr{C}_{2222} & \mathscr{C}_{2233} & \mathscr{C}_{2223} & \mathscr{C}_{2231} & \mathscr{C}_{2212} \\ \mathscr{C}_{3311} & \mathscr{C}_{3322} & \mathscr{C}_{3333} & \mathscr{C}_{3323} & \mathscr{C}_{3331} & \mathscr{C}_{3312} \\ \mathscr{C}_{2311} & \mathscr{C}_{2322} & \mathscr{C}_{2333} & \mathscr{C}_{2323} & \mathscr{C}_{2331} & \mathscr{C}_{2312} \\ \mathscr{C}_{3111} & \mathscr{C}_{3122} & \mathscr{C}_{3133} & \mathscr{C}_{3123} & \mathscr{C}_{3131} & \mathscr{C}_{3112} \\ \mathscr{C}_{1211} & \mathscr{C}_{1222} & \mathscr{C}_{1233} & \mathscr{C}_{1223} & \mathscr{C}_{1231} & \mathscr{C}_{1212} \end{bmatrix} :=$$

$$\begin{bmatrix} c_{11} & c_{12} & c_{13} & c_{14} & c_{15} & c_{16} \\ c_{12} & c_{22} & c_{23} & c_{24} & c_{25} & c_{26} \\ c_{13} & c_{23} & c_{33} & c_{34} & c_{35} & c_{36} \\ c_{14} & c_{24} & c_{34} & c_{44} & c_{45} & c_{46} \\ c_{15} & c_{25} & c_{35} & c_{45} & c_{55} & c_{56} \\ c_{16} & c_{26} & c_{36} & c_{46} & c_{56} & c_{66} \end{bmatrix}. \tag{10.2.8}$$

For an orthorhombic, poroelastic material, the equations of motion are given in Section (7.4.1) of Carcione [68]. It follows from his expression for kinetic energy that

$$T = \frac{1}{2}\Omega_b \left(q_{ij} v_i^{(m)} v_j^{(m)} + 2 r_{ij} v_i^{(m)} v_j^{(f)} + t_{ij} v_i^{(f)} v_j^{(f)} \right), \tag{10.2.9}$$

where $\mathbf{Q}$, $\mathbf{R}$ and $\mathbf{T}$ are 3×3 mass matrices, with $\mathbf{R}$ the induced mass matrix. For an orthorhombic material it is possible to find a coordinate system where each of these matrices can be diagonalized [68]. However, for our existence proof [157] to hold, it is not necessary to assume that the material is orthorhombic. Indeed, the general case with positive definite matrices suffices. The equations of motion then become [28, 30, 31, 51, 68]

$$\nabla \cdot \sigma^{(m)} = \left[(1-\beta)\rho_s \mathbb{I} - \underline{\underline{r}} \right] \partial_{tt} \mathbf{u}^{(m)} + \underline{\underline{r}} \partial_{tt} \mathbf{u}^{(f)} + \underline{\underline{b}} \left(\mathbf{v}^{(m)} - \mathbf{v}^{(f)} \right)(t) \tag{10.2.10}$$

and

$$-\nabla \left(\beta p^{(f)} \right) = \underline{\underline{r}} \partial_{tt} \mathbf{u}^{(m)} + \left(\beta \rho_f \mathbb{I} - \underline{\underline{r}} \right) \partial_{tt} \mathbf{u}^{(f)} - \underline{\underline{b}} \left(\mathbf{v}^{(m)} - \mathbf{v}^{(f)} \right)(t), \tag{10.2.11}$$

where $\mathbb{I}$ is the identity tensor, $\mathbf{v}^{(m)} = \partial_t \mathbf{u}^{(m)}$ and $\underline{\underline{r}}$ may or may not be $\mathrm{diag}(r_1, r_2, r_3)$; moreover, $\underline{\underline{b}}$ is the second-order, dissipation tensor

$$\underline{\underline{b}} = \beta^2 \eta \underline{\underline{K}}^{-1}, \tag{10.2.12}$$

where $\underline{\underline{K}}$ is the global permeability tensor and η the fluid viscosity.

Assuming the solutions are time harmonic, with frequency ω, equations take the form

$$\nabla \cdot \sigma^{(m)} = -\omega^2 \left[(1-\beta)\rho_s \mathbb{I} - \underline{\underline{r}}\right] \mathbf{u}^{(m)} - \omega^2 \underline{\underline{r}} \mathbf{u}^{(f)} - i\omega \mathbf{b} \left(\mathbf{u}^{(m)} - \mathbf{u}^{(f)}\right) \tag{10.2.13}$$

and likewise (10.2.11) may be written as

$$-\nabla \left(\beta p^{(f)}\right) + \omega^2 \underline{\underline{r}} \mathbf{u}^{(m)} + \omega^2 \left(\beta \rho_f \mathbb{I} - \underline{\underline{r}}_i\right) \mathbf{u}^{(f)} - i\omega \left(\mathbf{u}^{(m)} - \mathbf{u}^{(f)}\right) = 0. \tag{10.2.14}$$

This may be written in a more compact notation as

$$\nabla \cdot \sigma^{(m)} + \underline{\underline{p}}_{11} \mathbf{u}^{(m)} + \underline{\underline{p}}_{12} \mathbf{u}^{(f)} = 0$$

and

$$-\beta \nabla p^{(f)} + \underline{\underline{p}}_{12} \mathbf{u}^{(m)} + \underline{\underline{p}}_{22} \mathbf{u}^{(f)} = 0, \tag{10.2.15}$$

where the $\underline{\underline{p}}_{ij}$ tensors are defined as

$$\underline{\underline{p}}_{11} := -\omega^2((1-\beta)\rho_s \mathbb{I} - \underline{\underline{r}}) - i\omega \underline{\underline{b}}, \quad \underline{\underline{p}}_{12} := \omega^2 \underline{\underline{r}} + i\omega \underline{\underline{b}}, \quad \underline{\underline{p}}_{22} := \omega^2(\beta \rho_f \mathbb{I} - \underline{\underline{r}}) - i\omega \underline{\underline{b}}.$$

10.2.2 ELIMINATION OF THE FLUID DISPLACEMENTS

As in the isotropic case, there are too many unknowns. As before, we use $\mathbf{u}^{(m)}$ and $p^{(f)}$ and eliminate $\mathbf{u}^{(f)}$. Multiplying (10.2.15$_b$) by $\underline{\underline{p}}_{22}^{-1}$ and taking the divergence we have

$$-\beta \nabla \cdot \left(\underline{\underline{p}}_{22}^{-1} \nabla p^{(f)}\right) + \nabla \cdot \left(\underline{\underline{p}}_{22}^{-1} \underline{\underline{p}}_{12} \mathbf{u}^{(m)}\right) + \theta_f = 0, \tag{10.2.16}$$

with

$$\theta_f := \nabla \cdot \mathbf{u}^{(f)} = -\frac{1}{F}\left(\beta p^{(f)} + \mathscr{C} : \underline{\underline{e}}(\mathbf{u}^{(m)})\right)$$

from (10.2.2), where, as we recall, F is the multiplier of θ_f in the strain-energy (10.2.3). Now for (10.2.15$_a$), first multiplying it by $\underline{\underline{p}}_{12}^{-1}$ and substituting $\mathbf{u}^{(f)}$ from (10.2.15$_b$) such that

$$\mathbf{u}^{(f)} = -\underline{\underline{p}}_{22}^{-1} \left(-\beta \nabla p^{(f)} + \underline{\underline{p}}_{12} \mathbf{u}^{(m)}\right), \tag{10.2.17}$$

we obtain, in view of (10.2.1),

$$\nabla \cdot \left[\mathscr{C} : \underline{\underline{e}}(\mathbf{u}^{(m)}) + \theta_f \underline{\underline{C}}\right] + (\underline{\underline{p}}_{11} - \underline{\underline{p}}_{12} \underline{\underline{p}}_{22}^{-1} \underline{\underline{p}}_{12}) \mathbf{u}^{(m)} + \beta \underline{\underline{p}}_{12} \underline{\underline{p}}_{22}^{-1} \nabla p^{(f)} = \mathbf{0}. \tag{10.2.18}$$

It was shown using a variational approach that the Biot transmission problem was well-posed. A variation of the same arguments may be used for the above, fully anisotropic case. It is useful for us to pose several different Biot transmission problems.

Definition 10.3 (The transformed, non-homogeneous transmission problem (TP_f)). *The problem consists of finding the triplet $(\mathbf{u}^{(m)}, p^{(f)}, p^{(w)})$ such that its members satisfy the following equations:*

(E_u)

$$\nabla\cdot\left[\mathscr{C}:\underline{\underline{e}}(\mathbf{u}^{(m)})+\theta_f\underline{\underline{C}}\right]+(\underline{\underline{p}}_{11}-\underline{\underline{p}}_{12}\underline{\underline{p}}_{22}^{-1}\underline{\underline{p}}_{12})\mathbf{u}^{(m)}+\beta\underline{\underline{p}}_{12}\underline{\underline{p}}_{22}^{-1}\nabla p^{(f)}=\mathbf{0} \quad in \quad \Omega^b$$

$$(E_{p^{(f)}}) \qquad -\beta\nabla\cdot\left(\underline{\underline{p}}_{22}^{-1}\nabla p^{(f)}\right)+\nabla\cdot\left(\underline{\underline{p}}_{22}^{-1}\underline{\underline{p}}_{12}\mathbf{u}^{(m)}\right)+\theta_f=0 \quad in \quad \Omega^b$$

$$(E_{p^{(w)}}), \qquad -\left(\hat{p}^{(w)}+k_0^2p^{(w)}\right)=f \quad \text{in} \quad \Omega^w,\ f:=-div\,\mathbf{f}$$

with $p^{(w)}v$ having compact support in Ω^w, together with the transmission conditions:

$$(B_1) \qquad \left[\mathscr{C}:\underline{\underline{e}}(\mathbf{u}^{(m)})+\theta_f\underline{\underline{C}}-\beta p^{(f)}\mathbb{I}\right]\mathbf{n}=-p^{(w)}\mathbf{n} \quad \text{on} \quad \Gamma=\partial\Omega^b$$

with vanishing of the tangential frame stress

$$\left(\left[\mathscr{C}:\underline{\underline{e}}(\mathbf{u}^{(m)})+\theta_f\underline{\underline{C}}-\beta p^{(f)}\mathbb{I}\right]\mathbf{n}\right)\cdot\mathbf{t}=0,\ \mathbf{t},\ \textit{the unit tangent vector}, \ \text{on} \quad \Gamma=\partial\Omega^b$$

$$(B_2) \qquad \rho^w\omega^2\left(\left[\mathbb{I}-\beta\left(\mathbb{I}+\underline{\underline{p}}_{22}^{-1}\underline{\underline{p}}_{12}\right)\right]\mathbf{u}^m\right)\cdot\mathbf{n}+\beta^2\rho^w\omega^2\left[\underline{\underline{p}}_{22}^{-1}\nabla p^{(f)}\right]\cdot\mathbf{n}$$
$$=\left(\frac{\partial p^{(w)}}{\partial n}-\mathbf{n}\cdot\mathbf{f}\right) \quad \text{on} \quad \Gamma$$

and

$$(B_3). \qquad p^{(f)}=p^{(w)} \quad \text{on} \quad \Gamma.$$

In addition, we assume that the Sommerfeld radiation condition holds for $p^{(w)}$ when Ω^w is unbounded.

As before, the partial differential equation $(E_{p^{(w)}})$ for the pressure in the water tank, $p^{(w)}$, is

$$-\left(p^{(w)}+k_0^2p^{(w)}\right)=f \quad \text{in} \quad \Omega^w\subset\mathbb{R}^3 \tag{10.2.19}$$

This may be reformulated as a boundary integral equation for $p^{(w)}$ on Γ by seeking a solution in the form of a single-layer potential, namely

$$p^{(w)}=-\mathbf{S}\psi+P_{spec}^{(w)} \quad \text{in} \quad \Omega^w \tag{10.2.20}$$

where ψ is an unknown density function and $\mathbf{S}\psi$ is defined by

$$\mathbf{S}\psi(\mathbf{x}) := \int_\Gamma \frac{e^{ik_0|\mathbf{x}-\mathbf{y}|}}{4\pi|\mathbf{x}-\mathbf{y}|}\psi(\mathbf{y})ds_{\mathbf{y}}, \quad \mathbf{x} \in \Omega^w, \tag{10.2.21}$$

and $P^{(w)}_{spec}$ given by,

$$P^{(w)}_{spec} := \int\limits_{\text{supp}f} \frac{e^{ik_0|\mathbf{x}-\mathbf{y}|}}{4\pi|\mathbf{x}-\mathbf{y}|} f(\mathbf{y})\,d\mathbf{y}, \quad \mathbf{x} \in \Omega^w,$$

is a particular solution of (10.2.19), which is assumed to be known. Hence, if $p^{(w)}|_\Gamma$ is known, applying the trace operator γ_0 to (10.2.20), we then obtain a boundary integral equation for the unknown density ψ,

$$\hat{p}^{(w)}|_\Gamma = -\mathbf{V}\psi + \gamma_0 P^{(w)}_{spec}, \tag{10.2.22}$$

where $\mathbf{V} = \gamma_0\mathbf{S}$ is the single-layer boundary integral operator. Then from the transmission condition (B_3), we obtain the boundary integral equation

$$(E_{pb}) \qquad \mathbf{V}\psi + p^{(f)} = \gamma_0 P^{(w)}_{spec}.$$

Definition 10.4 (A nonlocal boundary value problem)**.** *The transmission problem* $\mathbf{TP}_f$ *is termed a nonlocal boundary value problem for the triple* $(\mathbf{u}^{(m)}, p^{(f)}, \psi)$ *if the triple* $(\mathbf{u}^{(m)}, p^{(f)}, \psi)$ *satisfies equations* (E_u)*,* $(E_{p^{(f)}})$*, and the boundary integral equation* (E_{pb}) *together with the transmission conditions*

$$(B_{1b}) \qquad \left[\mathscr{C} : \underline{\underline{e}}(\mathbf{u}^{(m)}) + \theta_f \underline{\underline{C}} - \beta p^{(f)}\mathbb{I}\right]\mathbf{n} = -p^{(w)}\mathbf{n} \quad \text{on} \quad \Gamma = \partial\Omega^b,$$

with

$$p^{(w)} = p^{(f)} = -\mathbf{V}\psi + \gamma_0 P^{(w)}_{spec}.$$

Here again we require vanishing of the tangential frame stress

$$\left\{\left[\mathscr{C} : \underline{\underline{e}}(\mathbf{u}^{(m)}) + \theta_f \underline{\underline{C}} - \beta p^{(f)}\mathbf{I}\right]\mathbf{n}\right\} \cdot \mathbf{t} = 0,$$

$$(B_{2b}) \qquad \left\{\rho^w\omega^2\left[\mathbb{I} - \beta\left(1 + \underline{\underline{P}}_{22}^{-1}\underline{\underline{P}}_{12}\right)\right]\mathbf{u}\right\}\cdot\mathbf{n} + \beta^2\rho^w\omega^2\left(\underline{\underline{P}}_{22}^{-1}\nabla p^{(f)}\right)\cdot\mathbf{n}$$

$$= \left(\frac{\partial p^{(w)}}{\partial n} - \mathbf{n}\cdot\mathbf{f}\right) \quad \text{on} \quad \Gamma,$$

with

$$\frac{\partial p^{(w)}}{\partial n} = \left(\frac{1}{2}\psi - \mathbf{K}'\psi\right) + \frac{\partial P^{(w)}_{spec}}{\partial n}.$$

The boundary integral operator $\mathbf{K}'$ in (B_{2b}) is defined by

$$\mathbf{K}'\psi(x) := \int_\Gamma \frac{\partial}{\partial n_x} \frac{e^{ik_0|\mathbf{x}-\mathbf{y}|}}{4\pi|\mathbf{x}-\mathbf{y}|}\psi(\mathbf{y})ds_{\mathbf{y}},\ \mathbf{x}\in\Gamma.$$

We note that condition (B_{2b}) can be explicitly written in terms of ψ

$$\underline{\underline{p}}_{22}^{-1}\nabla p^{(f)}\cdot\mathbf{n} = -\frac{1}{\beta^2}\left\{\left[\mathbb{I}-\beta\left(1+\underline{\underline{p}}_{22}^{-1}\underline{\underline{p}}_{12}\right)\right]\mathbf{u}\cdot\mathbf{n}-\frac{1}{\rho^w\omega^2}\left(\frac{1}{2}\psi-\mathbf{K}'\psi\right)\right\}$$
$$-\frac{1}{\beta^2\rho^w\omega^2}(\mathbf{n}\cdot\mathbf{f}-\frac{\partial}{\partial n}P^{(w)}_{spec}),$$

which will be needed for the variational formulation in the next section. In [157, 156] it was shown, using a variational approach, that a similar system of linear equations has at most one solution. It can be shown that this result extends to the present situation. In the next section we use this result to construct an iterative method for solving the non-linear system associated with an added fluid-fluid viscosity.

10.2.3 EXISTENCE THEOREM FOR THE ANISOTROPIC BIOT MODEL

It may be shown [157] that our boundary value problem may be posed in variational form.

Definition 10.5 (The variational formulation). *Given* $\mathbf{f}$*, find the triple* $(\mathbf{u}^{(m)}, p^{(f)}, \psi)$ $\in \left(H^1(\Omega^b)\right)^3 \times H^1(\Omega^b)\times H^{-1/2}(\Gamma)$ *such that*

$$\mathscr{A}\left(\mathbf{u}^{(m)}, p^{(f)}, \psi), (\overline{\mathbf{v}},\overline{\pi},\overline{\xi})\right) = \ell_f\,(\overline{\mathbf{v}},\overline{\pi},\overline{\xi}) \tag{10.2.23}$$

for all $(\overline{\mathbf{v}},\overline{\pi},\overline{\xi}) \in \left(H^1(\Omega^b)\right)^3 \times H^1(\Omega^b)\times H^{-1/2}(\Gamma)$*, where* $\mathscr{A}$ *and* ℓ_f *are respectively the sesquilinear form and linear functional defined by*

$$\mathscr{A}\left(\mathbf{u}^{(m)}, p^{(f)}, \psi), (\overline{\mathbf{v}},\overline{\pi},\overline{\xi})\right) := a(\mathbf{u}^{(m)},\overline{\mathbf{v}}) + b(p^{(f)},\pi) + \frac{1}{2\rho^w\omega^2}\left\langle \mathbf{V}\psi,\overline{\xi}\right\rangle_\Gamma$$

$$-\int\limits_{\Omega^b}\left(\underline{\underline{p}}_{11}-\underline{\underline{p}}_{12}\underline{\underline{p}}_{22}^{-1}\underline{\underline{p}}_{12}\right)\mathbf{u}^{(m)}\cdot\overline{\mathbf{v}}d\underline{x} - \frac{\beta^2}{F}\int\limits_{\Omega^b} p^{(f)}\overline{\pi}d\underline{x} + \beta\left\{\int\limits_{\Omega^b} p^{(f)}\nabla\cdot\left[\left(\underline{\underline{p}}_{11}\underline{\underline{p}}_{22}^{-1}\right)'\overline{\mathbf{v}}\right]d\underline{x}\right.$$

$$\left.+\int\limits_{\Omega^b}\overline{\pi}\nabla\cdot\left[\left(\underline{\underline{p}}_{22}^{-1}\underline{\underline{p}}_{12}\right)\mathbf{u}^{(m)}\right]d\underline{x}\right\} - \frac{\beta}{F}\left\{\int\limits_{\Omega^b} p^{(f)}\underline{\underline{C}}:\underline{\underline{e}}(\overline{\mathbf{v}})\,d\underline{x} + \int\limits_{\Omega^b}\overline{\pi}\underline{\underline{C}}:\underline{\underline{e}}(\mathbf{u}^{(m)})\,d\underline{x}\right\}.$$

$$-\left\langle\left[\mathbb{I}-\beta\left[\left(\mathbb{I}+\left(\underline{\underline{p}}_{12}\underline{\underline{p}}_{22}^{-1}\right)\right]'\right)\right]\mathbf{V}\psi,\mathbf{n}\cdot\overline{\mathbf{v}}\right\rangle_\Gamma + \left\langle\left[\mathbb{I}-\beta\left(\mathbb{I}+\underline{\underline{p}}_{22}^{-1}\underline{\underline{p}}_{12}\right)\right]\mathbf{u}^{(m)}\cdot\mathbf{n},\overline{\pi}\right\rangle_\Gamma.$$

$$-\frac{1}{\rho^{(w)}\omega^2}\left\{\left\langle\left(\frac{1}{2}\mathbb{I}-\mathbf{K}'\right)\psi,\overline{\pi}\right\rangle_\Gamma - \left\langle\frac{1}{2}p^{(f)},\overline{\xi}\right\rangle_\Gamma\right\}. \tag{10.2.24}$$

$$\ell_f\left(\overline{\mathbf{v}},\overline{\pi},\overline{\xi}\right) := -\left\langle\left[\mathbb{I}-\beta\left(\mathbb{I}+(\underline{\underline{p}}_{22}^{-1}\underline{\underline{p}}_{12})'\right)\right]\gamma_0 P_{spec}^{(w)}\mathbf{n},\overline{\mathbf{v}}\right\rangle_\Gamma$$
$$-\frac{1}{\rho^{(w)}\omega^2}\left\langle\mathbf{n}\cdot\mathbf{f}-\frac{\partial}{\partial n}P_{spec}^{(w)},\overline{\pi}\right\rangle_\Gamma+\frac{1}{2\rho^{(w)}\omega^2}\left\langle\gamma_0 P_{spec}^{(w)},\overline{\xi}\right\rangle_\Gamma. \qquad (10.2.25)$$

Here

$$a(\mathbf{u}^{(m)},\mathbf{v})-\int_{\Omega^b}\left(\underline{\underline{p}}_{11}-\underline{\underline{p}}_{12}\underline{\underline{p}}_{22}^{-1}\underline{\underline{p}}_{12}\right)\mathbf{u}^{(m)}\cdot\overline{\mathbf{v}}\,d\underline{x}-\frac{\beta}{F}\int_{\Omega^b}p^{(f)}\underline{\underline{C}}:\underline{\underline{e}}(\overline{\mathbf{v}})\,d\underline{x}$$

(BV_1)

$$+\beta\int_{\Omega^b}p^{(f)}\nabla\cdot\left(\left(\underline{\underline{p}}_{12}\underline{\underline{p}}_{22}\right)'\overline{\mathbf{v}}\right)d\underline{x}-\left\langle\mathbb{I}-\beta\left[\mathbb{I}+\left(\underline{\underline{p}}_{12}\underline{\underline{p}}_{22}^{-1}\right)'\right]\mathbf{V}\psi\mathbf{n},\overline{\mathbf{v}}\right\rangle_\Gamma$$
$$=-\left\langle\mathbb{I}-\beta\left[\mathbb{I}+\left(\underline{\underline{p}}_{12}\underline{\underline{p}}_{22}^{-1}\right)'\right]\gamma_0 P_{spec}^{(w)}\mathbf{n},\overline{\mathbf{v}}\right\rangle_\Gamma.$$

and

$$b(p^{(f)},\pi)=\beta^2\int_{\Omega^b}\left(\underline{\underline{p}}_{22}^{-1}\nabla p^{(f)}\right)\cdot\nabla\overline{\pi}\,d\mathbf{x}, \qquad (10.2.26)$$

Theorem 10.1

Assuming that the stiffness matrix $\mathscr{C}$ is uniformly positive definite, the sesquilinear form in (10.2.23) satisfies the Garding's inequality in the form

$$Re\mathscr{A}\left(\mathbf{u}^{(m)},p^{(f)},\psi,\mathbf{u}^{(m)},p^{(f)},\psi\right)$$
$$\geq\alpha\left\{\|\mathbf{u}^{(m)}\|^2_{(H^1(\Omega^b))^2}+\|p^{(f)}\|^2_{H^1(\Omega^b)}+\|\psi\|^2_{H^{-\frac{1}{2}}(\Gamma)}\right\}$$
$$-\delta\left\{\|\mathbf{u}^{(m)}\|^2_{(H^{1-\varepsilon}(\Omega^b))^2}+\|p^{(f)}\|^2_{H^{1-\varepsilon}(\Omega^b)}+\|\psi\|^2_{H^{-\frac{1}{2}-\varepsilon}(\Gamma)}\right\},$$

where $\alpha>0$ and $\delta\geq 0$ are constants and $\varepsilon>0$ is a small parameter. ■

As is well known, Garding's inequality implies the validity of the Fredholm alternative. Hence uniqueness implies the existence. For this purpose, in order to ensure the existence of a solution of the variational equation (10.2.23), we make the following additional assumptions.

Additional assumptions:

- The homogeneous transmission problem $\mathbf{TP}_f$ with $f=0$ has only the trial solution $p^{(w)}$ and there is no traction free solution $(\mathbf{u}^m,p^{(f)})$.

- The square of the wave number, k_0^2, is not an eigenvalue of the Dirichlet problem for the negative Laplacian in Ω^b.

We remark that the first assumption is needed in order to ensure that the homogeneous transmission problem (TP$_0$) has the only trivial solution $(\mathbf{u}^{(m)}, p^{(f)}, p^{(w)}) = (\mathbf{0}, 0, 0)$, while the second additional assumption is a guarantee for the invertibility of the single-layer operator $\mathbf{V}$. (See p. 30, Hsiao and Wendland [216]). We now summarize our results in the following theorem.

Theorem 10.2

Under the additional assumptions, there exists a unique solution of the problem $\mathbf{TP}_f$ in $\left(H^1(\Omega^b)\right)^3 \times H^1(\Omega^b) \times H^{-\frac{1}{2}}(\Gamma)$. ■

The above result is essential to the approximation scheme that follows in the next section.

10.3 THE CASE OF A NON-NEWTONIAN, INTERSTITIAL FLUID

In this section we include a fluid-fluid interaction in the isotropic Biot-ype equations (10.2.10), (10.2.11) by using a Lagrangian formulation for a dissipative system of equations of the form [151] is obtained

$$\nabla \cdot \underline{\underline{\sigma}}^{(m)} = \lambda^2 \left[(1-\beta)\rho_s \mathbb{I} - \underline{\underline{r}}\right] \mathbf{u}^{(m)} - \omega^2 \underline{\underline{r}} \mathbf{u}^{(f)} - i\omega \mathbf{b} \left(\mathbf{u}^{(m)} - \mathbf{u}^{(f)}\right)$$

$$-\nabla \left(\beta p^{(f)}\right) = -\omega^2 \underline{\underline{r}} \mathbf{u}^{(m)} - \omega^2 \left(\beta \rho_f \mathbb{I} - \underline{\underline{r}}_i\right) \mathbf{u}^{(f)}$$

$$+i\omega \left(\mathbf{u}^{(m)} - \mathbf{u}^{(f)}\right) - \nabla \eta \left(\underline{\underline{e}}(\mathbf{v}^{(f)})\right) \underline{\underline{e}}\left(\mathbf{v}^{(f)}\right), \tag{10.3.1}$$

where we take the fluid-fluid viscosity to be, for example, of Carreau type [385], namely,

$$\eta\left(\underline{\underline{e}}(\mathbf{v}^{(f)})\right) := (\eta_0 - \eta_\infty)\left(1 + \lambda |\underline{\underline{e}}(\mathbf{v}^{(f)})|^2\right)^{\frac{r-2}{2}} \quad 0 < r < 2.$$

In [151] a scheme is suggested for calculating the solution to the nonlinear problem by starting with the solution to the linear problem E_u, $\mathrm{E}_p^{(f)}$, $\mathrm{E}_p^{(w)}$ with transition conditions B_1,B_2, B_3 as the 0th approximation $(\mathbf{u}_0^{(m)}, p_0^{(f)})$. Using this $p_0^{(f)}$ solve the nonlinear system in the porous medium numerically and obtain as a result $\mathbf{u}_1^{(m)}$. We now introduce corrector terms $(\delta\mathbf{u}, \delta p^{(f)})$, which are defined as

$$\mathbf{u}^{(m)} := \mathbf{u}_0^{(m)} + \delta\mathbf{u}^{(m)}, \quad p^{(f)} := p_0^{(f)} + \delta p^{(f)}. \tag{10.3.2}$$

Again we require that $(\mathbf{u}^{(m)}, p^{(f)})$ satisfy the partial differential system and the usual transmission conditions (10.3.5). This leads to a set of equations for $(\delta\mathbf{u}^{(m)}, \delta p^{(f)}, \delta p^{(w)})$, namely,

$$\nabla\cdot\left[\mathscr{C}:\underline{\underline{e}}(\delta u^{(m)}+\delta\theta^{(f)}\underline{\underline{C}}\right]+\left(\underline{\underline{p}}_{11}-\underline{\underline{p}}_{12}\underline{\underline{p}}_{22}^{-1}\underline{\underline{p}}_{12}\right)\delta u^{(m)}+$$

$$\beta\underline{\underline{p}}_{12}\underline{\underline{p}}_{22}^{-1}\nabla p_0^{(f)}+\underline{\underline{b}}\left(\delta\mathbf{v}^{(m)}-\delta\mathbf{v}^{(f)}\right)=0, \tag{10.3.3}$$

$$\nabla\left(\beta\delta p^{(f)}\right)=\omega^2\underline{\underline{r}}\delta u^{(m)}+\omega^2\left(\beta\rho_f\mathbb{I}+\underline{\underline{r}}\right)\delta\mathbf{u}^{(f)}-i\omega\underline{\underline{b}}\left(\delta\mathbf{u}^{(m)}-\delta\mathbf{u}^{(f)}\right)$$

$$-\nabla\left(\frac{\partial\eta_r}{\partial\dot{\mathbf{u}}}\delta\dot{\mathbf{u}}\right)=0. \tag{10.3.4}$$

In the water tank, the pressure satisfies a homogeneous equation

$$\Delta p^{(w)}+k_0^2p^{(w)}=0 \quad \text{in } \Omega_w$$

We next list the transmission conditions for the corrector terms

$$\left[\mathscr{C}:\underline{\underline{e}}(\mathbf{u}^{(m)})+\delta\theta^f\underline{\underline{C}}-\beta p^{(f)}\mathbb{I}\right]\cdot\mathbf{n}=-p^{(w)}, \quad \partial\Omega_b$$

$$\left[\mathscr{C}:\underline{\underline{e}}(\mathbf{u}^{(m)})+\theta^f\underline{\underline{C}}-\beta p^{(f)}\mathbb{I}\right]\cdot\mathbf{t}=0, \quad \partial\Omega_b$$

$$\rho^{(w)}\omega^2\left(\left[\mathbb{I}-\beta(\mathbb{I}+\underline{\underline{p}}_{22}^{-1}\underline{\underline{p}}_{12})\right]\mathbf{u}^{(m)}\right)$$

$$+\beta^2\rho^{(w)}\omega^2\left[\underline{\underline{p}}_{22}^{-1}\nabla p^{(f)}\right]\cdot\mathbf{n}=\frac{\partial p^{(w)}}{\partial\mathbf{n}}, \quad \partial\Omega_b \tag{10.3.5}$$

10.4 SOME TIME-DEPENDENT SOLUTIONS TO THE BIOT SYSTEM

In this section we show how a standard numerical technique based on Lubich's approach [284, 283] may be used to construct time-dependent solutions of the Biot equations. One may apply Galerkin's semi-discretization in space and convolution quadrature (CQ) in time for treating the interaction problem numerically. See, for instance [370, 213]. We will pursue this direction for the fluid-bone, interaction problem in separate communications.

As usual, the elastic strains are related to the displacements by

$$e_{ii}=\frac{\partial u_i}{\partial x_i}, \quad e_{ij}=\frac{\partial u_i}{\partial x_j}+\frac{\partial u_j}{\partial x_i}, \quad i\neq j,\ i,j=1,2,3. \tag{10.4.1}$$

Equations (10.1.1) and (10.4.1) and an argument based upon Lagrangian dynamics are shown in [28, 30, 31] to lead to the following equations of motion for the displacements and dilatations

$$\mu\Delta\mathbf{u}^{(s)}+\nabla[(\lambda+\mu)e+Q\varepsilon] \quad = \quad \frac{\partial^2}{\partial t^2}(\rho_{11}\mathbf{u}^{(s)}+\rho_{12}\mathbf{u}^{(f)}) \tag{10.4.2}$$

$$-\nabla p^{(f)} := \nabla[Qe + R\varepsilon] = \frac{\partial^2}{\partial t^2}(\rho_{12}\mathbf{u}^{(s)} + \rho_{22}\mathbf{u}^{(f)}), \tag{10.4.3}$$

$$\frac{\partial^2 \mathbf{u}^{(f)}}{\partial t^2} = -\frac{1}{\rho_{22}}\left(\rho_{12}\frac{\partial^2 \mathbf{u}^{(s)}}{\partial t^2} + \nabla p^{(f)}\right) \tag{10.4.4}$$

Substituting (10.4.4) and (10.4.3) into (10.4.2) yields

$$\mu\Delta\mathbf{u}^{(s)} + \left(\lambda + \mu - \frac{Q^2}{R}\right)\nabla\nabla\cdot\mathbf{u}^{(s)} - \frac{Q}{R}\nabla p^{(f)}$$

$$+\frac{\rho_{11}}{\rho_{22}}\nabla p^{(f)} = \left(\rho_{11} - \frac{\rho_{12}^2}{\rho_{22}}\right)\frac{\partial^2 \mathbf{u}^{(s)}}{\partial t^2}. \tag{10.4.5}$$

Here $p^{(f)}$ is the fluid pressure; whereas, ρ_{11} and ρ_{22} are density parameters for the solid and fluid, ρ_{12} is a density coupling parameter. The Biot system may be written as

$$\left(\rho_{11} - \frac{\rho_{12}^2}{\rho_{22}}\right)\frac{\partial^2 \mathbf{u}^{(s)}}{\partial t^2} - \left[\mu\Delta\mathbf{u}^{(s)} + \left\{(\lambda - \frac{Q^2}{R} + \mu\right\}\nabla\nabla\cdot\mathbf{u}^{(s)}\right]$$

$$+\left(\frac{Q}{R} - \frac{\rho_{12}}{\rho_{22}}\right)\nabla p^{(f)} = 0 \tag{10.4.6}$$

$$\rho_{22}\left(\frac{Q}{R} - \frac{\rho_{12}}{\rho_{22}}\right)\frac{\partial^2\nabla\cdot\mathbf{u}^{(s)}}{\partial t^2} + \frac{\rho_{22}}{R}\frac{\partial^2 p^{(f)}}{\partial t^2} - \Delta p^{(f)} = 0 \tag{10.4.7}$$

All parameters in the above equations are assumed to be constant. We now define the additional parameters and operators

$$\rho := (\rho_{11} - \frac{\rho_{12}^2}{\rho_{22}}) > 0, \quad \lambda' := (\lambda - \frac{Q^2}{R}) > 0, \quad \eta := (\frac{Q}{R} - \frac{\rho_{11}}{\rho_{12}}) > 0; \tag{10.4.8}$$

moreover,

$$\Delta^*\mathbf{u}^{(s)} := \mu\Delta\mathbf{u}^{(s)} + \left((\lambda' + \mu)\nabla\nabla\cdot\mathbf{u}^{(s)}\right). \tag{10.4.9}$$

Then we may rewrite the system in a more compact form as:

$$\rho\frac{\partial^2 \mathbf{u}^{(s)}}{\partial t^2} - \Delta^*\mathbf{u}^{(s)} + \eta\nabla p^{(f)} = 0, \tag{10.4.10}$$

$$\rho_{22}\eta\frac{\partial\nabla\cdot\mathbf{u}^{(s)}}{\partial t} + \frac{\rho_{22}}{R}\frac{\partial^2 p^{(f)}}{\partial t} - \Delta p^{(f)} = 0. \tag{10.4.11}$$

For utilization of the transmission condition, we need to examine the fluid outside the bone specimen, i.e. in $\Omega^{(w)} := \mathbb{R}^3 \setminus \bar{\Omega}$. In the water $\Omega^{(w)}$, we consider the barotropic flow of an inviscid and compressible fluid. Let $\mathbf{v}^{(w)} := \mathbf{v}^{(w)}(\mathbf{x},t)$ be the velocity field and $\rho^{(w)} := \rho^{(w)}(\mathbf{x},t)$ and $p^{(w)} := p^{(w)}(\mathbf{x},t)$ be respectively the density and pressure of the fluid encompassing the bone sample. We assume that $\mathbf{v}^{(w)}$, $p^{(w)}$ and $\rho^{(w)}$ are small perturbations from the static state $\left\{\mathbf{v}^{(w)} = 0, p^{(w)} = \text{constant}, \rho^{(w)} = \text{constant}\right\}$. In this case, the governing fluid equations may be linearized to yield the linear Euler equation

$$\rho^{(w)}\frac{\partial \mathbf{v}^{(w)}}{\partial t} + \nabla p^{(w)} = 0,$$

the linear equation of continuity

$$\frac{\partial \rho^{(w)}}{\partial t} + \rho_0^{(w)} \nabla \cdot \mathbf{v}^{(w)} = 0$$

and the linearized state equation

$$p^{(w)} = c^2 \rho^{(w)}$$

in $\Omega^{(w)} \times (0,T]$, where c is the sound of speed defined by $c^2 = f'(\rho_0^{(w)})$ and $f(\rho^{(w)})$ is a function depending on the properties of the fluid [398, 212]. It is well known that for an irrotational flow, we may introduce a velocity potential [398] $\phi := \phi(\mathbf{x},t)$ such that

$$\mathbf{v}^{(w)} = -\nabla \phi, \quad \text{and } p^{(w)} = \rho_0^{(w)} \frac{\partial \phi}{\partial t} \tag{10.4.12}$$

From this it follows that the velocity potential satisfies the wave equation

$$\frac{\partial^2 \phi}{\partial t^2} - c^2 \Delta \phi = 0, \quad (\mathbf{x},t) \in \Omega^{(w)} \times (0,T]. \tag{10.4.13}$$

The time-dependent problem can be formulated as an initial-boundary-transmission problem consisting of the partial differential equations

$$(E_1) \qquad \rho \frac{\partial \mathbf{u}^{(s)}}{\partial t} - \Delta^* \mathbf{u}^{(s)} + \eta \nabla p^{(f)} = 0, \quad (\mathbf{x},t) \in \Omega \times (0,T]$$

$$(E_2) \qquad \rho_{22}\eta \frac{\partial^2(\nabla \cdot \mathbf{u}^{(s)})}{\partial t} + \frac{\rho_{22}}{R}\frac{\partial^2 p^{(f)}}{\partial t} - \Delta p^{(f)} = 0, \quad (\mathbf{x},t) \in \Omega \times (0,T]$$

$$(E_3) \qquad \frac{\partial^2 \phi}{\partial t^2} - c^2 \Delta \phi = 0, \quad (\mathbf{x},t) \in \Omega^{(w)} \times (0,T]$$

together with the transmission conditions:

Conservation of the normal component of stress

$$(TC_1) \quad \left(\underline{\underline{\sigma}}(\mathbf{u}^{(s)}) - \eta\, p^{(f)} \mathbb{I}\right)\mathbf{n} = -\rho_0^{(w)} \left(\frac{\partial \phi}{\partial t} + \frac{\partial \phi^{inc}}{\partial t}\right)\mathbf{n}, \quad (\mathbf{x},t) \in \Gamma \times (0,T]$$

Conservation of flux

$$(TC_2) \qquad \frac{\partial \mathbf{v}^{(w)}}{\partial t} \cdot \mathbf{n} = -\left(\frac{\partial \phi}{\partial \mathbf{n}} + \frac{\partial \phi^{inc}}{\partial \mathbf{n}}\right), \quad (\mathbf{x},t) \in \Gamma \times (0,T]$$

Continuity of pore pressure

$$(TC_3) \qquad p^{(f)} = \beta p^{(w)}, \quad (\mathbf{x},t) \in \Gamma \times (0,T]$$

and the initial conditions

$$(I_1) \qquad \mathbf{u}^{(s)}(\mathbf{x},0) = \frac{\partial \mathbf{u}^{(s)}}{\partial t} = 0, \quad \mathbf{x} \in \Omega$$

$$(I_2) \qquad p^{(f)}(\mathbf{x},0) = \frac{\partial p^{(f)}}{\partial t} = 0, \quad \mathbf{x} \in \Omega$$

$$(I_3) \qquad \phi(\mathbf{x},0) = \frac{\partial \phi}{\partial t}(\mathbf{x},0) = 0, \quad \mathbf{x} \in \Omega^{(w)}.$$

Here ϕ^{inc} and $\partial\phi^{inc}/\partial t$ are prescribed incident fields.

We are going to solve the problem using Lubisch's approach based on the Laplace transform methodology. To this end we introduce the notation

$$\mathbf{U}^{(s)}(\mathbf{x},s) = \int_0^\infty e^{-st}\mathbf{u}^{(s)}(\mathbf{x},t)\,dt, \quad \mathbf{U}^{(f)}(\mathbf{x},s) = \int_0^\infty e^{-st}\mathbf{u}^{(f)}(\mathbf{x},t)\,dt$$

$$P^{(f)}(\mathbf{x},s) = \int_0^\infty e^{-st}p^{(f)}(\mathbf{x},t)\,dt, \quad P^{(w)}(\mathbf{x},s) = \int_0^\infty e^{-st}p^{(w)}(\mathbf{x},t)\,dt$$

$$\Phi(\mathbf{x},s) = \int_0^\infty e^{-st}\phi(\mathbf{x},t)\,dt,$$

etc. The transformed equations now become

$$(\hat{E}_1) \qquad -\Delta^*\mathbf{U}^{(s)} + \rho s^2\mathbf{U}^{(s)} + \eta\nabla P^{(f)} = 0, \quad \text{in} \quad \Omega,$$

$$(\hat{E}_2) \qquad \rho_{22}\eta s^2(\nabla\cdot\mathbf{U}^{(s)}) - \Delta P^{(f)} + \frac{\rho_{22}}{R}s^2P^{(f)} = 0, \quad \text{in} \quad \Omega,$$

$$(\hat{E}_3) \qquad -\Delta\Phi + \frac{s^2}{c^2}\Phi = 0, \quad \text{in} \quad \Omega^{(w)}$$

$$(\hat{TC}_1) \qquad \left(\underline{\underline{\sigma}}(\mathbf{U}^{(s)}) - \eta P^{(f)}\mathbb{I}\right)\cdot\mathbf{n} = -s\rho_0^{w)}\left(\Phi + \Phi^{inc}\right)\mathbf{n}$$

$$(\hat{TC}_2) \qquad \left((1-\beta) + \beta\frac{\rho_{12}}{\rho_{22}}\right)s\mathbf{U}^{(s)}\cdot\mathbf{n} = -\left(\frac{\partial\Phi}{\partial n} + \frac{\partial\Phi^{inc}}{\partial n}\right)$$

$$(\hat{TC}_3). \qquad P^{(f)} = \beta P^{(w)}$$

10.4.1 THE NONLOCAL BOUNDARY VALUE PROBLEM

The next step is to reduce the above transmission problem consisting of $(\hat{E}_1)-(\hat{TC}_3)$ to a nonlocal boundary value problem in Ω. We begin with $(\hat{E}_3)$

$$-\Delta\Phi+\frac{s^2}{c^2}\Phi=0 \quad \text{in } \Omega^c \tag{10.4.1}$$

by seeking a solution of (10.4.1) in the form of simple and double layer potentials, i.e.

$$\Phi=\mathscr{D}(\psi)-\mathscr{S}(\zeta), \quad \text{in } \Omega^c, \tag{10.4.2}$$

where $\psi=\Phi|_\Gamma$ and $\zeta=\frac{\partial\Phi}{\partial n}|_\Gamma$ are the Cauchy data of the solution Φ of (10.4.1) on $\Gamma:=\partial\Omega$. Here

$$\mathscr{D}(\psi)(\mathbf{x}) := \int_\Gamma \frac{\partial}{\partial n_y}E_{s/c}(\mathbf{x};\mathbf{y})\psi(\mathbf{y})\,d\Gamma_y, \quad \mathbf{x}\in\mathbb{R}^3\setminus\Gamma \tag{10.4.3}$$

$$\mathscr{S}(\psi)(\mathbf{x}) := \int_\Gamma E_{s/c}(\mathbf{x};\mathbf{y})\zeta(\mathbf{y})\,d\Gamma_y, \quad \mathbf{x}\in\mathbb{R}^3\setminus\Gamma, \tag{10.4.4}$$

where

$$E_{s/c}(\mathbf{x};\mathbf{y}) := \frac{e^{-s/c}(\mathbf{x}-\mathbf{y})}{\|\mathbf{x}-\mathbf{y}\|} \tag{10.4.5}$$

is the fundamental solution of the shielded Coulomb potential operator, known also as the Yukowa potential operator [312]. The Cauchy data ψ and ζ satisfy the operator equation [216, 215]

$$\begin{pmatrix}\psi\\ \zeta\end{pmatrix}=\begin{pmatrix}\frac{1}{2}\mathbb{I}+\mathscr{K}(s) & -\mathscr{V}(s)\\ -\mathscr{W}(s) & (\frac{1}{2}\mathbb{I}+\mathscr{K}(s))'\end{pmatrix}\begin{pmatrix}\psi\\ \zeta\end{pmatrix}, \quad \mathbf{x}\in\Gamma \tag{10.4.6}$$

Here $\mathscr{V}$, $\mathscr{K}$, $\mathscr{K}'$ and $\mathscr{W}$ are the four basic boundary integral operators familiar from potential theory [216, 215]

$$\left\{\begin{aligned}
\mathscr{V}(s)\zeta(\mathbf{x}) &:= \int_\Gamma E_{s/c}(\mathbf{x},\mathbf{y})\zeta(\mathbf{y})d\Gamma_y, \quad \mathbf{x}\in\Gamma\\
\mathscr{K}(s)\psi(\mathbf{x}) &:= \int_\Gamma \frac{\partial}{\partial n_y}E_{s/c}(\mathbf{x},\mathbf{y})\psi(\mathbf{y})d\Gamma_y, \quad \mathbf{x}\in\Gamma\\
\mathscr{K}'(s)\psi(\mathbf{x}) &:= \int_\Gamma \frac{\partial}{\partial n_x}E_{s/c}(\mathbf{x},\mathbf{y})\zeta(\mathbf{y})d\Gamma_y, \quad \mathbf{x}\in\Gamma\\
\mathscr{W}(s)\psi(\mathbf{x}) &:= -\frac{\partial}{\partial n_x}\int_\Gamma \frac{\partial}{\partial n_y}E_{s/c}(\mathbf{x},\mathbf{y})\psi(\mathbf{y})d\Gamma_y, \quad \mathbf{x}\in\Gamma
\end{aligned}\right. \tag{10.4.7}$$

From the second equation of (10.4.6), we see that

$$\frac{\partial\Phi}{\partial n}|_\Gamma =: \zeta = -\mathscr{W}(s)\psi+(\frac{1}{2}\mathbb{I}+\mathscr{K}(s))'\zeta$$

and the transmission condition $(\hat{TC}_2)$ then leads to the BIE

$$-s\,q(\beta)\,\mathbf{U}^{(s)}\cdot\mathbf{n}+\mathscr{W}(s)\psi-\left(\frac{1}{2}\mathbb{I}-\mathscr{K}(s)\right)'\zeta=\frac{\partial\psi^{inc}}{\partial n}\quad\text{on}\quad\Gamma;\qquad(10.4.8)$$

while ψ and ζ are required to satisfy the first boundary integral equation in (10.4.6) as a constraint,

$$\left(\frac{1}{2}\mathbb{I}-\mathscr{K}(s)\right)\psi+\mathscr{V}(s)\zeta=0,\quad\text{on}\quad\Gamma.$$

To simplify the representation, the coefficient in $(\hat{TC}_2)$ has been denoted by

$$q(\beta):=\left((1-\beta)+\beta\frac{\rho_{12}}{\rho_{22}}\right)$$

in the BIE (10.4.8). We note that $q(\beta)>0$ for $\beta\in[0,1]$.

With the Cauchy data ψ and ζ as new unknowns, the partial differential equation (10.4.1) in $\Omega^{(w)}$ is eliminated. This reformulation leads to a nonlocal boundary problem in Ω for the unknowns $\left\{\mathbf{U}^{(s)},P^{(w)},\psi,\ \zeta\right\}$, consisting of the partial differential equations

$$-\Delta^*\mathbf{U}^{(s)}+\rho\,s^2\mathbf{U}^{(s)}+\eta\nabla P^{(f)}=0,\qquad(10.4.9)$$

$$\rho_{22}\,\eta\,s\nabla\cdot\mathbf{U}^{(s)})-\Delta P^{(f)}+s^2\frac{\rho_{22}}{R}P^{(f)},=0,\qquad(10.4.10)$$

and the boundary integral equations

$$-s^2\,q(\beta)\mathbf{U}^{(s)}\cdot\mathbf{n}+\mathscr{W}(s)\psi-\left(\frac{1}{2}\mathbb{I}-\mathscr{K}(s)\right)'\zeta=\frac{\partial\Phi^{inc}}{\partial n}\quad\text{on}\,\Gamma;\qquad(10.4.11)$$

$$\left(\frac{1}{2}\mathbb{I}-\mathscr{K}(s)\right)\psi+\mathscr{V}(s)\zeta=0,\quad\text{on}\,\Gamma,\qquad(10.4.12)$$

together with the transmission conditions

$$\left(\underline{\underline{\sigma}}(\mathbf{U}^{(s)})-\eta\,P^{(f)}\,\mathbb{I}\right)\mathbf{n}=-s\rho^{(w)}w_0\left(\Phi+\Phi^{inc}\right)\mathbf{n}\qquad(10.4.13)$$

and the condition $(\hat{TC}_3)$, from which we tacitly induce to an alternative condition

$$\frac{\partial}{\partial n}P^{(f)}=\beta s\rho^{(w)}w_0\frac{\partial\Phi^{inc}}{\partial n}.\qquad(10.4.14)$$

To be more precise, let us first consider the unknowns $\left(\mathbf{U}^{(s)},P^{(f)}\right)\in\mathbf{H}^1(\Omega)\times H^1(\Omega)$. Multiplying (10.4.9) by the test function $\mathbf{U}\in\mathbf{H}^1(\Omega)$ and integrating by parts, we obtain the weak formulation of (10.4.9), namely,

$$\mathscr{A}\left(\mathbf{U}^{(s)},\mathbf{U};s\right)-\int_\Gamma\underline{\underline{\sigma}}(\mathbf{U}^{(s)},P^{(f)})\mathbf{n}\cdot\gamma^-\overline{\mathbf{U}}d\Gamma-\eta\int_\Omega P^{(f)}\nabla\cdot\overline{\mathbf{U}}d\mathbf{x}=0,\qquad(10.4.15)$$

where

$$\mathscr{A}\left(\mathbf{U}^{(s)},\mathbf{U};s\right) := \int_{\Omega}\left(\underline{\underline{\sigma}}(\mathbf{U}^{(s)}) : \mathbf{u}^{\varepsilon}(\overline{\mathbf{U}}) + s^2\rho\mathbf{U}^{(s)}\cdot\overline{\mathbf{U}}\right) d\mathbf{x} \tag{10.4.16}$$

is the sesquilinear form with

$$\underline{\underline{\sigma}}(\mathbf{U}^{(s)}) := \lambda'\nabla\mathbf{U}^{(s)}\mathbb{I}+2\mu\,\varepsilon(\mathbf{U}^{(s)}),\quad \varepsilon(\mathbf{U}^{(s)}) := \frac{1}{2}\left(\nabla\mathbf{U}^{(s)}+\nabla\mathbf{U}^{(s)^{T}}\right). \tag{10.4.17}$$

Then from the transmission condition (10.4.13), we obtain

$$\mathscr{A}(\mathbf{U}^{(s)},\mathbf{U};s) - \eta\left(P^{(f)},\nabla\cdot\overline{\mathbf{U}}\right)_{\Omega} + \rho_0^{(w)}s\langle\Phi\mathbf{n},\gamma^{-}\overline{\mathbf{U}}\rangle_{\Gamma} = -\rho^{(w)}\rho_0^{(w)}s\langle\Phi^{inc}\mathbf{n},\gamma^{-}\overline{\mathbf{U}}\rangle_{\Gamma} \tag{10.4.18}$$

Similarly, multiplying (10.4.10) by the test function $\overline{P}$ yields

$$\rho_{22}\,\eta\,s^2(\nabla\cdot\mathbf{U}^{(s)},\overline{P})_{\Omega} + \mathscr{B}(P^{(f)},\overline{P};s) = 0, \tag{10.4.19}$$

where the sesqiulinear form $\mathscr{B}(P^{(f)},\overline{P};s)$ is defined by

$$\mathscr{B}(P^{(f)},\overline{P};s) := \int_{\Omega}\left(\nabla P^{(f)}\cdot\nabla\overline{P}+\frac{\rho_{22}}{R}s^2P^{(f)}\overline{P}\right) d\mathbf{x}. \tag{10.4.20}$$

Now for the unknowns $(\psi,\zeta)\in H^{1/2}(\Gamma)\times H^{-1/2}(\Gamma)$, we proceed in the same manner. Multiplying (10.4.11) and (10.4.12) by the test functions $\bar{\varphi}$ and $\bar{\xi}$, respectively, we obtain

$$-sq(\beta)\langle\mathbf{U}^{(s)}\cdot\mathbf{n},\bar{\varphi}\rangle_{\Gamma} + \langle\mathscr{W}(s)\psi,\bar{\varphi}\rangle_{\Gamma} - \langle\left(\frac{1}{2}\mathbb{I}-\mathscr{K}(s)\right)'\zeta,\bar{\varphi}\rangle_{\Gamma} = \langle\frac{\partial\Phi^{inc}}{\partial n},\bar{\varphi}\rangle_{\Gamma}, \tag{10.4.21}$$

$$\langle\left(\frac{1}{2}\mathbb{I}-K(s)\right)\psi,\bar{\xi}\rangle_{\Gamma} + \langle V(s)\zeta,\,\bar{\xi}\rangle_{\Gamma} = 0 \tag{10.4.22}$$

for $\bar{\varphi}\in H^{1/2}(\Gamma)$ and $\bar{\xi}\in H^{-1/2}(\Gamma)$. Finally we define the operators

$$\mathfrak{A}_s : \mathbf{H}^1(\Omega)\to\left(\mathbf{H}^1(\Omega)\right)',\quad \mathfrak{B}_s : H^1(\Omega)\to\left(H^1(\Omega)\right)' \tag{10.4.23}$$

by their sesquilinear forms:

$$\langle\mathfrak{A}_s\mathbf{U}^{(s)},\mathbf{U}\rangle_{\Omega} := \mathscr{A}\left(\mathbf{U}^{(s)},\mathbf{U};\mathbf{s}\right),\quad \langle\mathfrak{B}_sP^{(f)},P\rangle_{\Omega} := \mathscr{B}\left(P^{(f)},P;s\right) \tag{10.4.24}$$

Then from (10.4.18), (10.4.20), (10.4.21) and (10.4.22), the non-local boundary problem may be formulated as a system of operator equations:
Let $\mathbf{X} := \mathbf{H}^1(\Omega)\times H^1(\Omega)\times H^{1/2}(\Omega)\times H^{-1/2}(\Omega)$. Then given data $(d_1,d_2,d_3,d_4)\in\mathbf{X}'$, find $\left(\mathbf{U}^{(s)},P^{(f)},\psi,\zeta\right)$ $in\,\mathbf{X}$ satisfying

$$\mathcal{A}\begin{pmatrix}\mathbf{U}^{(s)}\\ P^{(f)}\\ \psi\\ \zeta\end{pmatrix} := \begin{pmatrix}\mathfrak{A}_s & -\eta(\nabla\cdot)' & s\rho_0^{(w)}\gamma^{-'}\mathbf{n} & 0\\ s^2\rho_{22}\,\eta\,\nabla\cdot & \mathfrak{B}_s & 0 & 0\\ -sq(\beta)\,\mathbf{n}^T\gamma^{-} & 0 & \mathscr{W}(s) & -\left(\frac{1}{2}\mathbb{I}-\mathscr{K}(s)\right)'\\ 0 & 0 & \left(\frac{1}{2}\mathbb{I}-\mathscr{K}(s)\right) & \mathscr{V}(s)\end{pmatrix}$$

$$\begin{pmatrix} \mathbf{U}^{(s)} \\ P^{(f)} \\ \psi \\ \zeta \end{pmatrix} = \begin{pmatrix} d_1 \\ d_2 \\ d_3 \\ d_4 \end{pmatrix}, \tag{10.4.25}$$

where the data (d_1, d_2, d_3, d_4) are given by

$$\begin{aligned} d_1 &= -s\rho_0^{(w)} \left(\gamma^+ \Phi^{inc} \mathbf{n}\right), \\ d_2 &= \beta\, s\, \rho_0^{(w)} \left(\partial_n^+ \Phi^{inc}\right), \\ d_3 &= \partial_n^+ \Phi^{inc}, \\ d_4 &= 0 \end{aligned} \tag{10.4.26}$$

Here, and in the sequel, $\gamma^{\pm}$ represents the trace operator on Γ from inside (-) and outside (+) of Ω and $\gamma^{\pm\prime}$ denotes the transpose of γ.; while $\partial_n^{\pm}$ represents limits of the corresponding normal derivatives. Our intention is now to show that (10.4.25) admits a unique solution $\left(\mathbf{U}^{(s)}, P^{(f)}, \psi, \zeta\right) \in \mathbf{X}$.

It is noteworthy to mention, that for the time-independent case, one may simply consider the weak formulation of the corresponding nonlocal boundary problem, since the boundary integral operators involved do not depend upon the complex parameter s. However, for the time-dependent case, such as (10.4.25), one cannot analyze the weak form of (10.4.25) directly. As will be seen, we need to consider the weak form of (10.4.25) indirectly. We will pursue the idea in the next section.

10.4.2 VARIATIONAL FORMULATION

We proceed much in the same way as we did for the isotropic case. For this we need the following energy norms:

$$|||\mathbf{U}^{(s)}|||^2_{|s|,\Omega} := \left(\underline{\underline{\sigma}}(\mathbf{U}^{(s)}), \varepsilon(\mathbf{U}^{(s)})\right)_{\Omega} + \rho \|s\mathbf{U}^{(s)}\|^2_{\Omega}, \quad \mathbf{U}^{(s)} \in \mathbf{H}^1(\Omega)$$

$$|||P^{(f)}|||^2_{|s|,\Omega} := \|\nabla P^{(f)}\|^2_{\Omega} + \frac{\rho_{22}}{R}\|sP^{(f)}\|^2_{\Omega}, \quad P^{(f)} \in H^1(\Omega)$$

$$|||\Phi|||_{|s|,\Omega^c} := \|\nabla\Phi\|^2_{\Omega^c} + c^{-2}\|s\Phi\|^2_{\Omega^c}, \quad \Phi \in H^1(\Omega^c)$$

For the complex Laplace parameter $s \in \mathbb{C}^+$, we will denote

$$\sigma := \Re e\, s > 0, \quad \underline{\sigma} := \min\{1, \sigma\}$$

and will make use of the following equivalence relations for the norms

$$\underline{\sigma}\, |||\mathbf{U}^{(s)}|||_{1,\Omega} \le |||\mathbf{U}^{(s)}|||_{|s|,\Omega} \le \frac{|s|}{\underline{\sigma}} |||\mathbf{U}^{(s)}|||_{1,\Omega}$$

and similar relations hold for norms of $P^{(f)}$ and Φ, which may be obtained from the inequalities

$$\min\{1,\sigma\} \le \min\{1,|s|\} \text{ and } \max\ \{1,|s|\} \times \min\{1,\sigma\} \le |s|, \forall s \in \mathbb{C}^+$$

We remark that the norms $|||P^{(f)}|||_{1,\Omega}$ and $|||\Phi|||_{H^1(\Omega)}$ are equivalent to $\|P^{(f)}\|_{H^1(\Omega)}$ and $\|\Phi\|_{H^1(\Omega)}$ respectively and so is energy norm $|||\mathbf{U}^{(s)}|||_{1,\Omega}$ equivalent to the $\mathbf{H}^1(\Omega)$ norm of $\mathbf{U}^{(s)}$ by the second Korn equality. Now suppose that $\left(\mathbf{U}^{(s)}, P^{(f)}, \psi, \zeta\right)$ *in* $\mathbf{X}$ is a solution of (10.4.25). Let

$$V(s) := \mathscr{D}(s)\psi - \mathscr{S}(s)\zeta \quad \text{in} \quad \mathbb{R}^3 \setminus \Gamma. \tag{10.4.1}$$

Then $V \in H^1(\mathbb{R}^3 \setminus \Gamma)$ is a solution of the equation

$$-\Delta V + \frac{s^2}{c^2} V = 0 \quad \text{in} \quad \mathbb{R}^3 \setminus \Gamma. \tag{10.4.2}$$

By the standard argument in potential theory (see, e.g. [216]), it can be shown that the following jump relations across Γ hold:

$$[\gamma V] := \gamma^+ V - \gamma^- V = \psi \in H^{\frac{1}{2}}(\Gamma)$$

$$[\partial_n V] := \partial_n^+ V - \partial_n^- V = \zeta \in H^{-\frac{1}{2}}(\Gamma),$$

It follows from (10.4.25) that

$$\begin{cases} \mathfrak{A}_s \mathbf{U}^{(s)} - \eta(\operatorname{div})' P^{(f)} + s\rho_0^{(w)} \gamma^{-\prime} [\gamma V] \mathbf{n} = d_1 \quad \text{in} \quad \Omega \\ s^2 \rho_{22}\, \eta\, (\operatorname{div} \mathbf{U}^{(s)}) + \mathfrak{B}_s P^{(f)} = d_2 \quad \text{in} \quad \Omega \\ -s q(\beta)\, \mathbf{n} \cdot \gamma^- \mathbf{U}^{(s)} + \mathscr{W}(s)[\gamma V] - \left(\frac{1}{2}\mathbb{I} - \mathscr{K}(s)\right)' [\partial_n V] = d_3 \quad \text{on} \quad \Gamma \\ (\frac{1}{2}\mathbb{I} - \mathscr{K}(s))[\gamma V] + \mathscr{V}(s)[\partial_n V] = d_4. \end{cases} \tag{10.4.3}$$

The 4th equation in (10.4.3), denoted $(10.4.3)_4$, implies that

$$-\gamma^- V = 0,$$

since $(\frac{1}{2}\mathbb{I} - \mathscr{K}(s))(\gamma^+ V) + \mathscr{V}(s)(\partial_n^+ V) = d_4$ from (10.4.25). This means that V given by (10.4.1) is a solution of (10.4.2) in Ω with homogeneous Dirichlet boundary condition on Γ. By the uniqueness of the solution to the interior Dirichlet problem of (10.4.2), we conclude that $V \equiv 0$ in $\overline{\Omega}$. Consequently, we have

$$[\gamma V] = \gamma^+ V = \psi \quad \text{and} \quad [\partial_n V] = \partial_n^+ V = \xi \tag{10.4.4}$$

and the system (10.4.3) reduces to the following simple form

$$\begin{cases} \mathfrak{A}_s \mathbf{U}^{(s)} - \eta\, (\operatorname{div})' P^{(f)} + s\rho_0^{(w)} \gamma^{-\prime} (\gamma^+ V) \mathbf{n} = d_1 \quad \text{in} \quad \Omega \\ s^2 \rho_{22}\, \eta\, (\operatorname{div} \mathbf{U}^{(s)}) + \mathfrak{B}_s P^{(f)} = d_2 \quad \text{in} \quad \Omega \\ -s q(\beta)\, (\mathbf{n} \cdot \gamma^- \mathbf{U}^{(s)}) - (\partial_n^+ V) = d_3 \quad \text{on} \quad \Gamma. \end{cases} \tag{10.4.5}$$

To derive a variational formulation of (10.4.5), let us first consider the third equation $(10.4.5)_3$ in (10.4.5). Multiplying the last term in $(10.4.5)_3$ by the trace of test function $Z \in H^1(\Omega^{(w)})$, we see that

$$-\langle \partial_n^+ V, \gamma^+ \overline{Z}\rangle_\Gamma = \int_{\Omega^{(w)}} \left(\nabla V \cdot \overline{\nabla Z} + (\frac{s}{c})^2 V \overline{Z} \right) d\mathbf{x}$$
$$=: \mathscr{C}(V, \overline{Z}; s) \tag{10.4.6}$$

In the same way, we may define the operator $\mathfrak{C}_s$

$$\mathfrak{C}_s : \mathbf{H}^1(\Omega^{(w)}) \to \mathbf{H}^1(\Omega^{(w)})', \tag{10.4.7}$$

by the sesquilinear form:

$$(\mathfrak{C}_s V,\, Z)_{\Omega^{(w)}} := \mathscr{C}(V, Z; s). \tag{10.4.8}$$

Together with the operators $\mathfrak{A}_s$ and $\mathfrak{B}_s$, we arrive at the following variational formulation: Find $(\mathbf{U}^{(s)}, P^{(f)}, V) \in \mathbf{H} := \mathbf{H}^1(\Omega) \times H^1(\Omega) \times H^1(\Omega^{(w)})$ for given $(d_1, d_2, d_3) \in \mathbf{H}'$ such that

$$\begin{cases} \frac{\bar{s}}{|s|} \left\{ \left(\mathfrak{A}_s \mathbf{U}^{(s)}, \overline{\mathbf{U}} \right)_\Omega - \eta \left(P^{(f)}, \nabla \cdot \overline{\mathbf{U}} \right)_\Omega + s \rho_0^{(w)} \left\langle (\gamma^+ V)\mathbf{n}, \gamma^- \overline{\mathbf{U}} \right\rangle_\Gamma \right\} = \frac{\bar{s}}{|s|} \left\{ (d_1, \overline{\mathbf{U}})_\Omega \right\}, \\ \frac{\bar{s}}{|s|^3} \left\{ s^2 \eta \left(\nabla \cdot \mathbf{U}^{(s)}, \overline{P} \right)_\Omega + \frac{1}{\rho_{22}} \left(\mathfrak{B}_s P^{(f)}, \overline{P} \right)_\Omega \right\} = \frac{\bar{s}}{|s|^3} \frac{1}{\rho_{22}} \left\{ (d_2, \overline{P})_\Omega \right\}, \\ \frac{\bar{s}}{|s|} \left\{ -s \rho_0^{(w)} \left\langle \mathbf{n} \cdot \gamma^- \mathbf{U}^{(s)}, \gamma^+ \overline{Z} \right\rangle_\Gamma + a(\beta) \left(\mathfrak{C}_s V, \overline{Z} \right)_{\Omega^{(w)}} \right\} = \frac{\bar{s}}{|s|} a(\beta) \left\{ \left\langle d_3, \gamma^+ \overline{Z} \right\rangle_\Gamma \right\}, \end{cases} \tag{10.4.9}$$

for all $(\mathbf{U}, V, Z) \in \mathbf{H}$, where $a(\beta) := \rho_0^{(w)}/q(\beta)$. We notice that in the formulation (10.4.9), each of the equations in (10.4.5) has been multiplied by appropriate weight factors. It is necessary and will be transparent later.

10.4.3 EXISTENCE AND UNIQUENESS

We remark that this variational problem (10.4.9) is equivalent to the nonlocal problem (10.4.25), from which we have the following basic results.

Theorem 10.3

The variational problem (10.4.9) has a unique solution $\left\{ \mathbf{U}^{(s)}, P^{(f)}, V \right\} \in \mathbf{H}$ for given $(d_1, d_2, d_3) \in \mathbf{H}'$. Moreover, the following estimate.

$$|||(\mathbf{U}^{(s)}, P^{(f)}, V)|||_{\mathbf{H}} \leq c_0 \frac{|s|^3}{\sigma \underline{\sigma}^6} ||(d_2, d_2, d_3)||_{\mathbf{H}'}, \tag{10.4.1}$$

where c_0 is a constant depending only on the physical parameters $\beta, \rho_0^{(w)}, \eta, \rho_{22}$ and where $\mathbf{H}' := \mathbf{H}^1(\Omega)' \times H^1(\Omega)' \times H^1(\Omega^{(w)})'$. ■

Proof

From the system of equations in (10.4.9), we notice that

$$\Re e \left\{ \frac{\bar{s}}{|s|} \left(-\eta \left(P^{(f)}, \nabla \cdot \overline{\mathbf{U}^{(s)}} \right)_\Omega + s\rho_0^{(w)} \left\langle (\gamma^+ V)\mathbf{n}, \gamma^- \overline{\mathbf{U}^{(s)}} \right\rangle_\Gamma \right) \right.$$
$$\left. + \frac{\bar{s}}{|s|^3} \left(s^2 \eta \left(\nabla \cdot \mathbf{U}^{(s)}, \overline{P^{(f)}} \right)_\Omega \right) + \frac{\bar{s}}{|s|} \left(-s\rho_0^{(w)} \left\langle \mathbf{n} \cdot \gamma^- \mathbf{U}^{(s)}, \gamma^+ \overline{V} \right\rangle_\Gamma \right) \right\} = 0.$$

Consequently, from (10.4.9) we have

$$\Re e \left\{ \frac{\bar{s}}{|s|} \left(\mathfrak{A}_s \mathbf{U}^{(s)}, \overline{\mathbf{U}^{(s)}} \right)_\Omega + \frac{\bar{s}}{\rho_{22}|s|^3} \left(\mathfrak{B}_s P^{(f)}, \overline{P^{(f)}} \right)_\Omega + \frac{\bar{s}}{|s|} a(\beta) \left(\mathfrak{C}_s, V, \overline{V} \right)_{\Omega^{(w)}} \right\}$$
$$= \Re e \left\{ \frac{\bar{s}}{|s|} \left(d_1, \overline{\mathbf{U}^{(s)}} \right)_\Omega + \frac{\bar{s}}{\rho_{22}|s|^3} \left(d_2, P^{\bar{(f)}} \right)_\Omega + \frac{\bar{s}}{|s|} a(\beta) \langle d_3, \gamma^+ \overline{V} \rangle_{\Gamma^{(w)}} \right\} \tag{10.4.2}$$

Now let us examine each term on the LHS of (10.4.2): For the first term, we see that

$$\frac{\bar{s}}{|s|} \left(\mathfrak{A}_s \mathbf{U}^{(s)}, \overline{\mathbf{U}^{(s)}} \right) := \frac{\bar{s}}{|s|} \left\{ \left(\underline{\underline{\sigma}}(\mathbf{U}^{(s)}), \varepsilon(\overline{\mathbf{U}^{(s)}}) \right)_\Omega + s^2 \rho \|\mathbf{U}^{(s)}\| \right\}$$
$$= \frac{\bar{s}}{|s|} \left\{ \left(\underline{\underline{\sigma}}(\mathbf{U}^{(s)}), \varepsilon(\overline{\mathbf{U}^{(s)}}) \right)_\Omega + \frac{\bar{s}}{|s|} \|s\mathbf{U}^{(s)}\|_\Omega^2 \right\}.$$

Then,

$$\Re e \left\{ \frac{\bar{s}}{|s|} \left(\mathfrak{A}_s \mathbf{U}^{(s)}, \overline{\mathbf{U}^{(s)}} \right) \right\} = \frac{\sigma}{|s|} |||\mathbf{U}^{(s)}|||_{|s|,\Omega}^2 \tag{10.4.3}$$

Similarly, for the second term, we have

$$\frac{\bar{s}}{|s|^3} \frac{1}{\rho_{22}} \left(\mathfrak{B}_s P^{(f)}, \overline{P^{(f)}} \right)_\Omega = \frac{\bar{s}}{|s|^3} \frac{1}{\rho_{22}} \left\{ \|\nabla P^{(f)}\|_\Omega^2 + \frac{\rho_{22}}{R} s^2 \|P^{(f)}\|_\Omega^2 \right\}$$
$$= \frac{1}{\rho_{22}} \left\{ \frac{\bar{s}}{|s|^3} \|\nabla P^{(f)}\|_\Omega^2 + \frac{s}{|s|^3} \frac{\rho_{22}}{R} \|sP^{(f)}\|_\Omega^2 \right\}$$

Thus,

$$\Re e \left\{ \frac{\bar{s}}{|s|^3} \frac{1}{\rho_{22}} \left(\mathfrak{B}_s P^{(f)}, \overline{P^{(f)}} \right)_\Omega \right\} = \frac{\sigma}{|s|^3} \frac{1}{\rho_{22}} |||P^{(f)}|||_{|s|,\Omega}^2 \tag{10.4.4}$$

For the third term, we obtain similarly as the first term, namely,

$$\Re e \left\{ \frac{\bar{s}}{|s|} a(\beta) \left(\mathfrak{C}_s, V, \overline{V} \right)_{\Omega^{(w)}} \right\} = \frac{\sigma}{|s|} a(\beta) \, |||V|||_{|s|,\Omega^{(w)}}^2 \tag{10.4.5}$$

Therefore, combining (10.4.3)- (10.4.5) and substituting into (10.4.2) yields

$$\begin{aligned}
&\frac{\sigma}{|s|}|||\mathbf{U}^{(s)}|||^2_{|s|,\Omega}+\frac{\sigma}{|s|^3}\frac{1}{\rho_{22}}|||P^{(f)}|||^2_{|s|,\Omega}+\frac{\sigma}{|s|}a(\beta)\,|||V|||^2_{|s|,\Omega^c}\\
&\quad=\Re e\left\{\frac{\bar{s}}{|s|}\left(d_1,\overline{\mathbf{U}^{(s)}}\right)_\Omega+\frac{\bar{s}}{|s|^3}\frac{1}{\rho_{22}}\left(d_2,\overline{P^{(f)}}\right)_\Omega+\frac{\bar{s}}{|s|}a(\beta)\left\langle d_3,\overline{\gamma^+V}\right\rangle_\Gamma\right\}\\
&\quad\le\left|\left(d_1,\overline{\mathbf{U}^{(s)}}\right)_\Omega+\frac{1}{|s|^2}\frac{1}{\rho_{22}}\left(d_2,\overline{P^{(f)}}\right)_\Omega+a(\beta)\left\langle d_3,\overline{\gamma^+V}\right\rangle_\Gamma\right|
\end{aligned} \tag{10.4.6}$$

However, from the definition of $\underline{\sigma}=\min\{1,\sigma\}$, we see that $1/|s|>\underline{\sigma}/|s|$ and $1>\underline{\sigma}/|s|$. This implies that the LHS of (10.4.6) satisfies the estimate, namely,

$$\begin{aligned}
&\frac{\sigma}{|s|}|||\mathbf{U}^{(s)}|||^2_{|s|,\Omega}+\frac{\sigma}{|s|^3}\frac{1}{\rho_{22}}|||P^{(f)}|||^2_{|s|,\Omega}+\frac{\sigma}{|s|}a(\beta)\,|||V|||^2_{|s|,\Omega^c}\\
&\quad\ge\frac{\sigma}{|s|}\left(|||\mathbf{U}^{(s)}|||^2_{|s|,\Omega}+\left(\frac{\underline{\sigma}}{|s|}\right)^2\frac{1}{\rho_{22}}|||P^{(f)}|||^2_{|s|,\Omega}+a(\beta)\,|||V|||^2_{|s|,\Omega^c}\right)\\
&\quad\ge\frac{\sigma\underline{\sigma}^2}{|s|^3}\left(|||\mathbf{U}^{(s)}|||^2_{|s|,\Omega}+\frac{1}{\rho_{22}}|||P^{(f)}|||^2_{|s|,\Omega}+a(\beta)\,|||V|||^2_{|s|,\Omega^c}\right)
\end{aligned} \tag{10.4.7}$$

Consequently, from (10.4.6), we obtain the estimates

$$\begin{aligned}
&\left(|||\mathbf{U}^{(s)}|||^2_{1,\Omega}+|||P^{(f)}|||^2_{1,\Omega}+|||V|||^2_{1,\Omega^{(w)}}\right)\\
&\le\quad c\,\frac{|s|^3}{\sigma\underline{\sigma}^4}\left(\left|\left(d_1,\overline{\mathbf{U}^{(s)}}\right)_\Omega+\frac{1}{|s|^2}\frac{1}{\rho_{22}}\left(d_2,\overline{P^{(f)}}\right)_\Omega+a(\beta)\left\langle d_3,\overline{\gamma^+V}\right\rangle_\Gamma\right|\right)\\
&\le\quad c_0\,\frac{|s|^3}{\sigma\underline{\sigma}^6}\left(\left|\left(d_1,\overline{\mathbf{U}^{(s)}}\right)_\Omega\right|+\left|\left(d_2,\overline{P^{(f)}}\right)_\Omega\right|+\left|\left\langle d_3,\overline{\gamma^+V}\right\rangle_\Gamma\right|\right),
\end{aligned} \tag{10.4.8}$$

from which we obtain finely the desired estimate

$$|||(\mathbf{U}^{(s)},P^{(f)},V)|||_{\mathbf{H}}\le c_0\,\frac{|s|^3}{\sigma\underline{\sigma}^6}\,|||(d_1,d_2,d_3)|||_{\mathbf{H}'}, \tag{10.4.9}$$

where c_0 is a constant depending only on the physical parameters ρ_0,η,ρ_{22}.

An alternative proof shows the following sharper estimate for the norm of $|||(\mathbf{U}^{(s)},P^{(f)},V)|||_{\mathbf{H}}$ can be established.

Corollary 10.1. *The estimate of* (10.4.1) *in* Theorem 10.3 *can be improved as*

$$|||(\mathbf{U}^{(s)},P^{(f)},V)|||_{\mathbf{H}}\le c_0\,\frac{|s|^2}{\sigma\underline{\sigma}^4}\,|||(d_1,d_2,d_3)|||_{\mathbf{H}'}, \tag{10.4.10}$$

where c_0 is a constant depending only on the physical parameters ρ_0,η,ρ_{22}.

Proof

From (10.4.6) in the proof of Theorem 10.3, we see that

$$\frac{\sigma}{|s|}|||\mathbf{U}^{(s)}|||^2_{|s|,\Omega}+\frac{\sigma}{|s|^3}\frac{1}{\rho_{22}}|||P^{(f)}|||^2_{|s|,\Omega}+\frac{\sigma}{|s|}a(\beta)\,|||V|||^2_{|s|,\Omega^c}$$
$$\leq\left|\left(d_1,\overline{\mathbf{U}^{(s)}}\right)_\Omega+\frac{1}{|s|^2}\frac{1}{\rho_{22}}\left(d_2,\overline{P^{(f)}}\right)_\Omega+a(\beta)\left\langle d_3,\overline{\gamma^+V}\right\rangle_\Gamma\right| \tag{10.4.11}$$

Then

$$\frac{\sigma\underline{\sigma}^2}{|s|}\left(|||\mathbf{U}^{(s)}|||^2_{1,\Omega}+\frac{1}{|s|^2}\frac{1}{\rho_{22}}|||P^{(f)}|||^2_{1,\Omega}+a(\beta)\,|||V|||^2_{1,\Omega^c}\right)$$
$$\leq\left|\left(d_1,\overline{\mathbf{U}^{(s)}}\right)_\Omega+\frac{1}{|s|^2}\frac{1}{\rho_{22}}\left(d_2,\overline{P^{(f)}}\right)_\Omega+a(\beta)\left\langle d_3,\overline{\gamma^+V}\right\rangle_\Gamma\right|$$
$$\leq\frac{1}{\underline{\sigma}}\left\{\left|\left(d_1,\overline{\mathbf{U}^{(s)}}\right)_\Omega\right|+\frac{1}{|s|}\frac{1}{\rho_{22}}\left|\left(d_2,\overline{P^{(f)}}\right)_\Omega\right|+a(\beta)\left|\left\langle d_3,\overline{\gamma^+V}\right\rangle_\Gamma\right|\right\}, \tag{10.4.12}$$

from which we obtain the improved estimate immediately

$$|||(\mathbf{U}^{(s)},P^{(f)},V)|||_{\mathbf{H}}\leq c_0\,\frac{|s|^2}{\sigma\underline{\sigma}^4}\,|||(d_1,d_2,d_3)|||_{\mathbf{H}'}, \tag{10.4.13}$$

where c_0 is a constant depending only on the physical parameters ρ_0,η,ρ_{22}.

Theorem 10.4

Let

$$\begin{cases}\mathbf{X}:=\mathbf{H}^1(\Omega)\times H^1(\Omega)\times H^{1/2}(\Gamma)\times H^{-1/2}(\Gamma),\\ \mathbf{X}':=\mathbf{H}^1(\Omega)'\times H^1(\Omega)'\times H^{-1/2}(\Gamma)\times H^{1/2}(\Gamma),\\ \mathbf{X_0}':=\{(d_1,d_2,d_3,d_4)\in\mathbf{X}'\mid d_4=0.\}\end{cases}$$

Then $\mathcal{A}:\mathbf{X}\to\mathbf{X_0}'$ is invertible. Moreover, we have the estimate

$$\|\mathcal{A}^{-1}|_{\mathbf{X_0}}\|_{\mathbf{X'X}}\leq c_0\,\frac{|s|^{3\frac{1}{2}}}{\sigma\underline{\sigma}^{5\frac{1}{2}}}, \tag{10.4.14}$$

where c_0 is a constant independent of s and $\sigma:=\Re e\, s>0$. ■

Proof

Recall

$$\gamma^+V:=[\gamma V]=\psi\in H^{1/2}(\Gamma),\quad \partial_n^+V=\left[\partial_n^+V\right]=\zeta\in H^{-1/2}(\Gamma).$$

Then we have the estimates (see, e.g. [212]).

$$\|\psi|^2_{H^{1/2}(\Gamma)}=\|\gamma^+V\|^2_{H^{1/2}(\Gamma)}\leq c_1|||V|||_{1,|\Omega^{(w)}} \tag{10.4.15}$$

Similarly, an application of the Bamberger and Ha-Duong's optimal lifting leads to the estimate

$$\|\zeta\|_{H^{-1/2}(\Gamma)} = \|\partial_n^+ V\|_{H^{-1/2}(\Gamma)} \le \left(\frac{|s|}{\underline{\sigma}}\right)^{1/2} |||V|||_{|s|,\Omega^c}.$$

Hence,

$$\|\zeta\|^2_{H^{1/2}(\Gamma)} \le c_2^2 \left(\frac{|s|}{\underline{\sigma}}\right) |||V|||^2_{|s|,\Omega^c} \le c_2^2 \frac{|s|^3}{\underline{\sigma}^3} |||V|||^2_{1,\Omega^c} \tag{10.4.16}$$

From (10.4.15) and (10.4.16), we obtain the estimates

$$\frac{1}{2}\left\{\frac{1}{c_1}\|\psi\|^2_{H^{1/2}(\Gamma)} + \frac{1}{c_2^2}\frac{\underline{\sigma}^3}{|s|^3}\|\zeta\|^2_{H^{-1/2}(\Gamma)}\right\} \le |||V|||^2_{1,\Omega^c}.$$

As a consequence of (10.4.10), it follows that

$$\left\{|||\mathbf{U}^{(s)}|||^2_{1,\Omega} + |||P^{(f)}|||^2_{1,\Omega} + C\left(\|\psi\|^2_{H^{1/2}(\Gamma)} + \frac{\underline{\sigma}^3}{|s|^3}\|\zeta\|^2_{H^{-1/2}(\Gamma)}\right)\right\}^{\frac{1}{2}} \le c_0 \frac{|s|^2}{\sigma\underline{\sigma}^4}\|(d_1, 0, d_3, 0)\|_{\mathbf{X}'}$$

Thus we have the estimate

$$|||(\mathbf{U}^{(s)}, P^{(f)}, \psi, \zeta)|||_{\mathbf{X}} \le c_0' \frac{|s|^{3\frac{1}{2}}}{\sigma\underline{\sigma}^{5\frac{1}{2}}}\|(d_1, d_2, d_3, 0)\|_{\mathbf{X}'}$$

from which the desired estimate (10.4.14) follows. This completes the proof.

10.4.4 THE TIME DOMAIN

A major advantage of this approach is that estimates of solutions in the time domain are obtained without the need for applying the inverse Laplace transform. Indeed Proposition (10.1) is employed to retrieve time-domain estimates from estimates obtained in the Laplace transformed domain.

In order to explain our results, first we need some notation. Let $\mathfrak{B}(\mathbf{X}, \mathbf{Y})$ denote the set of bounded linear operators from the Banach space $\mathbf{X}$ to the Banach space $\mathbf{Y}$. An analytic function

$$\mathscr{A} : \mathbb{C}_+ \longrightarrow \mathfrak{B}(\mathbf{X}, \mathbf{Y})$$

is said to be an element of the class of symbols $\mathrm{Sym}(\mu, \mathfrak{B}(\mathbf{X}, \mathbf{Y}))$, if there exists a $\mu \in \mathbb{R}$ and $m \ge 0$ such that

$$\|\mathscr{A}(s)\|_{\mathbf{X},\mathbf{Y}} \le C_{\mathscr{A}}(\Re e\, s)|s|^\mu \text{ for } s \in \mathbb{C}_+ := \{s \in \mathbb{C} \mid \Re e\, s > 0\},$$

where $C_{\mathscr{A}} : (0, \infty) \to (0, \infty)$ is a non-increasing function such that

$$C_{\mathscr{A}}(\sigma) \le \frac{c}{\sigma^m}, \ \forall \sigma \in (0, 1].$$

In order to make the formulation of the time-domain estimates more compact, we will make use of the regularity spaces

$$W_+^h(\mathscr{H}) := \left\{\omega \in C^{k-1}(\mathbb{R};\mathscr{H}) : \omega \equiv 0 \in (-\infty,0),\ \omega^{(k)} \in L^1(\mathbb{R},\mathscr{H})\right\},$$

where $\mathscr{H}$ denotes a Banach space. The following proposition has been established in [368, 254]

Proposition 10.1. *Let* $\mathscr{A} = \mathscr{L}\{\mathfrak{A}\} \in \mathrm{Sym}(\mu, \mathfrak{B}(\mathbf{X},\mathbf{Y}))$*, with* $\mu \geq 0$*; furthermore, let* $k = \lfloor \mu+2 \rfloor$*, be the largest integer less than or equal to* $\mu+2$*, and let* $\varepsilon : k-(\mu+1) \in (0,1]$.

If $g \in W_+^k(\mathbb{R},\mathbf{X})$*, then* $\mathfrak{A} \star g \in \mathscr{C}(\mathbb{R}), \mathbf{Y}$ *is causal and*

$$\|(\mathfrak{A}\star g)(t)\|_{\mathbf{Y}} \leq 2^{n+1} C_\varepsilon(t)\, C_{\mathscr{A}}(t^{-1}) \int_0^1 \|(\mathscr{P}_k g)(\tau)\|_{\mathbf{X}}\, d\tau$$

where

$$C_{\varepsilon(t)} = \frac{t^\varepsilon}{\pi\,\varepsilon}, \quad \textit{and} \quad (\mathscr{P}_k g)(t) = \sum_{\ell=0}^{k} \binom{k}{\ell} q^{(\ell)}(t).$$

As an immediate consequence of Proposition (10.1), we see from Theorem (10.4) that

$$\mathcal{A}^{-1}|_{\mathbf{X}_0'} \in \mathrm{Sym}\left(3\frac{1}{2}, \mathfrak{B}(\mathbf{X}',\mathbf{X})\right).$$

Moreover $\mu = 3\frac{1}{2}$, $k = \lfloor 3\frac{1}{2}+2 \rfloor = 5$ and $\varepsilon = 5-(3\frac{1}{2}+1) = \frac{1}{2} \in (0,1]$, and we have the following estimate.

Theorem 10.5

Let $\mathscr{D}(t) := \mathscr{L}^{-1}\left\{(d_1,d_2,d_3,0)^T\right\}$ belong $W_+^5(\mathbb{R},\tilde{\mathbf{X}}')$. Then

$$\left(\mathbf{u}^{(s)}),\ p^{(f)}, \phi,\ \partial_n\phi)\right)^T \in \mathscr{C}([0,T],\mathbf{X})$$

and there exists a constant $c > 0$ depending only on the geometry such that

$$\begin{aligned}\|\left(\mathbf{u}^{(s)}, p^{(f)},\ \phi,\ \partial_n\phi\right)^T\|_{\mathbf{X}} &\leq c\,t^{\frac{1}{2}}\left(\sigma\underline{\sigma}^{5\frac{1}{2}}\right)|_{\sigma=\frac{1}{t}} \int_0^1 \|(\mathscr{P}_6\mathscr{D})(\tau)\|_{\mathbf{X}'}\, d\tau \\ &= c\,t^{\frac{1}{2}+1}\max\left\{1, t^{5\frac{1}{2}}\right\} \int_0^1 \|(\mathscr{P}_6\mathscr{D})(\tau)\|_{\mathbf{X}'}\, d\tau.\end{aligned}$$

■

Similarly, applying Proposition 10.1 to Corollary 10.1, we have

$$\mu = 2,\ k = \lfloor \mu+2 \rfloor = 4,\ \varepsilon := k-(\mu+1) = 1,$$

$$\left(\frac{1}{\sigma\underline{\sigma}^4}\right)|_{\sigma=\frac{1}{t}} = t\max\left\{1,t^4\right\}$$

we have the result.

Theorem 10.6

Let $\mathbf{H} := \mathbf{H}^1(\Omega) \times H^1(\Omega) \times H^1(\Omega^{(w)})$ and

$$\mathbf{D}(t) := \mathscr{L}^{-1}\left\{(d_1,d_2,d_3)^T\right\}(t) \in W_+^4(\mathbb{R},\mathbf{H}').$$

Then

$$\left(\mathbf{u}^{(s)}, p^{(f)}, \mathscr{L}^{-1}\{V\}\right)^T \in \mathscr{C}([0,T],\mathbf{H})$$

and there holds the estimate

$$\|\left(\mathbf{u}^{(s)}, p^{(f)},\ \mathscr{L}^{-1}\{V\}\right)^T(t)\|_{\mathbf{H}} \leq c_0\ t^2 \max\left\{1,t^4\right\} \int_0^1 \|(\mathscr{P}_4\mathscr{D})(\tau)\|_{\mathbf{H}'}\, d\tau,$$

where $c_0 > 0$ is a constant. ∎

11 Creation of RVE for Bone Microstructure

11.1 THE RVE MODEL

We assume some Voigt dissipation in the trabeculae, which makes matching with the fluid phase numerically more stable; consequently the constitutive equations in the trabeculae take the form

$$\tau^{(s)} = A^{(s)}\mathbf{e}(\mathbf{u}) + B^{(s)}\mathbf{e}(\mathbf{v}),$$

where, as usual, $\mathbf{e}(\mathbf{u})$ is the strain tensor; $\mathbf{u}$ and $\mathbf{v}$ are the displacement and velocity fields, respectively.

These constitutive equations may be written in component form as

$$\tau_{ij}^{(s)} = A_{ijkl}^{(s)} e(\mathbf{u})_{kl} + B_{ijkl}^{(s)} e(\mathbf{v})_{kl}, \qquad (11.1.1)$$

where the $A_{ijkl}^{(s)}$ are the elasticity coefficients for the solid and are assumed to have the classical symmetry and positivity properties, i.e.

$$A_{ijkl}^{(s)} e_{ij} e_{kl} \geq 0 \quad \text{and} \quad A_{ijkl}^{(s)} = A_{klij}^{(s)} = A_{jikl}^{(s)} = A_{ijlk}^{(s)},$$

while the $B_{ijkl}^{(s)}$ would correspond to instantaneous (Voigt) viscosity terms where we assume that these are isotropic in form. In the isotropic elastic case, we have

$$A_{ijkl}^{(s)} e_{kl} = \left(\lambda^{(1)} \delta_{ij} \delta_{kl} + 2\mu^{(1)} \delta_{ik} \delta_{jl}\right) e_{kl} = \lambda^{(1)} \delta_{ij} e_{kk} + 2\mu^{(1)} e_{ij},$$

where $\{\lambda^{(1)}, \mu^{(1)}\}$ are the elastic Lamé parameters, and

$$B_{ijkl}^{(s)} e_{kl} = \left(\lambda^{(2)} \delta_{ij} \delta_{kl} + 2\mu^{(2)} \delta_{ik} \delta_{jl}\right) e_{kl} = \lambda^{(2)} \delta_{ij} e_{kk} + 2\mu^{(2)} e_{ij}, \qquad (11.1.2)$$

where $\{\lambda^{(2)}, \mu^{(2)}\}$ are the viscosity parameters. Rewriting the solid stress in tensor form one gets

$$\tau^{(s)} = \lambda^{(1)} \mathbb{I} e(\mathbf{u}) + 2\mu^{(1)} \mathbf{e}(\mathbf{u}) + \lambda^{(2)} \mathbb{I} e(\mathbf{v}) + 2\mu^{(2)} \mathbf{e}(\mathbf{v}), \qquad (11.1.3)$$

where e is the dilatation, and $\mathbf{e}$ is the strain tensor and $\mathbb{I}$ the identity tensor. The equations of motion for the solid part are given by

$$\frac{\partial \mathbf{v}}{\partial t} = b^{(s)} \operatorname{div}\left[A^{(s)} \mathbf{e}(\mathbf{u}) + B^{(s)} \mathbf{e}(\mathbf{v})\right] + \mathbf{f}^{(s)}, \qquad (11.1.4)$$

in $\Omega_s \times [0,T]$, where $b^{(s)} = 1/\rho^{(s)}$ is the solid buoyancy. As we discuss the isotropic case, we have

$$\begin{aligned}\frac{\partial \mathbf{v}}{\partial t} &= \mathbf{f}^{(s)} + b^{(s)}\left[\frac{\partial}{\partial x_j}\left(\lambda^{(1)}\delta_{ij}e_{kk}(\mathbf{u}) + 2\mu^{(1)}e_{ij}(\mathbf{u})\right)\right. \\ &\quad \left. + \frac{\partial}{\partial x_j}\left(\lambda^{(2)}\delta_{ij}e_{kk}(\mathbf{v}) + 2\mu^{(2)}e_{ij}(\mathbf{v})\right)\right]. \end{aligned} \tag{11.1.5}$$

In the fluid part, $\Omega_f \times [0,T]$, we assume that the small compressibility approximation holds, in which case, the constitutive equations may be written as (11.1.1), namely

$$\tau^{(f)} = A^{(f)}\mathbf{e}(\mathbf{u}) + B^{(f)}\mathbf{e}(\mathbf{v}),$$

with

$$A^{(f)} = c^2\rho^{(f)}\mathbb{I}, \quad B^{(f)} = 2\eta\mathbb{I},$$

where c is the speed of sound in the fluid and η is the fluid viscosity.

The equations of motion for this Stokes system are

$$\frac{\partial \mathbf{v}}{\partial t} = b^{(f)}\mathrm{div}\left[c^2\rho^f\mathrm{div}(\mathbf{u}\mathbb{I}) + 2\eta\, e(\mathbf{v})\right] + \mathbf{f}^{(f)}, \tag{11.1.6}$$

which hold in $\Omega_f \times [0,T]$. Here $b^{(f)} = 1/\rho^{(f)}$ is the fluid buoyancy.

In both fluid and solid parts, the system of equations for τ, $\mathbf{v}$ and $\mathbf{u}$ is completed by adding

$$\frac{\partial \mathbf{u}}{\partial t} = \mathbf{v}. \tag{11.1.7}$$

11.2 REFORMULATION AS A GRAVES-LIKE SCHEME

Since we assume that the trabeculae are viscoelastic and isotropic, this leads to a system of equations similar to that in Graves [165] for seismic waves. Hence, in three-dimensional, Cartesian coordinates, the stresses in the trabeculae are written as

$$\begin{aligned}\tau^{(s)}_{xx} &= \left(\lambda^{(1)} + 2\mu^{(1)}\right)\partial_x u_x + \lambda^{(1)}\left(\partial y u_y + \partial_z u_z\right) \\ &\quad + \left(\lambda^{(2)} + 2\mu^{(2)}\right)\partial_x v_x + \lambda^{(2)}\left(\partial y v_y + \partial_z v_z\right), \\ \tau^{(s)}_{yy} &= \left(\lambda^{(1)} + 2\mu^{(1)}\right)\partial y u_y + \lambda^{(1)}\left(\partial_x u_x + \partial_z u_z\right) \\ &\quad + \left(\lambda^{(2)} + 2\mu^{(2)}\right)\partial y v_y + \lambda^{(2)}\left(\partial_x v_x + \partial_z v_z\right), \\ \tau^{(s)}_{zz} &= \left(\lambda^{(1)} + 2\mu^{(1)}\right)\partial_z u_z + \lambda^{(1)}\left(\partial_x u_x + \partial y u_y\right) \\ &\quad + \left(\lambda^{(2)} + 2\mu^{(2)}\right)\partial_z v_z + \lambda^{(2)}\left(\partial_x v_x + \partial y v_y\right), \\ \tau^{(s)}_{xy} &= \mu^{(1)}\left(\partial_x u_y + \partial y u_x\right) + \mu^{(2)}\left(\partial_x v_y + \partial y v_x\right), \end{aligned} \tag{11.2.1}$$

$$
\begin{aligned}
\tau_{xz}^{(s)} &= \mu^{(1)}(\partial_x u_z + \partial_z u_x) + \mu^{(2)}(\partial_x v_z + \partial_z v_x),\\
\tau_{yz}^{(s)} &= \mu^{(1)}(\partial y u_z + \partial_z u_y) + \mu^{(2)}(\partial y v_z + \partial_z v_y),
\end{aligned}
$$

and their evolutionary equations become

$$
\begin{aligned}
\partial_t \tau_{xx}^{(s)} &= \left(\lambda^{(1)} + 2\mu^{(1)}\right)\partial_x v_x + \lambda^{(1)}(\partial y v_y + \partial_z v_z)\\
&\quad + \left(\lambda^{(2)} + 2\mu^{(2)}\right)\partial_x \dot{v}_x + \lambda^{(2)}(\partial y \dot{v}_y + \partial_z \dot{v}_z),\\
\partial_t \tau_{yy}^{(s)} &= \left(\lambda^{(1)} + 2\mu^{(1)}\right)\partial y v_y + \lambda^{(1)}(\partial_x v_x + \partial_z v_z)\\
&\quad + \left(\lambda^{(2)} + 2\mu^{(2)}\right)\partial y \dot{v}_y + \lambda^{(2)}(\partial_x \dot{v}_x + \partial_z \dot{v}_z),\\
\partial_t \tau_{zz}^{(s)} &= \left(\lambda^{(1)} + 2\mu^{(1)}\right)\partial_z v_z + \lambda^{(1)}(\partial_x v_x + \partial y v_y)\\
&\quad + \left(\lambda^{(2)} + 2\mu^{(2)}\right)\partial_z \dot{v}_z + \lambda^{(2)}(\partial_x \dot{v}_x + \partial y \dot{v}_y),\\
\partial_t \tau_{xy}^{(s)} &= \mu^{(1)}(\partial_x v_y + \partial y v_x) + \mu^{(2)}(\partial_x \dot{v}_y + \partial y \dot{v}_x),\\
\partial_t \tau_{xz}^{(s)} &= \mu^{(1)}(\partial_x v_z + \partial_z v_x) + \mu^{(2)}(\partial_x \dot{v}_z + \partial_z v_x),\\
\partial_t \tau_{yz}^{(s)} &= \mu^{(1)}(\partial y v_z + \partial_z v_y) + \mu^{(2)}(\partial y \dot{v}_z + \partial_z \dot{v}_y).
\end{aligned}
\tag{11.2.2}
$$

Here the dot over a variable is meant to represent differentiation with respect to t (e.g. $\dot{v}_x = \partial_t v_x$). In (11.2.2), the acceleration field is given by

$$
\begin{aligned}
\partial_t v_x &= b^{(s)}\left[\partial_x \tau_{xx}^{(s)} + \partial y \tau_{xy}^{(s)} + \partial_z \tau_{xz}^{(s)}\right],\\
\partial_t v_y &= b^{(s)}\left[\partial_x \tau_{xy}^{(s)} + \partial y \tau_{yy}^{(s)} + \partial_z \tau_{yz}^{(s)}\right],\\
\partial_t v_z &= b^{(s)}\left[\partial_x \tau_{xz}^{(s)} + \partial y \tau_{yz}^{(s)} + \partial_z \tau_{zz}^{(s)}\right].
\end{aligned}
\tag{11.2.3}
$$

In the interstitial fluid, the evolutionary equations for the stress field are

$$
\begin{aligned}
\partial_t v_x &= b^{(f)}\left[\partial_x \tau_{xx}^{(f)} + \partial y \tau_{xy}^{(f)} + \partial_z \tau_{xz}^{(f)}\right],\\
\partial_t v_y &= b^{(f)}\left[\partial_x \tau_{xy}^{(f)} + \partial y \tau_{yy}^{(f)} + \partial_z \tau_{yz}^{(f)}\right],\\
\partial_t v_z &= b^{(f)}\left[\partial_x \tau_{xz}^{(f)} + \partial y \tau_{yz}^{(f)} + \partial_z \tau_{zz}^{(f)}\right],
\end{aligned}
\tag{11.2.4}
$$

combined with

$$
\begin{aligned}
\partial_t \tau_{xx}^{(f)} &= c^2\rho^{(f)}(\partial_x v_x + \partial y v_y + \partial_z v_z) + 2\eta\,\partial_x \dot{v}_x,\\
\partial_t \partial_t \tau_{yy}^{(f)} &= c^2\rho^{(f)}(\partial_x v_x + \partial y v_y + \partial_z v_z) + 2\eta\,\partial y \dot{v}_y,\\
\partial_t \tau_{zz}^{(f)} &= c^2\rho^{(f)}(\partial_x v_x + \partial y v_y + \partial_z v_z) + 2\eta\,\partial_z \dot{v}_x,\\
\partial_t \tau_{xy}^{(f)} &= \eta\,(\partial_x \dot{v}_y + \partial y \dot{v}_x),
\end{aligned}
\tag{11.2.5}
$$

$$\begin{aligned}
\partial_t \tau_{xz}^{(f)} &= \eta\left(\partial_x \dot{v}_z + \partial_z \dot{v}_x\right), \\
\partial_t \tau_{yz}^{(f)} &= \eta\left(\partial y \dot{v}_z + \partial_z \dot{v}_y\right).
\end{aligned}$$

We note that both the fluid and solid phases are described by the generalized system

$$\begin{aligned}
\tau_{xx}^{(a)} &= \left(\lambda^{(k)} + 2\mu^{(k)}\right)\partial_x u_x + \lambda^{(k)}\left(\partial y u_y + \partial_z u_z\right) \\
&\quad + \left(\lambda^{(k+1)} + 2\mu^{(k+1)}\right)\partial_x v_x + \lambda^{(k+1)}\left(\partial y v_y + \partial_z v_z\right), \\
\tau_{yy}^{(a)} &= \left(\lambda^{(k)} + 2\mu^{(k)}\right)\partial y u_y + \lambda^{(k)}\left(\partial_x u_x + \partial_z u_z\right) \\
&\quad + \left(\lambda^{(k+1)} + 2\mu^{(k+1)}\right)\partial y v_y + \lambda^{(k+1)}\left(\partial_x v_x + \partial_z v_z\right), \\
\tau_{zz}^{(a)} &= \left(\lambda^{(k)} + 2\mu^{(1)}\right)\partial_z u_z + \lambda^{(k)}\left(\partial_x u_x + \partial y u_y\right) \\
&\quad + \left(\lambda^{(k+1)} + 2\mu^{(k+1)}\right)\partial_z v_z + \lambda^{(k+1)}\left(\partial_x v_x + \partial y v_y\right), \\
\tau_{xy}^{(a)} &= \mu^{(k)}\left(\partial_x u_y + \partial y u_x\right) + \mu^{(k+1)}\left(\partial_x v_y + \partial y v_x\right), \\
\tau_{xz}^{(a)} &= \mu^{(k)}\left(\partial_x u_z + \partial_z u_x\right) + \mu^{(k+1)}\left(\partial_x v_z + \partial_z v_x\right), \\
\tau_{yz}^{(a)} &= \mu^{(k)}\left(\partial y u_z + \partial_z u_y\right) + \mu^{(k+1)}\left(\partial y v_z + \partial_z v_y\right),
\end{aligned} \tag{11.2.6}$$

together with the evolutionary equations

$$\begin{aligned}
\partial_t \tau_{xx}^{(a)} &= \left(\lambda^{(k)} + 2\mu^{(k)}\right)\partial_x v_x + \lambda^{(k)}\left(\partial y v_y + \partial_z v_z\right) \\
&\quad + \left(\lambda^{(k+1)} + 2\mu^{(k+1)}\right)\partial_x \dot{v}_x + \lambda^{(k+1)}\left(\partial y \dot{v}_y + \partial_z \dot{v}_z\right), \\
\partial_t \tau_{yy}^{(a)} &= \left(\lambda^{(k)} + 2\mu^{(k)}\right)\partial y v_y + \lambda^{(k)}\left(\partial_x v_x + \partial_z v_z\right) \\
&\quad + \left(\lambda^{(k+1)} + 2\mu^{(k+1)}\right)\partial y \dot{v}_y + \lambda^{(k+1)}\left(\partial_x \dot{v}_x + \partial_z \dot{v}_z\right), \\
\partial_t \tau_{zz}^{(a)} &= \left(\lambda^{(k)} + 2\mu^{(1)}\right)\partial_z v_z + \lambda^{(k)}\left(\partial_x v_x + \partial y v_y\right) \\
&\quad + \left(\lambda^{(k+1)} + 2\mu^{(k+1)}\right)\partial_z \dot{v}_z + \lambda^{(k+1)}\left(\partial_x \dot{v}_x + \partial y \dot{v}_y\right), \\
\partial_t \tau_{xy}^{(a)} &= \mu^{(k)}\left(\partial_x v_y + \partial y v_x\right) + \mu^{(k+1)}\left(\partial_x \dot{v}_y + \partial y \dot{v}_x\right), \\
\partial_t \tau_{xz}^{(a)} &= \mu^{(k)}\left(\partial_x v_z + \partial_z v_x\right) + \mu^{(k+1)}\left(\partial_x \dot{v}_z + \partial_z \dot{v}_x\right), \\
\partial_t \tau_{yz}^{(a)} &= \mu^{(k)}\left(\partial y v_z + \partial_z v_y\right) + \mu^{(k+1)}\left(\partial y \dot{v}_z + \partial_z \dot{v}_y\right),
\end{aligned} \tag{11.2.7}$$

where $(a,k) = \{(s,1),(f,3)\}$, $\lambda^{(3)} = c^2\rho^{(f)}$, $\lambda^{(4)} = 0$, $\mu^{(3)} = 0$ and $\mu^{(4)} = \eta$. The acceleration field also has the same form for both phases,

$$\begin{aligned}
\partial_t v_x &= b^{(a)}\left[\partial_x \tau_{xx}^{(a)} + \partial y \tau_{xy}^{(a)} + \partial_z \tau_{xz}^{(a)}\right], \\
\partial_t v_y &= b^{(a)}\left[\partial_x \tau_{xy}^{(a)} + \partial y \tau_{yy}^{(a)} + \partial_z \tau_{yz}^{(a)}\right],
\end{aligned} \tag{11.2.8}$$

$$\partial_t v_z = b^{(a)} \left[\partial_x \tau_{xz}^{(a)} + \partial y \tau_{yz}^{(a)} + \partial_z \tau_{zz}^{(a)} \right].$$

This is completed by

$$\partial_t u_x = v_x, \quad \partial_t u_y = v_y, \quad \partial_t u_z = v_z, \tag{11.2.9}$$

which yields the displacements in both phases as well.

11.3 ABSORBING BOUNDARY CONDITION: A PERFECTLY MATCHED LAYER

For computational purposes we assume the bone sample is a rectangular parallelepiped and the ultrasound transponder produces on the bone surface a sinusoidal plane wave signal. The transponder and receiver are situated on the opposite sides of the sample. Two kinds of boundary conditions are assumed on different sides of the parallelogram; for example, a free-surface boundary condition is applied on all sides except the receiver side, as in [153, 154]. In order to make a good impedance match where the receiver is, an absorbing layer is attached at the boundary there. There are various options of absorbing layers. The one we used is called a perfectly matched layers. It was first introduced by Berenger (1994) for a two-dimensional, time-domain electromagnetic field, and then widely used for both finite-difference and finite element methods in electromagnetic, acoustic and elastodynamic wave propagation simulations. Hasting et al. (1996) first applied it to elastic waves. Collino and Tsogka [92] incorporated it into the stress-velocity formulation (2001).

The perfectly matched layer (**PML**) is an artificial region attached to the boundary of the domain. Theoretically, the interior area and the absorbing region are perfectly matched and no reflection is generated on the interface between the two media. As a result, we can use very high damping parameters in the region and generally use a thin layer to save computational cost.

The idea of a PML can be viewed as an equation-splitting procedure. Consider equations (11.2.7)-(11.2.8); each equation is split into a parallel and a perpendicular component. The perpendicular component contains the spatial derivative in the direction that we are interested in, and an added damping term; the parallel component contains all the rest of the spatial derivatives. Assuming the wave propagates in the x-direction, then a damping term $d(x)$ is introduced. Equation (11.2.7)-(11.2.8) can now be rewritten as the following split equations.

First of all, the variables are split as perpendicular and parallel components,

$$\begin{aligned}
\tau_{xx}^{(a)} &= \tau_{xx}^{(a)\perp} + \tau_{xx}^{(a)\|}, \\
\tau_{yy}^{(a)} &= \tau_{yy}^{(a)\perp} + \tau_{yy}^{(a)\|}, \\
\tau_{zz}^{(a)} &= \tau_{zz}^{(a)\perp} + \tau_{zz}^{(a)\|}, \\
\tau_{xy}^{(a)} &= \tau_{xy}^{(a)\perp} + \tau_{xy}^{(a)\|}, \\
\tau_{xz}^{(a)} &= \tau_{xz}^{(a)\perp} + \tau_{xz}^{(a)\|},
\end{aligned} \tag{11.3.1}$$

$$\begin{aligned}
\tau_{yz}^{(a)} &= \tau_{yz}^{(a)\perp} + \tau_{yx}^{(a)\|}, \\
v_x &= v_x^{\perp} + v_x^{\|}, \\
v_y &= v_y^{\perp} + v_y^{\|}, \\
v_z &= v_z^{\perp} + v_z^{\|}.
\end{aligned}$$

Hence the equations can be written in the form

$$\begin{aligned}
\partial_t \tau_{xx}^{(a)\perp} + d(x)\tau_{xx}^{(a)\perp} &= \left(\lambda^{(k)} + 2\mu^{(k)}\right)\partial_x v_x + \left(\lambda^{(k+1)} + 2\mu^{(k+1)}\right)\partial_x \dot{v}_x, \\
\partial_t \tau_{yy}^{(a)\perp} + d(x)\tau_{yy}^{(a)\perp} &= \lambda^{(k)}\left(\partial_x v_x\right) + \lambda^{(k+1)}\left(\partial_x \dot{v}_x\right), \\
\partial_t \tau_{zz}^{(a)\perp} + d(x)\tau_{zz}^{(a)\perp} &= \lambda^{(k)}\left(\partial_x v_x\right) + \lambda^{(k+1)}\left(\partial_x \dot{v}_x\right), \\
\partial_t \tau_{xy}^{(a)\perp} + d(x)\tau_{xy}^{(a)\perp} &= \mu^{(k)}\left(\partial_x v_y\right) + \mu^{(k+1)}\left(\partial_x \dot{v}_y\right), \\
\partial_t \tau_{xz}^{(a)\perp} + d(x)\tau_{xz}^{(a)\perp} &= \mu^{(k)}\left(\partial_x v_z\right) + \mu^{(k+1)}\left(\partial_x \dot{v}_z\right), \qquad (11.3.2) \\
\partial_t \tau_{yz}^{(a)\perp} + d(x)\tau_{yz}^{(a)\perp} &= 0, \\
\partial_t v_x^{\perp} + d(x)v_x^{\perp} &= b^{(a)}\left(\partial_x \tau_{xx}^{(a)}\right), \\
\partial_t v_y^{\perp} + d(x)v_y^{\perp} &= b^{(a)}\left(\partial_x \tau_{xy}^{(a)}\right), \\
\partial_t v_z^{\perp} + d(x)v_z^{\perp} &= b^{(a)}\left(\partial_x \tau_{xz}^{(a)}\right),
\end{aligned}$$

for perpendicular components which contain all x derivatives. And

$$\begin{aligned}
\partial_t \tau_{xx}^{(a)\|} &= \lambda^{(k)}\left(\partial y v_y + \partial_z v_z\right) + \lambda^{(k+1)}\left(\partial y \dot{v}_y + \partial_z \dot{v}_z\right), \\
\partial_t \tau_{yy}^{(a)\|} &= \left(\lambda^{(k)} + 2\mu^{(k)}\right)\partial y v_y + \lambda^{(k)}\left(\partial_z v_z\right) \\
&\quad + \left(\lambda^{(k+1)} + 2\mu^{(k+1)}\right)\partial y \dot{v}_y + \lambda^{(k+1)}\left(\partial_z \dot{v}_z\right), \\
\partial_t \tau_{zz}^{(a)\|} &= \left(\lambda^{(k)} + 2\mu^{(1)}\right)\partial_z v_z + \lambda^{(k)}\left(\partial y v_y\right) \\
&\quad + \left(\lambda^{(k+1)} + 2\mu^{(k+1)}\right)\partial_z \dot{v}_z + \lambda^{(k+1)}\left(\partial y \dot{v}_y\right), \\
\partial_t \tau_{xy}^{(a)\|} &= \mu^{(k)}\left(\partial y v_x\right) + \mu^{(k+1)}\left(\partial y \dot{v}_x\right), \qquad (11.3.3) \\
\partial_t \tau_{xz}^{(a)\|} &= \mu^{(k)}\left(\partial_z v_x\right) + \mu^{(k+1)}\left(\partial_z \dot{v}_x\right), \\
\partial_t \tau_{yz}^{(a)\|} &= \mu^{(k)}\left(\partial y v_z + \partial_z v_y\right) + \mu^{(k+1)}\left(\partial y \dot{v}_z + \partial_z \dot{v}_y\right), \\
\partial_t v_x^{\|} &= b^{(a)}\left(\partial y \tau_{xy}^{(a)} + \partial_z \tau_{xz}^{(a)}\right), \\
\partial_t v_y^{\|} &= b^{(a)}\left(\partial y \tau_{yy}^{(a)} + \partial_z \tau_{yz}^{(a)}\right), \\
\partial_t v_z^{\|} &= b^{(a)}\left(\partial y \tau_{yz}^{(a)} + \partial_z \tau_{zz}^{(a)}\right),
\end{aligned}$$

for parallel components which contain all the y, z derivatives.

The damping parameter $d(x)$ is zero everywhere except in the PML absorbing regions. So the equations stay exactly the same as (11.2.7)-(11.2.8) in the interior region. Similar equations can be formed for absorption in the y and z directions by using damping parameters $d(y)$ and $d(z)$.

11.4 DISCRETIZED SYSTEMS

We use a staggered-grid finite difference scheme to discretize the system (11.3.2)–(11.3.3) in each phase, where the different components of displacement, velocity and stress are defined at different grid points in the computational domain and the variables are also staggered temporally. A unit cell showing the staggered-grid is found in Figure (11.4.1). In our notation, the subscripts refer to the spatial indices while the superscripts refer to the time index. For example, the expression $v^{n+\frac{1}{2}}_{xi+\frac{1}{2},j,k}$, represents the x-component of the velocity at point $x_{i+1/2} = (i+1/2)\Delta x$, $y_j = j\Delta y$, $z_k = k\Delta z$ and at time $t_{n+1/2} = (n+1/2)\Delta t$, where Δx, Δy, Δz are the mesh sizes in the three spatial directions and Δt is the time step. To avoid overly cumbersome expressions, D_j denotes the difference operator for the discretization of the partial derivative ∂_j in space. The superscript (a) refers to fluid or solid depending on which phase the grid point is in. We briefly present the discretized equations below and refer the reader to [165, 153] for further details.

The discretization of (11.3.2) reads

$$
\begin{aligned}
\frac{(\tau_{xx}^{(a)\perp})^{n+1}_{i,j,k} - (\tau_{xx}^{(a)\perp})^{n}_{i,j,k}}{\Delta t} &+ d(x_i)\frac{(\tau_{xx}^{(a)\perp})^{n+1}_{i,j,k} + (\tau_{xx}^{(a)\perp})^{n}_{i,j,k}}{2} \\
&= \left[\left(\lambda^{(k)} + 2\mu^{(k)}\right) D_x v_x + \left(\lambda^{(k+1)} + 2\mu^{(k+1)}\right) D_x \dot{v}_x\right]^{n+\frac{1}{2}}_{i,j,k}, \\
\frac{(\tau_{yy}^{(a)\perp})^{n+1}_{i,j,k} - (\tau_{yy}^{(a)\perp})^{n}_{i,j,k}}{\Delta t} &+ d(x_i)\frac{(\tau_{yy}^{(a)\perp})^{n+1}_{i,j,k} + (\tau_{yy}^{(a)\perp})^{n}_{i,j,k}}{2} \\
&= \left[\lambda^{(k)} (D_x v_x) + \lambda^{(k+1)} (D_x \dot{v}_x)\right]^{n+\frac{1}{2}}_{i,j,k}, \\
\frac{(\tau_{zz}^{(a)\perp})^{n+1}_{i,j,k} - (\tau_{zz}^{(a)\perp})^{n}_{i,j,k}}{\Delta t} &+ d(x_i)\frac{(\tau_{zz}^{(a)\perp})^{n+1}_{i,j,k} + (\tau_{zz}^{(a)\perp})^{n}_{i,j,k}}{2} \\
&= \left[\lambda^{(k)} (D_x v_x) + \lambda^{(k+1)} (D_x \dot{v}_x)\right]^{n+\frac{1}{2}}_{i,j,k}, \\
\frac{(\tau_{xy}^{(a)\perp})^{n+1}_{i+\frac{1}{2},j+\frac{1}{2},k} - (\tau_{xy}^{(a)\perp})^{n}_{i+\frac{1}{2},j+\frac{1}{2},k}}{\Delta t} &+ d(x_{i+\frac{1}{2}})\frac{(\tau_{xy}^{(a)\perp})^{n+1}_{i+\frac{1}{2},j+\frac{1}{2},k} + (\tau_{xy}^{(a)\perp})^{n}_{i+\frac{1}{2},j+\frac{1}{2},k}}{2} \\
&= \left[\mu^{(k)} (D_x v_y) + \mu^{(k+1)} (D_x \dot{v}_y)\right]^{n+\frac{1}{2}}_{i+\frac{1}{2},j+\frac{1}{2},k}, \\
\frac{(\tau_{xz}^{(a)\perp})^{n+1}_{i+\frac{1}{2},j,k+\frac{1}{2}} - (\tau_{xz}^{(a)\perp})^{n}_{i+\frac{1}{2},j,k+\frac{1}{2}}}{\Delta t} &+ d(x_{i+\frac{1}{2}})\frac{(\tau_{xz}^{(a)\perp})^{n+1}_{i+\frac{1}{2},j,k+\frac{1}{2}} + (\tau_{xz}^{(a)\perp})^{n}_{i+\frac{1}{2},j,k+\frac{1}{2}}}{2} \\
&= \left[\mu^{(k)} (D_x v_z) + \mu^{(k+1)} (D_x \dot{v}_z)\right]^{n+\frac{1}{2}}_{i+\frac{1}{2},j,k+\frac{1}{2}}, \qquad (11.4.1) \\
\frac{(\tau_{yz}^{(a)\perp})^{n+1}_{i,j+\frac{1}{2},k+\frac{1}{2}} - (\tau_{yz}^{(a)\perp})^{n}_{i,j+\frac{1}{2},k+\frac{1}{2}}}{\Delta t} &+ d(x_i)\frac{(\tau_{yz}^{(a)\perp})^{n+1}_{i,j+\frac{1}{2},k+\frac{1}{2}} + (\tau_{yz}^{(a)\perp})^{n}_{i,j+\frac{1}{2},k+\frac{1}{2}}}{2} \\
&= 0, \\
\frac{(v_x^{\perp})^{n+\frac{1}{2}}_{i+\frac{1}{2},j,k} - (v_x^{\perp})^{n-\frac{1}{2}}_{i+\frac{1}{2},j,k}}{\Delta t} &+ d(x_{i+\frac{1}{2}})\frac{(v_x^{\perp})^{n+\frac{1}{2}}_{i+\frac{1}{2},j,k} + (v_x^{\perp})^{n-\frac{1}{2}}_{i+\frac{1}{2},j,k}}{2} \\
&= b^{(a)}\left[D_x \tau_{xx}^{(a)}\right]^{n}_{i+\frac{1}{2},j,k}, \\
\frac{(v_y^{\perp})^{n+\frac{1}{2}}_{i,j+\frac{1}{2},k} - (v_y^{\perp})^{n-\frac{1}{2}}_{i,j+\frac{1}{2},k}}{\Delta t} &+ d(x_i)\frac{(v_y^{\perp})^{n+\frac{1}{2}}_{i,j+\frac{1}{2},k} + (v_y^{\perp})^{n-\frac{1}{2}}_{i,j+\frac{1}{2},k}}{2} \\
&= b^{(a)}\left[D_x \tau_{xy}^{(a)}\right]^{n}_{i,j+\frac{1}{2},k}, \\
\frac{(v_z^{\perp})^{n+\frac{1}{2}}_{i,j,k+\frac{1}{2}} - (v_z^{\perp})^{n-\frac{1}{2}}_{i,j,k+\frac{1}{2}}}{\Delta t} &+ d(x_i)\frac{(v_z^{\perp})^{n+\frac{1}{2}}_{i,j,k+\frac{1}{2}} + (v_z^{\perp})^{n-\frac{1}{2}}_{i,j,k+\frac{1}{2}}}{2} \\
&= b^{(a)}\left[D_x \tau_{xz}^{(a)}\right]^{n}_{i,j,k+\frac{1}{2}}.
\end{aligned}
$$

The discretization of (11.3.3) reads

$$
\begin{aligned}
\frac{(\tau_{xx}^{(a)\parallel})^{n+1}_{i,j,k} - (\tau_{xx}^{(a)\parallel})^{n}_{i,j,k}}{\Delta t} &= \left[\lambda^{(k)} (D_y v_y + D_z v_z) + \lambda^{(k+1)} (D_y \dot{v}_y + D_z \dot{v}_z)\right]^{n+\frac{1}{2}}_{i,j,k}, \\
\frac{(\tau_{yy}^{(a)\parallel})^{n+1}_{i,j,k} - (\tau_{yy}^{(a)\parallel})^{n}_{i,j,k}}{\Delta t} &= \left[\left(\lambda^{(k)} + 2\mu^{(k)}\right) D_y v_y + \lambda^{(k)} (D_z v_z)\right. \\
&+ \left.\left(\lambda^{(k+1)} + 2\mu^{(k+1)}\right) D_y \dot{v}_y + \lambda^{(k+1)} (D_z \dot{v}_z)\right]^{n+\frac{1}{2}}_{i,j,k},
\end{aligned}
$$

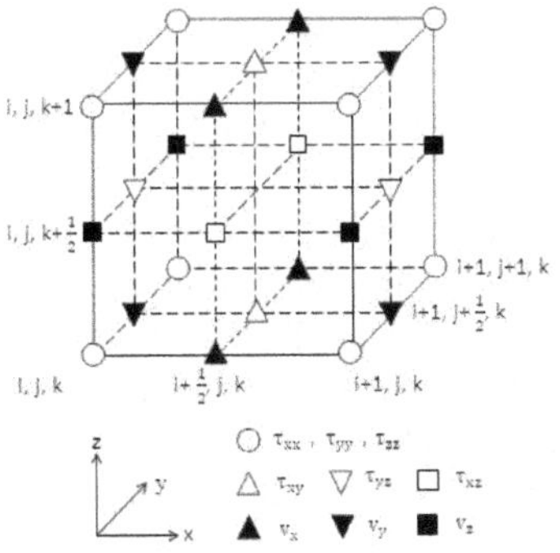

FIGURE 11.4.1 Unit cell of staggered grid lay-out.

$$
\begin{aligned}
\frac{(\tau_{zz}^{(a)\|})_{i,j,k}^{n+1} - (\tau_{zz}^{(a)\|})_{i,j,k}^{n}}{\Delta t} &= \Big[\left(\lambda^{(k)} + 2\mu^{(k)}\right) D_z v_z + \lambda^{(k)} \left(D_y v_y\right) \\
&+ \left(\lambda^{(k+1)} + 2\mu^{(k+1)}\right) D_z \dot{v}_z + \lambda^{(k+1)} \left(D_y \dot{v}_y\right)\Big]_{i,j,k}^{n+\frac{1}{2}}, \\
\frac{(\tau_{xy}^{(a)\|})_{i+\frac{1}{2},j+\frac{1}{2},k}^{n+1} - (\tau_{xy}^{(a)\|})_{i+\frac{1}{2},j+\frac{1}{2},k}^{n}}{\Delta t} &= \left[\mu^{(k)} \left(D_y v_x\right) + \mu^{(k+1)} \left(D_y \dot{v}_x\right)\right]_{i+\frac{1}{2},j+\frac{1}{2},k}^{n+\frac{1}{2}}, \\
\frac{(\tau_{xz}^{(a)\|})_{i+\frac{1}{2},j,k+\frac{1}{2}}^{n+1} - (\tau_{xz}^{(a)\|})_{i+\frac{1}{2},j,k+\frac{1}{2}}^{n}}{\Delta t} &= \left[\mu^{(k)} \left(D_z v_x\right) + \mu^{(k+1)} \left(D_z \dot{v}_x\right)\right]_{i+\frac{1}{2},j,k+\frac{1}{2}}^{n+\frac{1}{2}}, \\
\frac{(\tau_{yz}^{(a)\|})_{i,j+\frac{1}{2},k+\frac{1}{2}}^{n+1} - (\tau_{yz}^{(a)\|})_{i,j+\frac{1}{2},k+\frac{1}{2}}^{n}}{\Delta t} &= \left[\mu^{(k)} \left(D_z v_y + D_y v_z\right) + \mu^{(k+1)} \left(D_z \dot{v}_y + D_y \dot{v}_z\right)\right]_{i,j+\frac{1}{2},k+\frac{1}{2}}^{n+\frac{1}{2}}, \\
\frac{(v_x^{\|})_{i+\frac{1}{2},j,k}^{n+\frac{1}{2}} - (v_x^{\|})_{i+\frac{1}{2},j,k}^{n-\frac{1}{2}}}{\Delta t} &= b^{(a)} \left[D_y \tau_{xy}^{(a)} + D_z \tau_{xz}^{(a)}\right]_{i+\frac{1}{2},j,k}^{n}, \\
\frac{(v_y^{\|})_{i,j+\frac{1}{2},k}^{n+\frac{1}{2}} - (v_y^{\|})_{i,j+\frac{1}{2},k}^{n-\frac{1}{2}}}{\Delta t} &= b^{(a)} \left[D_y \tau_{yy}^{(a)} + D_z \tau_{yz}^{(a)}\right]_{i,j+\frac{1}{2},k}^{n}, \\
\frac{(v_z^{\|})_{i,j,k+\frac{1}{2}}^{n+\frac{1}{2}} - (v_z^{\|})_{i,j,k+\frac{1}{2}}^{n-\frac{1}{2}}}{\Delta t} &= b^{(a)} \left[D_y \tau_{yz}^{(a)} + D_z \tau_{zz}^{(a)}\right]_{i,j,k+\frac{1}{2}}^{n}.
\end{aligned}
\tag{11.4.2}
$$

Upon discretization, the PML is no longer perfectly matched. However, the reflection has been proved to be sufficiently small by Collino and Tsogka [92]. For the choice of quadratic damping parameter $d(x)$

$$
d(x) = d_0 \left(\frac{x}{\delta}\right)^2 \tag{11.4.3}
$$

with

$$
d_0 = \log\left(\frac{1}{R}\right) \frac{3V}{2\delta} \tag{11.4.4}
$$

where R is the theoretical reflection coefficient, δ is the width of the PML layer in terms of grid points, and V is a representative velocity. The reflection coefficients are about 1% for a $5h$ PML width, 0.1% for a $10h$ PML width and 0.01% for a $20h$ PML width.

In the present formulation, Eqs. (11.4.1) and (11.4.2) form a closed system of equations for the stress and velocity fields. Free-surface conditions are imposed at the boundaries of the domain. An auxiliary computation determines the displacements from the velocities at every time step, assuming their respective components are defined at the same grid points but staggered temporally. Using centered finite differences in time, the discretization of (11.2.9) is given by

$$\begin{aligned} u^{n+1}_{xi+\frac{1}{2},j,k} &= u^{n}_{xi+\frac{1}{2},j,k} + \Delta t\, v^{n+\frac{1}{2}}_{xi+\frac{1}{2},j,k}, \\ u^{n+1}_{yi,j+\frac{1}{2},k} &= u^{n}_{yi,j+\frac{1}{2},k} + \Delta t\, v^{n+\frac{1}{2}}_{yi,j+\frac{1}{2},k}, \\ u^{n+1}_{zi,j,k+\frac{1}{2}} &= u^{n}_{zi,j,k+\frac{1}{2}} + \Delta t\, v^{n+\frac{1}{2}}_{zi,j,k+\frac{1}{2}}. \end{aligned} \tag{11.4.5}$$

In [154] we investigated randomly created bone examples, and compared them with CT scans as we were interested in the comparison between such simulated domain and real bone samples. Considering the fact that real bone is orthotropic, the results were not bad at all.

The real bone sample is obtained from microCT images of human calcaneus, as shown in the left portion of Figure 11.4.2. The image was taken in the transverse plane, so that the top of the image is the anterior side, the bottom of the image is the posterior aspect of the bone, the right hand side of the image is the lateral aspect of the bone, and the left hand side of the image is the medial aspect of the bone. See our paper [154] for numerical simulations. From the picture, we can see that the solid frame shows strong anisotropy in the vertical direction. The bone frame shrinks as β increases, showing more and more serious osteoporosis, and the anisotropy property became less obvious for high-porosity scans [154].

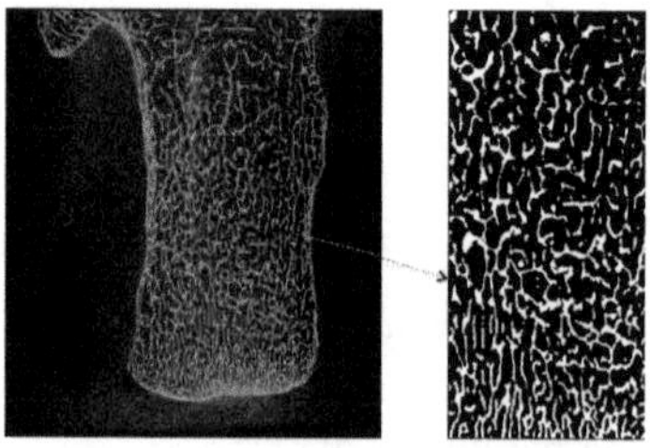

FIGURE 11.4.2 Left: microCT image for human calcaneus. Right: binary mapping of a rectangle cut from the calcanei sample.

The random simulated domain is obtained using the efficient and accurate, 'turning-band' method indexturning-band method of Mantoglou and Wilson [288] (see also [386]). It is a simulated multidimensional stationary random field with a prescribed covariance structure. The idea of the method is, instead of generating a high-dimensional field directly, several independent one-dimensional processes are simulated and superimposed. Details of the method can be found in [288] (see also [386],[153]). With this method, we have an example of a random two-dimensional field in Figure (11.4.3). Each point of the field is normally distributed, and the entire field has an exponential covariance structure.

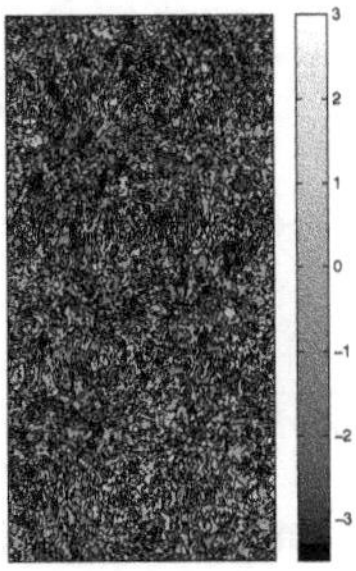

FIGURE 11.4.3 An example of a two-dimensioinal random field.

To turn the field into random 'fluid' or 'solid' points, we created a procedure, which selects whether a point is either fluid or solid. Since the field value at each point is normally distributed, this categorization can be achieved by using the error function.

11.4.1 ORTHOTROPIC RANDOM BONE

For a given point $\mathbf{x} \in \mathbb{R}^n, n = 2,3$ the random function $f(\mathbf{x})$ is a random field. If E is the expectation, the mean of $f(\mathbf{x})$ is defined as $m(\mathbf{x}) := E[f(\mathbf{x})]$. The covariance function is thereby defined as

$$C(\mathbf{x}_1,\mathbf{x}_2) := E\left[(f(\mathbf{x}_1) - m(\mathbf{x}_1))(f(\mathbf{x}_2) - m(\mathbf{x}_2))\right], \quad \forall \mathbf{x}_1,\mathbf{x}_2 \in \mathbb{R}^n.$$

For simplicity we shall only consider second-order stationary random fields, which assumes the following:

1. The mean is independent of the position of each point, i.e.

$$E\left[f(\mathbf{x})\right] = m(\mathbf{x}) = m, \in, \forall \mathbf{x} \in \mathbb{R}^n$$

2. The covariance function depends only on the vector difference between two points, i.e.

$$C\left(\mathbf{x}_1,\mathbf{x}_2\right) = C\left(\mathbf{x}_1 - \mathbf{x}_2\right) = C\left(\mathbf{h}\right), \forall \mathbf{x}_1,\mathbf{x}_2 \in \mathbb{R}^n,$$

   where $\mathbf{h} := \mathbf{x}_1 - \mathbf{x}_2$.

In her thesis Jing Li [271], see also [153, 154], considers the special case where the covariance depends only on the positive distance between the two points. In this case the covariance is said to be isotropic. We shall indicate here how that situation might be extended to orthotropic cases. A zero-mean, real-valued, multi-dimensional, random function $Z(\mathbf{x})$ can be viewed as the real part of a complex random function Z(x) [391]

$$z(\mathbf{x}) = \Re\{Z(\mathbf{x})\} = \Re\left\{\sum_k Z_k e^{i\omega_k \mathbf{x}}\right\} \approx \int E^{i\omega\cdot\mathbf{x}} Z(d\omega) \quad (11.4.6)$$

The random amplitudes $Z(d\omega)$ are uncorrelated random processes with zero mean in the frequency domain, such that

$$E[Z(d\omega)] = 0$$

and

$$E\left[\overline{Z[d\omega_1)} Z[d\omega_2)\right] = 0$$

for non-overlapping regions $d\omega_1)$ and $d\omega_2)$ and where the complex conjugate of $Z[d\omega)$ is indicated by $\overline{Z[d\omega)}$ The connection between the covariance function and the spectral density are given by [391]

$$C(\mathbf{h}) = \int_{\mathbb{R}^n} S(\omega) e^{i\omega\cdot\mathbf{h}}\, d\mathbf{h}$$

$$S(\omega) = \frac{1}{(2\pi)^n} \int_{\mathbb{R}^n} C(\mathbf{h}) e^{i\omega\cdot\mathbf{h}} d\omega. \quad (11.4.7)$$

In the paper by Montoglou and Wilson [288], the two-dimensional case, where the field is isotropic and $|\mathbf{h}| = r$, the positive distance is considered. This means that when performing the indicated integrations in 11.4.7 one integrates over circles of radius r and then over the angle from 0 to 2π, i.e. the Cartesian representation

$$S(\omega_1, \omega_2) = \frac{1}{(2\pi)^2} \int_{-\infty}^{\infty}\int_{-\infty}^{\infty} C(h_1, h_2) \cos(\omega_1 h_1 + \omega_2 h_2)\, dh_1\, dh_2 \quad (11.4.8)$$

becomes, using polar coordinates of integration,

$$h_1 = r\cos(\theta), \quad h_2 = r\sin(\theta)$$

$$\omega_1 = r\cos(\phi), \quad h_2 = \omega\sin(\phi)$$

Equation (11.4.8) becomes

$$S(\omega, \phi) = \frac{1}{(2\pi)^2} \int_0^{\infty}\int_0^{2\pi} C(r, \theta) (\cos(\omega r\cos(\theta)\cos(\phi) + \omega r\sin(\theta)\sin(\phi))\, r\, dr\, d\theta$$

$$= \frac{1}{(2\pi)^2} \int_0^\infty \int_0^{2\pi} C(r,\theta) \cos(\omega r \cos(\theta - \phi))\, r dr d\theta$$

If now $C(r,\theta) = C(r)$, then we may perform the θ integration first and obtain

$$= \frac{1}{(2\pi)^2} \int_0^\infty C(r) J_0(\omega r)\, r dr d\theta \tag{11.4.9}$$

Hence, in the special case of isotropy the spectral density is a Hankel transform of the covariance [271]. Since the Hankel transform is reciprocal we also have [111]

$$C(r) = \frac{1}{(2\pi)^2} \int_0^\infty S(r) J_0(\omega r)\, r dr d\theta. \tag{11.4.10}$$

Mantoglou and Wilson [288] make a list of some two-dimensional covariance functions. In particular, they are able to compute the corresponding radial spectral density functions. We shall try to emulate some of these calculations for the trans-isotropic case in $\mathbb{R}^2$. To emulate the trans-isotropic case we assume that the correlation function differs from the case of isotropy in that the points do not lie on a circle from the *center* point but instead lie on an ellipse, say

$$x^2 + \frac{y^2}{\alpha^2} = \rho^2,$$

where α is a fixed constant defining the eccentricity of the ellipse. In this case of trans-isotropy, $C(r,\theta)$ is not a function of the positive distance between two points, but lies on the distance between two points, one lying on an ellipse and the other at the ellipse center, in other words

$$C(r,\theta) = C(r^2[\cos^2(\theta) + \frac{1}{\alpha^2}\sin^2(\theta)]) \tag{11.4.11}$$

Consider

$$S(\omega,\phi) = \frac{1}{(2\pi)^2} \int_0^\infty \int_0^{2\pi} C\left(r^2[\cos^2(\theta) + \frac{1}{\alpha^2}\sin^2(\theta)]\right) \cos(\omega r \cos(\theta - \phi))\, r dr d\theta$$

If, for instance, $C(\rho) = \exp(-b\rho^2)$ we have integrals of the form

$$\frac{1}{(2\pi)^2} \int_0^\infty \int_0^{2\pi} e^{-b\left(\left(r^2[\cos^2(\theta) + \frac{1}{\alpha^2}\sin^2(\theta)\right)\right)} \cos(\omega r \cos(\theta - \phi))\, r dr d\theta$$

Integrals of this type may be done using infinite series expansions and integrating the trigonometric terms directly.

$$\mathrm{erf}(z) = \frac{2}{\sqrt{\pi}} \int_0^z e^{-t^2} dt,$$

which is related to the cumulative density function of a standard normal distribution by

$$\Phi(z) = \frac{1}{2} + \frac{1}{2}\mathrm{erf}\left(\frac{z}{\sqrt{2}}\right). \tag{11.4.12}$$

Suppose z_0 is the critical point such that the probability

$$P(-z_0 < z < z_0) = \Phi(z_0) - \Phi(-z_0) = \beta,$$

then z_0 can be found by using (11.4.12) as

$$z_0 = \sqrt{2}\,\mathrm{erf}^{-1}(\beta).$$

Therefore a point in the domain is fluid if its field value is in $[-z_0, z_0]$, otherwise it is a solid point. By this selection, we can guarantee that a proportion of β points out of the total number are fluid. If the probability distribution is not standard normal, then we can always make it so by a change of variables. Based on the field realization of Figure 11.4.3, we showed in [154] examples of random spatial distributions of fluid-solid points for $\beta = 0.7$, 0.75, 0.8, 0.85, using the categorization described above. As expected, the proportion of fluid points also increases with β, showing the same tendency as real bone scans. The difference from real samples is that the solid patch in the simulated domain is isotropic and more like little islands, rather than strings.

Comparing a real bone sample and the simulated domain, we see clearly that the real sample has a strong anisotropic structure while the simulated domain is isotropic according to its formulation. However, anisotropy, or more simply transverse isotropy, can be included in the random realizations by varying the correlation function to reflect this. Work is in progress concerning whether we can construct randomly distributed bone with orthotropic structure. For details concerning the numerical results we refer you to the papers [152, 153, 154]. In particular for a detailed description of the turning- band method and the numerical results of this procedure see Jing Li's PhD thesis [271].

ACKNOWLEDGMENTS

Thanks to Dr. Luis Cardoso from the City University of New York for providing the micro-CT images.

12 Bone Growth and Adaptive Elasticity

In the 1970s, Cowin and Hegedus proposed a mathematical model of bone deposition and reabsorption based on Wolff's Law [96] [97] [388]. According to Wolff's Law, bone is either deposited or reabsorbed according to the loading of the bone or the lack of loading. The theory of adaptive elasticity grew out of the works by [96], [97] and [388]. There are force balance equations for the domain Ω and for the total boundary Γ, as well as a rate equation for the remodeling (deposition and reabsorption) of bone. From them arise two unknowns: the displacement $\mathbf{u}$ and the change in volume fraction β of elastic bone material from some reference configuration. The two are coupled due to their presence together in the rate remodeling equation. In what follows, we consider the special case of a poro-elastic plate of thickness $\varepsilon \leq 1$. The thickness $\varepsilon \ll 1$ of the plate induces the use of asymptotic expansions for both the displacement $\mathbf{u}$ and the change in the volume fraction of elastic material β.[1]

In this chapter we also consider the adaptive elasticity model for thin plates, since medical imaging techniques (e.g. ultrasound and micro-computer tomography analysis) show that the trabeculae of healthy bone may be approximated by a mixture of rods [388] and plates. What is significant is that in the Trabucho study, as well as in what follows, the plate equations contain the time t only as a parameter, i.e. not as a differential in the usual elastic formation.

12.1 THE MODEL

The domain we consider is a plate, and in three-dimensional, Cartesian geometry, we visualize this as a cylinder whose axis coincides with the x_3 direction. The intersection of the domain with a plane perpendicular to the x_3 axis is a closed planar region ω. We assume the plate is of thickness 2ε and lies between $x_3 \pm \varepsilon$. We also assume that the boundary of ω is a smooth curve γ, see Figure 12.1.1. Hence, $\Omega^\varepsilon = \omega \times (-\varepsilon, \varepsilon)$ and $\Omega = \omega \times (-1, 1)$. The lateral boundary of Ω^ε, $\Gamma_0^\varepsilon = \gamma \times [-\varepsilon, \varepsilon]$; whereas, the lateral boundary of Ω, $\Gamma^0 = \gamma \times [-1, 1]$. The upper and lower boundaries of Ω^ε are $\Gamma_+^\varepsilon = \omega \times \{\varepsilon\}$ and $\Gamma_-^\varepsilon = \omega \times \{-\varepsilon\}$, respectively. For the top and bottom of Ω, there are $\Gamma^+ = \omega \times \{1\}$ and $\Gamma^- = \omega \times \{-1\}$. It is assumed that the constant $\varepsilon > 0$ and ε is much smaller than the dimensions of ω. The plate is assumed to be clamped so that a displacement vector $\mathbf{u}$ vanishes on the lateral boundary.

[1] Trabucho in his work with Viaño [388] had done an exhaustive characterization of linearly elastic rods. Then, later in a paper with Figueiredo [125], he extended his study to rods with a remodeling rate equation that is nonlinear with respect to the strain tensor, while the stress tensor itself is still linear with respect to the strain.

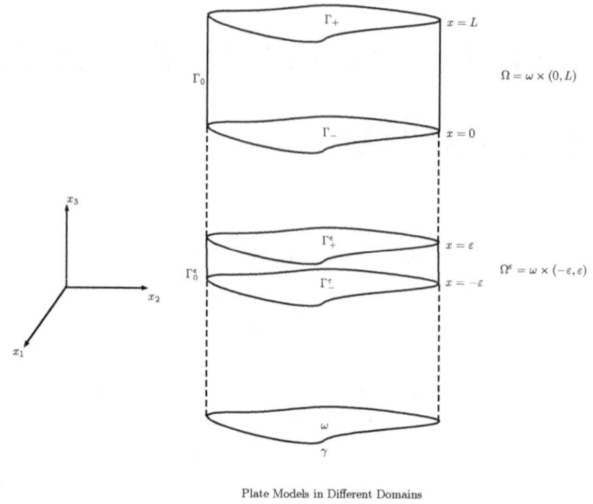

FIGURE 12.1.1 The plate of thickness 2ε

The notation for subscripts is the usual for tensor notation, namely, Latin indices take on values of –1, 2, 3″ whereas Greek indices take on values of –1,2″. For repeated indices, the summation convention will be employed. The plate Ω^ε is composed of a linearly elastic, homogeneous and isotropic material whose Lamè constants λ^ε and μ^ε are related to Young's modulus E^ε and Poisson's ratio ν^ε as follows:

$$\lambda^\varepsilon = \frac{\nu^\varepsilon E^\varepsilon}{(1+\nu^\varepsilon)(1-2\nu^\varepsilon)}$$

$$\mu^\varepsilon = \frac{E^\varepsilon}{2(1+\nu^\varepsilon)}$$

Volume forces, denoted by $\mathbf{f}^\varepsilon = (f_i^\varepsilon) : \Omega^\varepsilon \Rightarrow \mathbf{R}^3$, act in the interior of Ω^ε and surface forces, denoted by $\mathbf{g}^\varepsilon = (g_i^\varepsilon) : \Omega^\varepsilon \Rightarrow \mathbf{R}^3$, act on the boundaries $\Gamma^\varepsilon = \Gamma_0^\varepsilon \cup \Gamma_+^\varepsilon \cup \Gamma_-^\varepsilon$.

The equilibrium equations may be used to find the displacement $\mathbf{u}^\varepsilon = (u_i^\varepsilon) : \bar{\Omega} \Rightarrow \mathbf{R}^3$:

$$-\partial_j^\varepsilon[\lambda e_{pp}^\varepsilon(\mathbf{u}^\varepsilon)\delta_{ij} + 2\mu^\varepsilon e_{ij}^\varepsilon(\mathbf{u}^\varepsilon)] = f_i^\varepsilon, \; on \; \Omega^\varepsilon$$

$$[\lambda^\varepsilon e_{pp}^\varepsilon(\mathbf{u}^\varepsilon)\delta_{ij} + 2\mu^\varepsilon e_{ij}^\varepsilon(\mathbf{u})]n_j^\varepsilon = g_i^\varepsilon, \; on \; \Gamma^\varepsilon$$

where Γ^ε is the boundary of the plate on the side and on the top and bottom,

$$(\sigma^\varepsilon(\mathbf{u}^\varepsilon))_{ij} := \sigma_{ij}^\varepsilon(\mathbf{u}^\varepsilon) = \lambda^\varepsilon e_{pp}(\mathbf{u}^\varepsilon)\delta_{ij} + 2\mu^\varepsilon e_{ij}^\varepsilon(\mathbf{u}^\varepsilon)$$

$$e_{ij}^\varepsilon(\mathbf{u}^\varepsilon) = \frac{1}{2}(\partial_i^\varepsilon u_j^\varepsilon + \partial_j^\varepsilon u_i^\varepsilon).$$

Furthermore, the boundary is assumed to be clamped; i.e. $\mathbf{u}^\varepsilon = \mathbf{0}$ on the lateral boundary Γ_0^ε. The variational formulation of this boundary value problem may be formulated as

$$\mathbf{u}^\varepsilon \in V(\Omega^\varepsilon) = \{\mathbf{v}^\varepsilon = (v_i^\varepsilon) \in [H^1(\Omega^\varepsilon)]^3 : \mathbf{v}^\varepsilon = \mathbf{0} \in \Gamma_+^\varepsilon \cup \Gamma_-^\varepsilon\},$$

$$\int_{\Omega^\varepsilon} \sigma_{ij}^\varepsilon(\mathbf{u}^\varepsilon) e_{ij}^\varepsilon(\mathbf{v}^\varepsilon) d\mathbf{x}^\varepsilon = \int_{\Omega^\varepsilon} f_i^\varepsilon v_i^\varepsilon d\mathbf{x}^\varepsilon + \int_{\Gamma_0^\varepsilon} g_i^\varepsilon v_i^\varepsilon ds^\varepsilon + \int_{\Gamma_+^\varepsilon \cup \Gamma_-^\varepsilon} h_i^\varepsilon v_i d\omega$$

for all $\mathbf{v}^\varepsilon \in V(\Omega^\varepsilon)$.

If $f_i^\varepsilon \in L^2(\Omega^\varepsilon)$ and $g_i^\varepsilon \in L^2(\Gamma^\varepsilon)$, then the problem is well-posed. The left side of the above equation is symmetric and continuous. The right side is continuous as well.

Finally, there is a rate equation for the deposition and reabsorption of bone:

$$\dot{\beta}^\varepsilon = a^\varepsilon(\beta^\varepsilon) + A_{kl}^\varepsilon(\beta^\varepsilon) e_{kl}^\varepsilon(u^\varepsilon) + B_{ijkm}(\beta^\varepsilon) e_{ij}^\varepsilon(u^\varepsilon) e_{km}^\varepsilon(u^\varepsilon)$$

where $\dot{\beta}^\varepsilon$ is the time rate of change in the volume fraction of elastic material and A_{kl}^ε and B_{ijkm}^ε are remodeling rate coefficients. The right-hand side may be thought of as a second-order, polynomial approximation to $\dot{\beta}^\varepsilon$ in the u^ε. Higher-order, approximations could also be considered. Since there is symmetry about the Ox_3 axis:

$$B_{ijkm}(\beta^\varepsilon) = B_{jikm}(\beta^\varepsilon) = B_{kmij}(\beta^\varepsilon) = B_{ijkm}(\beta^\varepsilon),$$

and

$$B_{\alpha\beta\gamma 3}(\beta^\varepsilon) = 0, \quad B_{\alpha 333}(\beta^\varepsilon) = 0.$$

Moreover, we require an elliptic condition

$$\exists c > 0 : B_{ijkm}(\beta^\varepsilon) \tau_{ij} \tau_{km} \geq c \|\tau_{ij}\|^2, \ \forall \tau_{ij} \in \mathbb{R}^{3\times 3}.$$

12.2 SCALINGS OF UNKNOWNS

For the scaling of unknowns, their derivatives, and force fields $f_i^\varepsilon(\mathbf{x}^\varepsilon), g_i^\varepsilon(\mathbf{x}^\varepsilon)$, and $h_i^\varepsilon(\mathbf{x}^\varepsilon)$, we follow the ideas of Ciarlet [87]. Functions in the reference domain, Ω, are related to those in Ω^ε and follow the rules

$\mathbf{x}^\varepsilon := (x_i^\varepsilon) = (x_1, x_2, \varepsilon x_3)$

$\partial_\alpha^\varepsilon = \partial_\alpha$ and $\partial_3^\varepsilon = \frac{1}{\varepsilon}\partial_3$

$v_\alpha^\varepsilon(\mathbf{x}^\varepsilon) = v_\alpha(\mathbf{x})$ and $v_3^\varepsilon(\mathbf{x}^\varepsilon) = \frac{1}{\varepsilon} v_3(\mathbf{x}), \ \forall x^\varepsilon \in \overline{\Omega}$

$f_\alpha^\varepsilon(\mathbf{x}^\varepsilon) = f_\alpha(\mathbf{x})$ and $f_3^\varepsilon(\mathbf{x}^\varepsilon) = \varepsilon f_3(\mathbf{x}), \ \forall \mathbf{x}^\varepsilon \in \Omega^\varepsilon$

$g_\alpha^\varepsilon(\mathbf{x}^\varepsilon) = \varepsilon g_\alpha(\mathbf{x})$ and $g_3^\varepsilon(\mathbf{x}^\varepsilon) = \varepsilon^2 g_3(\mathbf{x}), \ \forall \mathbf{x}^\varepsilon \in \Gamma_0^\varepsilon$

$h_\alpha^\varepsilon(\mathbf{x}^\varepsilon) = h_\alpha(\mathbf{x})$ and $h_3^\varepsilon(\mathbf{x}^\varepsilon) = \varepsilon h_3(\mathbf{x}), \ \forall \mathbf{x}^\varepsilon \in \Gamma_+^\varepsilon \cup \Gamma_-^\varepsilon$

$$\sigma^{\varepsilon}_{\alpha\beta}(x^{\varepsilon}) = \sigma_{\alpha\beta}(\mathbf{u}(\varepsilon))(x),\ \sigma^{\varepsilon}_{\alpha 3}(x^{\varepsilon}) = \varepsilon\sigma_{\alpha\beta}(\mathbf{u}(\varepsilon))(x),\ \sigma^{\varepsilon}_{33}(x^{\varepsilon})(x^{\varepsilon}) = \varepsilon^2\sigma_{33}(\mathbf{u}(\varepsilon))(x)$$

$$\mathbf{e}^{\varepsilon}_{\alpha\beta}(\mathbf{u}(\varepsilon)) = \mathbf{e}_{\alpha\beta}(\mathbf{u}(\varepsilon))(x),\ \mathbf{e}^{\varepsilon}_{\alpha\beta}(\mathbf{u}^{\varepsilon}) = \tfrac{1}{\varepsilon}\mathbf{e}_{\alpha 3}(\mathbf{u}(\varepsilon))(x),\ \mathbf{e}^{\varepsilon}_{33}(\mathbf{u}(\varepsilon)) = \tfrac{1}{\varepsilon^2}\mathbf{e}_{33}(\mathbf{u}(\varepsilon))(x)$$

where Γ^{ε}_{+} and Γ^{ε}_{-} represent the top and the bottom of the plate respectively. This leads to the following equation which can be found in [87] (p. 29):

$$\begin{aligned}
&\int_{\Omega}\{\lambda e_{\sigma\sigma}(\mathbf{u}(\varepsilon))e_{\tau\tau}(\mathbf{v}) + 2\mu e_{\alpha\beta}(\mathbf{u}(\varepsilon))e_{\alpha\beta}(\mathbf{v})\}d\mathbf{x} \\
&+\frac{1}{\varepsilon^2}\int_{\Omega}\{\lambda e_{\sigma\sigma}(\mathbf{u}(\varepsilon))e_{33}(\mathbf{v}) + \lambda e_{33}(\mathbf{u}(\varepsilon))e_{\tau\tau}(\mathbf{v}) + 4\mu e_{\alpha 3}(\mathbf{u}(\varepsilon))e_{\alpha 3}(\mathbf{v})\}d\mathbf{x} \\
&+\frac{1}{\varepsilon^4}\int_{\Omega}(\lambda + 2\mu)e_{33}(\mathbf{u}(\varepsilon))e_{33}(\mathbf{v})d\mathbf{x} = \int_{\Omega} f_i v_i d\mathbf{x} + \int_{\Gamma^0} g_i v_i ds \\
&+\int_{\Gamma_+\cup\Gamma_-} h_i v_i d\omega,\ \forall\mathbf{v}\in\mathbf{V}(\Omega).
\end{aligned}$$

The scalings of the previous quantities will then be substituted into the remodeling rate equation:

$$\begin{aligned}
\beta(\varepsilon) &= a^{\varepsilon}(\beta^{\varepsilon}) + A_{\alpha\beta}(\beta^{\varepsilon})e_{\alpha\beta}(\varepsilon) + B_{\alpha\beta\gamma\mu}(\beta^{\varepsilon})e_{\gamma\mu}(\varepsilon)e_{\alpha\beta}(\varepsilon) + \frac{2}{\varepsilon}A_{3\beta}(\beta^{\varepsilon})e_{3\beta}(\varepsilon) \\
&+\frac{1}{\varepsilon^2}[A_{33}(\beta^{\varepsilon})e_{33}(\varepsilon) + 4B_{3\alpha 3\beta}(\beta^{\varepsilon})e_{3\beta}(\varepsilon)e_{3\alpha}(\varepsilon) + B_{\alpha\beta 33}(\beta^{\varepsilon})e_{33}(\varepsilon)e_{\alpha\beta}(\varepsilon)] \\
&+\frac{1}{\varepsilon^4}B_{3333}(\beta^{\varepsilon})e_{33}(\varepsilon)e_{33}(\varepsilon).
\end{aligned}$$

12.3 ASYMPTOTIC SOLUTIONS

Following Ciarlet [88] the unknowns are expanded in ascending powers of ε.

$$\sigma_{\alpha\beta}(\varepsilon) = \varepsilon^{-2}\sigma^{-2}_{\alpha\beta} + \varepsilon^{-1}\sigma^{-1}_{\alpha\beta} + \sigma^{0}_{\alpha\beta} + \dots$$

$$\sigma_{\alpha 3}(\varepsilon) = \varepsilon^{-2}\sigma^{-2}_{\alpha 3} + \varepsilon^{-1}\sigma^{-1}_{\alpha 3} + \dots$$

$$\sigma_{33}(\varepsilon) = \varepsilon^{-4}\sigma^{-4}_{33} + \varepsilon^{-3}\sigma^{-3}_{33} + \dots$$

$$\mathbf{u}(\varepsilon) = \mathbf{u}^0 + \varepsilon\mathbf{u}^1 + \varepsilon^2\mathbf{u}^2 + \varepsilon^3\mathbf{u}^3 + \dots$$

Matching like powers of ε leads to the following:

$$\sigma^{-2}_{\alpha\beta} = \lambda e_{33}(\mathbf{u}^0)\delta_{\alpha\beta} \tag{12.3.1}$$

$$\sigma^{-1}_{\alpha\beta} = \lambda e_{33}(\mathbf{u}^1)\delta_{\alpha\beta} \tag{12.3.2}$$

$$\sigma^{p}_{\alpha\beta} = \lambda e_{\mu\mu}(\mathbf{u}^p)\delta_{\alpha\beta} + 2\mu e_{\alpha\beta}(\mathbf{u}^p) + \lambda e_{33}(\mathbf{u}^{p+2})\delta_{\alpha\beta},\ p \geq 0 \tag{12.3.3}$$

$$\sigma^{p}_{\alpha 3} = 2\mu e_{\alpha 3}(\mathbf{u}^{p+2}),\ p \geq -2 \tag{12.3.4}$$

$$\sigma_{33}^{-4} = (\lambda + 2\mu)e_{33}(\mathbf{u}^0) \tag{12.3.5}$$

$$\sigma_{33}^{-3} = (\lambda + 2\mu)e_{33}(\mathbf{u}^1) \tag{12.3.6}$$

$$\sigma_{33}^{p} = (\lambda + 2\mu)e_{33}(\mathbf{u}^{p+4}) + \lambda e_{\mu\mu}(\mathbf{u}^{p+2}),\ p \geq -2. \tag{12.3.7}$$

With these, the equation of equilibrium (12.2.1) now reads:

$$\varepsilon^{-4}\int^{\Omega} \sigma_{33}^{-4} e_{33}^{-4} e_{33}(\mathbf{v})d\mathbf{x} + \varepsilon^{-3}\int_{\Omega} \sigma_{33}^{-3} e_{33}(\mathbf{v})d\mathbf{x} + \varepsilon^{-2}\int_{\Omega} \sigma_{ij}^{-2} e_{ij}(\mathbf{v})d\mathbf{x} +$$
$$+ \varepsilon^{-1}\int_{\Omega} \sigma_{ij}^{-1} e_{ij}(\mathbf{v})d\mathbf{x} + \int_{\Omega} \sigma_{ij}^{p} e_{ij}(\mathbf{v})d\mathbf{x} = \int_{\Omega} f_i v_i d\mathbf{x} + \int_{\Gamma^0} g_i v_i ds + \int_{\Gamma^+\cup\Gamma^-} h_i v_i d\omega$$
$$+ \sum_{p\geq 1} \varepsilon^{p}\int_{\Omega} \sigma_{ij}^{p} e_{ij}(\mathbf{v})d\mathbf{x},\ \forall \mathbf{v} \in V(\Omega)$$

Setting like powers of ε on each side equal to one another gives:

$$\varepsilon^{-4} : \int_{\Omega} \sigma_{33}^{-4} e_{33}(\mathbf{v})d\mathbf{x} = 0 \tag{12.3.8}$$

$$\varepsilon^{-3} : \int_{\Omega} \sigma_{33}^{-3} e_{33}(\mathbf{v})d\mathbf{x} = 0 \tag{12.3.9}$$

$$\varepsilon^{-2} : \int_{\Omega} \sigma_{ij}^{-2} e_{ij}(\mathbf{v})d\mathbf{x} = 0 \tag{12.3.10}$$

$$\varepsilon^{-1} : \int_{\Omega} \sigma_{ij}^{-1} e_{ij}(\mathbf{v})d\mathbf{x} = 0 \tag{12.3.11}$$

$$\varepsilon^{0} : \int_{\Omega} \sigma_{ij}^{0} e_{ij}(\mathbf{v})d\mathbf{x} = \int_{\Omega} f_i v_i d\mathbf{x} + \int_{\Gamma^0} g_i v_i ds + \int_{\Gamma^+\cup\Gamma^-} h_i v_i d\omega \tag{12.3.12}$$

$$\varepsilon^{p} : \int_{\Omega} \sigma_{ij}^{p} e_{ij}(\mathbf{v})d\mathbf{x} = 0,\ p \geq 1. \tag{12.3.13}$$

The following results follow.

$$\sigma_{33}^{-4} = 0,\ \sigma_{\alpha 3}^{-2} = 0,\ \sigma_{\alpha\beta}^{-2} = 0,\ \int_{\Omega} \sigma_{33}^{-2} e_{33}(\mathbf{v})d\mathbf{x} = 0, \forall \mathbf{v} \in V(\Omega)$$
$$\mathbf{u}_3^0 = \mathbf{u}_3^0(x_1, x_2),\ \mathbf{u}_\beta^0 = \zeta_\beta^0(x_1, x_2) - x_3\frac{\partial u_3^0}{\partial x_\beta}$$
$$\sigma_{33}^{-3} = 0,\ \sigma_{3\beta}^{-1} = 0,\ \sigma_{\alpha\beta}^{-1} = 0,\ \int_{\Omega} \sigma_{33}^{-1} e_{33}(\mathbf{v})d\mathbf{x} = 0, \forall \mathbf{v} \in V(\Omega)$$

To find out more about σ_{33}^{-2} and σ_{33}^{-1}, we repeat an argument by Ciarlet and Destuynder [88].

Theorem 12.1

For $n = \{-2, -1\}$, if $\sigma_{\alpha\beta}^{n} = 0 = \sigma_{3\beta}^{n}$, then $\sigma_{33}^{n} = 0$. ■

Proof. For $n = \{-2, -1\}$,

$$\int_{\Omega} \sigma_{ij}^n e_{ij} d\mathbf{x} = 0 = \int_{\Omega} (\sigma_{\alpha\beta}^n e_{\alpha\beta} + \sigma_{3\beta}^n e_{3\beta} + \sigma_{33}^n e_{33}) d\mathbf{x}$$

Using Green's formula, we have

$$\begin{aligned} \int_{\Omega} (\sigma_{33}^n \frac{\partial v_3}{\partial x_3} + \sigma_{3\beta}^n \frac{\partial v_3}{\partial x_\beta}) d\mathbf{x} &= -\int_{\Omega} (\frac{\partial \sigma_{33}^n}{\partial x_3} v_3 + \frac{\partial \sigma_{3\beta}^n}{\partial x_\beta} v_3) d\mathbf{x} \\ &+ \int_{\Gamma^+ \cup \Gamma^-} \sigma_{33}^n d\mathbf{x} + \int_{\Gamma^+ \cup \Gamma^-} \sigma_{3\beta}^n v_3 d\mathbf{x} \end{aligned}$$

Since $\int_{\Omega} (\sigma_{33}^n e_{33} + \sigma_{3\beta}^n e_{3\beta}) d\mathbf{x} = 0$, now $\sigma_{3\beta} = 0$, so $0 = \frac{\partial \sigma_{3\beta}^n}{\partial x_3} = \frac{\partial \sigma_{3\beta}^n}{\partial x_\beta} = \frac{\partial \sigma_{33}^n}{\partial x_3}$. Thus, σ_{33}^n is a constant with respect to x_3; hence, it follows that

$$0 = \int_{\Gamma^+ \cup \Gamma^-} \sigma_{33}^n v_3 d\mathbf{x} = \int_{\Gamma^+} \sigma_{33}^n v_3 d\mathbf{x} - \int_{\Gamma^-} \sigma_{33}^n v_3 d\mathbf{x} \ \forall v_3 (\Gamma^+ \cup \Gamma^-).$$

So $\sigma_{33}^n = 0$ on $\Gamma^+ \cup \Gamma^-$. However, $\frac{\partial \sigma_{33}^n}{\partial x_3} = 0$, so $\sigma_{33}^n = 0$ in Ω. □

Therefore, all negative powers of σ_{ij} are zero.

Corollary 12.1. *For any positive odd integer n, if* $\sigma_{3\beta}^n = 0 = \sigma_{\alpha\beta}^n$, *then* $\sigma_{33}^n = 0$.

Proof. For any positive odd integer n, $\int_{\Omega} \sigma_{ij}^n e_{ij}(\mathbf{v}) d\mathbf{x} = 0$. Given that $\sigma_{\alpha\beta}^n = 0$, this simplifies to $\int_{\Omega} (\sigma_{33}^n e_{33} + \sigma_{3\beta}^n e_{3\beta}) d\mathbf{x} = 0$. Now, Green's formula can be applied to this equation just as it was in the cases of $n = \{-2, -1\}$. □

Theorem 12.2

The following properties hold:

$$\sigma^{-2} = 0, \ e_{33}(\mathbf{u}^2) = -\frac{\lambda e_{\mu\mu}(\mathbf{u}^0)}{\lambda + 2\mu}, \ \sigma_{\alpha\beta}^0 = \frac{2\mu\lambda}{\lambda + 2\mu} e_{\mu\mu}(\mathbf{u}^0) \delta_{\alpha\beta} + 2\mu e_{\alpha\beta}(\mathbf{u}^0),$$

where $\mathbf{u}^0$ is a unique solution to the ε^0-order equation (12.3.12). ■

Proof. From the previous theorem, $\int_{\Omega} \sigma_{33}^{-2} e_{33}(\mathbf{v}) d\mathbf{x} = 0$. Since $\sigma_{33}^{-2} = 0, \sigma^{-2} = 0$, and there exists a $\mathbf{u}^2$ such that $e_{33}(\mathbf{u}^2) = -\frac{\lambda e_{\mu\mu}(\mathbf{u}^0)}{\lambda + 2\mu}$. Inserting this into equation (12.3.3) with $p = 0$,

$$\sigma_{\alpha\beta}^0 = \lambda e_{\mu\mu}(\mathbf{u}^0) \delta_{\alpha\beta} + 2\mu e_{\alpha\beta}(\mathbf{u}^0) - \frac{\lambda^2}{\lambda + 2\mu} e_{\mu\mu}(\mathbf{u}^0) \delta_{\alpha\beta}$$

$$\sigma^0_{\alpha\beta} = \frac{2\mu\lambda}{\lambda+2\mu}e_{\mu\mu}(\mathbf{u}^0)\delta_{\alpha\beta} + 2\mu e_{\alpha\beta}(\mathbf{u}^0)$$

Now $\int_\Omega \sigma^0_{ij}e_{ij}(\mathbf{v})d\mathbf{x} = \int_\Omega f\mathbf{v}d\mathbf{x} + \int_{\Gamma^0} g\mathbf{v}ds + \int_{\Gamma^+\cup\Gamma^-} h\mathbf{v}d\omega$.

Set $\mathbf{v} = \mathbf{u}^0$ and expand:

$$\int_\Omega \sigma^0_{\alpha\beta}e_{\alpha\beta}(\mathbf{u}^0)d\mathbf{x} + \int_\Omega \sigma^0_{3\beta}e_{3\beta}(\mathbf{u}^0)d\mathbf{x} + \int_\Omega \sigma^0_{33}e_{33}(\mathbf{u}^0)$$
$$= \int_\Omega f\mathbf{u}^0 d\mathbf{x} + \int_{\Gamma^0} g\mathbf{u}^0 ds + \int_{\Gamma^+\cup\Gamma^-} h\mathbf{u}^0 d\omega.$$

Since $e_{3\beta}(\mathbf{u}^0) = 0 = e_{33}(\mathbf{u}^0)$, the following ensues:

$$\int_\Omega [\frac{2\mu\lambda}{\lambda+2\mu}e_{\mu\mu}(\mathbf{u}^0)\delta_{\alpha\beta} + 2\mu e_{\alpha\beta}(\mathbf{u}^0)]e_{\alpha\beta}(\mathbf{u}^0)]d\mathbf{x}$$
$$= \int_\Omega f\mathbf{u}^0 d\mathbf{x} + \int_{\Gamma^0} g\mathbf{u}^0 ds + \int_{\Gamma^+\cup\Gamma^-} h\mathbf{u}^0 d\omega$$

$$\int_\Omega [\frac{2\mu\lambda}{\lambda+2\mu}e_{\mu\mu}(\mathbf{u}^0)e_{\rho\rho}(\mathbf{u}^0)d\mathbf{x} + 2\mu e_{\alpha\beta}(\mathbf{u}^0)e_{\alpha\beta}(\mathbf{u}^0)]d\mathbf{x}$$
$$= \int_\Omega f\mathbf{u}^0 d\mathbf{x} + \int_{\Gamma^0} g\mathbf{u}^0 ds + \int_{\Gamma^+\cup\Gamma^-} h\mathbf{u}^0 d\omega.$$

Using the *Lax-Milgram lemma*, one can show that there is one and only one solution to the above equation [87] (p. 39). □

Now, one comes upon a major simplification of the terms in the asymptotic expansions of the displacement vector and of the stress tensor.

Theorem 12.3

The odd powers of ε in the asymptotic expansions of the displacement vector and of the stress tensor are all zero. ■

Proof. From a previous theorem [87], we have $\int_\Omega \sigma^{-1}_{33}e_{33}(\mathbf{v})d\mathbf{x} = 0$.
From Corollary 12.1, $\sigma^{-1}_{33} = 0$, $e_{33}(\mathbf{u}^3) = -\frac{\lambda e_{\mu\mu}(\mathbf{u}^1)}{\lambda+2\mu}$, and $\sigma^{-1} = 0$

$$\sigma^1_{\alpha\beta} = \lambda e_{\mu\mu}(\mathbf{u}^1)\delta_{\alpha\beta} + 2\mu e_{\alpha\beta}(\mathbf{u}^1) - \frac{\lambda^2}{\lambda+2\mu}e_{\mu\mu}(\mathbf{u}^1)\delta_{\alpha\beta} \quad (12.3.14)$$
$$= \frac{2\mu\lambda}{\lambda+2\mu}e_{\mu\mu}(\mathbf{u}^1)\delta_{\alpha\beta} + 2\mu e_{\alpha\beta}(\mathbf{u}^1) \quad (12.3.15)$$

Now, $\int_\Omega \sigma^1_{ij}e_{ij}(\mathbf{v})d\mathbf{x} = 0$. Set $\mathbf{v} = \mathbf{u}^1$ and expand:

$$\int_{\Omega}\sigma^1_{\alpha\beta}e_{\alpha\beta}(\mathbf{u}^1)d\mathbf{x}+\int_{\Omega}\sigma^1_{3\beta}e^{\varepsilon}_{3\beta}(\mathbf{u}^1)d\mathbf{x}+\int_{\Omega}\sigma^1_{33}e_{33}(\mathbf{u}^1)d\mathbf{x}=0.$$

Since $e_{3\beta}(\mathbf{u}^1)=0=e_{33}(\mathbf{u}^1)$, the following ensues:

$$\int_{\Omega}\left[\frac{2\mu\lambda}{\lambda+2\mu}e_{\mu\mu}(\mathbf{u}^1)\delta_{\alpha\beta}+2\mu e_{\alpha\beta}(\mathbf{u}^1)\right]e_{\alpha\beta}(\mathbf{u}^1)d\mathbf{x}=0$$

$$\int_{\Omega}\left[\frac{2\mu\lambda}{\lambda+2\mu}e_{\mu\mu}(\mathbf{u}^1)e_{\rho\rho}(\mathbf{u}^1)+2\mu e_{\alpha\beta}(\mathbf{u}^1)e_{\alpha\beta}(\mathbf{u}^1)\right]d\mathbf{x}=0$$

Noting that $\lambda\geq 0$ and $\mu\geq 0$, one sees that $e_{\mu\mu}(\mathbf{u}^1)=0=e_{\alpha\beta}(\mathbf{u}^1)\Rightarrow\mathbf{u}^1=0$ and $e_{33}(\mathbf{u}^3)=0\Rightarrow\sigma^1_{\alpha\beta}=0$.

From Corollary 12.1, $\sigma^1_{33}=0\Rightarrow e_{33}(\mathbf{u}^5)=-\frac{\lambda e_{\mu\mu}(\mathbf{u}^3)}{\lambda+2\mu}$. Using this in equation (12.3.3) with $p=3$ and setting $\mathbf{v}=\mathbf{u}^3$, the integral $\sigma^3_{ij}e_{ij}(\mathbf{v})d\mathbf{x}=0$ becomes:

$$\int_{\Omega}\left[\frac{2\mu\lambda}{\lambda+2\mu}e_{\mu\mu}(\mathbf{u}^3)e_{\mu\mu}(\mathbf{u}^3)+2\mu e_{\alpha\beta}(\mathbf{u}^3)e_{\alpha\beta}(\mathbf{u}^3)\right]d\mathbf{x}=0$$

$\Rightarrow e_{\mu\mu}(\mathbf{u}^3)=0$ and $e_{\alpha\beta}(\mathbf{u}^3)=0\Rightarrow e_{33}(\mathbf{u}^5)=0$.
Setting $\mathbf{v}=\mathbf{u}^3$ into $\int_{\Omega}\sigma^1_{ij}e_{ij}(\mathbf{v})d\mathbf{x}=0$;

$$0=\int_{\Omega}\sigma^1_{3\beta}e_{3\beta}(\mathbf{u}^3)d\mathbf{x}=\int_{\Omega}4\mu e_{3\beta}(\mathbf{u}^3)e_{3\beta}(\mathbf{u}^3)d\mathbf{x}$$

$\Rightarrow e_{3\beta}(\mathbf{u}^3)=0\Rightarrow\mathbf{u^3}=\mathbf{0}$ and $\sigma^1_{3\beta}=0\Rightarrow\boldsymbol{\sigma}^{\mathbf{1}}=\mathbf{0}$.
By replacing the superscripts on σ and $\mathbf{u}$ with odd exponents starting with some odd integer $m\geq -1$ and using induction, one finds that positive odd powers of both $\mathbf{u}$ and σ are all zero. □

Next, the second order term, $\mathbf{u}^2$ in the approximant is considered. We present the approach of Ciarlet [87] (p. 35).

Theorem 12.4

$$u_3^2=-\frac{\lambda}{\lambda+2\mu}(x_3[\frac{\partial\zeta_1^0}{\partial x_1}+\frac{\partial\zeta_2^0}{\partial x_2}]-\frac{(x_3)^2}{2}[\frac{\partial^2u_3^0}{\partial x_1^2}+\frac{\partial^2u_3^0}{\partial x_2^2}])+\zeta_3^2(x_1,x_2)$$ ■

Proof. From Theorem 12.2, $e_{33}(\mathbf{u}^2)=-\frac{\lambda}{\lambda+2\mu}e_{\mu\mu}(\mathbf{u}^0)$

$$\frac{\partial u_3^2}{\partial x_3}=-\frac{\lambda}{\lambda+2\mu}(\frac{\partial u_1^0}{\partial x_1}+\frac{\partial u_2^0}{\partial x_2})\tag{12.3.16}$$

$$= -\frac{\lambda}{\lambda+2\mu}\{\frac{\partial}{\partial x_1}(\zeta_1^0(x_1,x_2) - x_3\frac{\partial u_3^0}{\partial x_1}) + \frac{\partial}{\partial x_2}(\zeta_2^0(x_1,x_2) - x_3\frac{\partial u_3^0}{\partial x_2})\} \tag{12.3.17}$$

$$= -\frac{\lambda}{\lambda+2\mu}(\frac{\partial \zeta_1^0}{\partial x_1} - x_3\frac{\partial^2 u_3^0}{\partial x_1^2} + \frac{\partial \zeta_2^0}{\partial x_2} - x_3\frac{\partial^2 u_3^0}{\partial x_2}) \tag{12.3.18}$$

$$u_3^2 = -\frac{\lambda}{\lambda+2\mu}(x_3[\frac{\partial \zeta_1^0}{\partial x_1} + \frac{\partial \zeta_2^0}{\partial x_2}] - \frac{(x_3)^2}{2}[\frac{\partial^2 u_3^0}{\partial x_1^2} + \frac{\partial^2 u_3^0}{\partial x_2^2}]) + \zeta_3^2(x_1,x_2) \tag{12.3.19}$$

This implies that $u_3^0 \in H^2(\Omega)$. □

Other expressions with $\sigma_{\alpha\beta}^0$ can be obtained from the force balance equation as follows:

Lemma 12.1

For $v_\beta^0 \in C^1(\Omega)$ and $v_3^0 \in C^1(\Omega)$,

$$\int_\Omega \sigma_{\alpha\beta}\partial_\alpha v_\beta^0 d\mathbf{x} = \int_\Omega \gamma' f_\beta v_\beta^0 d\mathbf{x} + \int_{\Gamma^0} g_\beta v_\beta^0 d\omega + \int_{\Gamma^+\cup\Gamma^-} h_\beta v_\beta^0 d\omega$$

$$0 = \int_\Omega \gamma' f_3 v_3^0 d\mathbf{x} + \int_{\Gamma^0} g_3 v_3^0 d\omega + \int_{\Gamma^+\cup\Gamma^-} h_3 v_3^0 d\omega$$

■

Proof. Let $\gamma' = \gamma(\xi_0 + P_\eta(\beta_0))$. In the ε^0 equation, one can set $\mathbf{v} = \mathbf{u^0}$. Since $e_{3\beta}^0 = 0 = e_{33}^0$, one can get $\int_\Omega \sigma_{3\beta} e_{3\beta}^0 d\mathbf{x} = 0$, $\int_\Omega \sigma_{33} e_{33}^0 d\mathbf{x} = 0$, and $\int_\Omega \sigma_{\alpha\beta} e_{\alpha\beta} d\mathbf{x}$ remains on the left side of the equation.

Now, from the force balance equation one has:

$$\int_\Omega (\sigma_{\alpha\beta}\partial_\alpha v_\beta + \sigma_{3\beta}\partial_3 v_\beta) d\mathbf{x} = \int_\Omega \gamma' f_\beta v_\beta d\mathbf{x} + \int_{\Gamma^0} g_\beta v_\beta d\omega + \int_{\Gamma^+\cup\Gamma^-} h_\beta v_\beta d\omega$$

$$\int_\Omega (\sigma_{33}\partial_3 v_3 + \sigma_{3\beta}\partial_\beta v_3) d\mathbf{x} = \int_\Omega \gamma' f_3 v_3 d\mathbf{x} + \int_{\Gamma^0} g_3 v_3 d\omega + \int_{\Gamma^+\cup\Gamma^-} h_3 v_3 d\omega$$

The two equations can be added together:

$$\int_\Omega (\sigma_{\alpha\beta}\partial_\alpha v_\beta + \sigma_{3\beta} e_{3\beta} + \sigma_{33} e_{33}) d\mathbf{x} = \int_\Omega \gamma'(f_\beta v_\beta + f_3 v_3) d\mathbf{x} + \int_{\Gamma^0} (g_\beta v_\beta + g_3 v_3) d\omega + \int_{\Gamma^+\cup\Gamma^-} (h_\beta v_\beta + h_3 v_3) d\omega.$$

If v is replaced by v^0, then the second and third terms of the left side will be zero. The v_β^0 terms on the right side can be grouped with $\int_\Omega \sigma_{\alpha\beta}\partial_\alpha v_\beta^0 d\mathbf{x}$ to form one equation. The v_3^0 terms on the right side can be grouped with zero to form another. □

Theorem 12.5

For $v^0_\beta \in C^1(\Omega)$ and $\sigma_{\alpha\beta} \in C^1(\Omega)$,

$$-\int_\omega \partial_\alpha \sigma_{\alpha\beta} d\omega = \int_\omega \gamma' f_\beta d\omega + \int_{\partial\omega} g_3 ds$$

with boundary conditions at $\chi = \{-1,1\}$ given as: $-\int_\omega \partial_\alpha \sigma_{\alpha\beta}(\chi) d\omega = \int_{\partial\omega} h_\beta v^0_\beta(\chi) d\omega$. ■

Proof. From the above lemma, the first of the resulting equations is:

$$\int_\Omega \sigma_{\alpha\beta} \partial_\alpha v^0_\beta d\mathbf{x} = \int_\Omega \gamma' f_\beta v^0_\beta d\mathbf{x} + \int_{\Gamma^0} g_\beta v^0_\beta d\omega + \int_{\Gamma^+ \cup \Gamma^-} h_\beta v^0_\beta d\omega$$

Integrate by parts on the left side noting that $v^0_3 = 0$ on Γ^0 to get $-\int_\Omega \partial_\alpha \sigma_{\alpha\beta} v^0_\beta d\mathbf{x}$. Decompose each integral as follows:

$$-\int_{-1}^{1} \left[\int_\omega \partial_\alpha \sigma_{\alpha\beta} d\omega\right] v^0_\beta dx_3 = \int_{-1}^{1} \left[\int_\omega \gamma' f_\beta d\omega\right] v^0_\beta dx_3 + \int_0^L \left[\int_{\partial\omega} g_\beta ds\right] v^0_\beta dx_3$$

with boundary conditions given at $\chi = \{-1,1\}$ as:

$$\int_\omega \partial_\alpha \sigma_{\alpha\beta}(\chi) d\omega = \int_\omega h_\beta v^0_\beta(\chi) d\omega.$$

Take the strong formulation to obtain the boundary value problem in the domain Ω. □

The theorem that follows may be found in [139].

Theorem 12.6

For $v^0_\beta \in C^1(\Omega)$, $f_3 \in C^1(\Omega)$, $g_3 \in C^1(\Omega)$, $h_3 \in C^1(\Omega)$,

$$0 = \int_\omega \gamma' \partial_3 f_3 d\omega + \int_{\partial\omega} \partial_3 g_3 d\omega$$

with boundary conditions at $\chi = \{-1,1\}$ given as: $0 = \int_{\Gamma^+ \cup \Gamma^-} \partial_3 h_3 v^0_\beta(\chi) d\omega$. ■

Proof. From the above lemma, the second of the resulting equations is:

$$0 = \int_\Omega \gamma' f_3 v^0_3 d\mathbf{x} + \int_{\Gamma^0} g_3 v^0_3 d\omega + \int_{\Gamma^+ \cup \Gamma^-} h_3 v^0_3 d\omega.$$

Choose $v_3^0 = \partial_3 v_\beta^0$.

$$0 = \int_\Omega \gamma f_3 \partial_3 v_\beta^0 d\mathbf{x} + \int_{\Gamma^0} g_3 \partial_3 v_\beta^0 ds + \int_{\Gamma^+ \cup \Gamma^-} h_3 \partial_3 v_\beta^0 ds.$$

Integrate by parts noting that $v_\beta^0 = 0$ on Γ^0.

$$0 = \int_\Omega \gamma \partial_3 f_3 v_\beta^0 d\mathbf{x} + \int_{\Gamma^0} \partial_3 g_3 v_\beta^0 d\omega + \int_{\Gamma^+ \cup \Gamma^-} \partial_3 h_3 v_\beta^0 d\omega$$

Decompose each integral as follows:

$$0 = \int_{-1}^{1} \left[\int_\omega \gamma \partial_3 f_3 d\omega \right] v_\beta^0 dx_3 + \int_{-1}^{1} \left[\int_{\partial\omega} \partial_3 g_3 ds \right] v_\beta^0 dx_3$$

with boundary conditions at $\chi = \{-1, 1\}$ given as: $0 = \int_{\Gamma^+ \cup \Gamma^-} \partial_3 h_3 v_\beta^0(\chi) d\omega$. Take the strong formulation to obtain the equation in the domain Ω. □

Theorem 12.7

The terms $\dot\beta^0, \dot\beta^1$, and $\dot\beta^2$ of the asymptotic expansion satisfy the following ordinary differential equations:

$$\begin{aligned}
\dot\beta^0 &= a + A_{\alpha\beta} e_{\alpha\beta}^0 + \lambda e_{\mu\mu}^0 e_{\rho\rho}^0 + 2\mu e_{\alpha\beta}^0 e_{\alpha\beta}^0 + A_{33} e_{33}^2 + 2\lambda e_{33}^2 e_{\rho\rho}^0 + (\lambda + 2\mu) e_{33}^2 e_{33}^2 \\
\dot\beta^1 &= A_{3\beta} e_{3\beta} \\
\dot\beta^2 &= A_{\alpha\beta} e_{\alpha\beta}^2 + 2\lambda e_{\mu\mu}^0 e_{\rho\rho}^2 + 4\mu e_{\alpha\beta}^0 e_{\alpha\beta}^2 + A_{33} e_{33}^4 + 2\lambda e_{33}^2 e_{\rho\rho}^2 + 2\lambda e_{33}^4 e_{\rho\rho}^0 \\
&\quad + 2\mu e_{3\beta}^2 e_{3\beta}^2 + 2(\lambda + 2\mu) e_{33}^2 e_{33}^4.
\end{aligned}$$

■

Proof. In the domain Ω^ε, the time rate change in the volume fraction of elastic material is given by:

$$\dot\beta^\varepsilon = a^\varepsilon + A_{kl}^\varepsilon e_{kl}^\varepsilon + c_{ijkl}^\varepsilon e_{kl}^\varepsilon e_{ij}^\varepsilon.$$

a^ε is a constitutive function. A_{ij}^ε is a remodeling rate coefficient. c_{ijkl}^ε is a modified elasticity coefficient that can be approximated by $\sigma_{ij}^\varepsilon e_{ij}^\varepsilon$ [88](p. 4). In turn, the linearized stress tensor can be replaced as follows:

$$\sigma_{\alpha\beta}^\varepsilon = \lambda e_{\mu\mu}^\varepsilon \delta_{\alpha\beta} + 2\mu e_{\alpha\beta}^\varepsilon + \lambda e_{33}^\varepsilon \delta_{\alpha\beta}$$

$$\sigma_{3\beta}^\varepsilon = 2\mu e_{3\beta}^\varepsilon, \quad \sigma_{33} = \lambda e_{\mu\mu}^\varepsilon + (\lambda + 2\mu) e_{33}^\varepsilon.$$

The equation for $\dot\beta^\varepsilon$ now appears as:

$$\dot\beta^\varepsilon \;=\; a^\varepsilon + A_{\alpha\beta}^\varepsilon e_{\alpha\beta}^\varepsilon + A_{3\beta}^\varepsilon e_{3\beta}^\varepsilon + (\lambda e_{\mu\mu}^\varepsilon \delta_{\alpha\beta} + 2\mu e_{\alpha\beta}^\varepsilon + \lambda e_{33}^\varepsilon \delta_{\alpha\beta}) e_{\alpha\beta}^\varepsilon$$

$$+2\mu e^{\varepsilon}_{3\beta}e^{\varepsilon}_{3\beta}+(\lambda e^{\varepsilon}_{\mu\mu}+(\lambda+2\mu)e^{\varepsilon}_{33})e^{\varepsilon}_{33}.$$

Furthermore, one can replace the strain tensors in Ω^{ε} with their equivalents in Ω due to the scalings:

$$e^{\varepsilon}_{\alpha\beta}=e_{\alpha\beta}(\mathbf{u}(\varepsilon)),\;\; e^{\varepsilon}_{3\beta}=\frac{1}{\varepsilon}e_{3\beta}(\mathbf{u}(\varepsilon)),\;\; e^{\varepsilon}_{33}=\frac{1}{\varepsilon^2}e_{33}(\mathbf{u}(\varepsilon)).$$

This results in:

$$\begin{aligned}\dot{\beta} &= a+A_{\alpha\beta}e_{\alpha\beta}+\lambda e_{\mu\mu}e_{\rho\rho}+2\mu e_{\alpha\beta}e_{\alpha\beta}+\frac{1}{\varepsilon}A_{3\beta}e_{3\beta}\\ &\quad+\frac{1}{\varepsilon^2}[A_{33}e_{33}+\lambda e_{33}e_{\rho\rho}+2\mu e_{3\beta}e_{3\beta}+\lambda e_{\mu\mu}e_{33}]+\frac{1}{\varepsilon^4}(\lambda+2\mu)e_{33}e_{33}.\end{aligned}$$

Now, the strain tensors are replaced with their asymptotic expansions found below, noting that $e^0_{3\beta}=0=e^0_{33}$ and the odd powers are equal to zero:

$$e_{\alpha\beta}=e^0_{\alpha\beta}+\varepsilon^2e^2_{\alpha\beta}+\varepsilon^4e^4_{\alpha\beta}+\ldots$$

$$e_{3\beta}=\varepsilon^2e^2_{3\beta}+\varepsilon^4e^4_{3\beta}+\ldots,\qquad e_{33}=\varepsilon^2e^2_{33}+\varepsilon^4e^4_{33}+\ldots$$

or

$$\varepsilon^4\dot{\beta}=\varepsilon^4\dot{\beta}^0+\varepsilon^5\dot{\beta}^1+\varepsilon^6\dot{\beta}^2+\ldots$$

on the left side and terms with coefficients of various powers of ε, starting with ε^4, on the right. Equate like powers of ε to get the statement of the theorem. □

Those interested in more details are suggested to see the references [139, 141, 140, 148] and, in particular, [389, 125, 388]. For numerical simulations we refer the reader to [141].

A Appendix

A.1 MOVING INTERFACE IN THE INERTIAL TERMS AND FROZEN INTERFACE IN THE CONSTITUTIVE EQUATIONS

In this section we present formal calculations leading to the weak formulation of the momentum balance equation.

1. Inertial terms.

$$\int_{I_T}\int_{V^\varepsilon} \partial_t(\rho^\varepsilon \mathbf{v}^\varepsilon)\cdot \psi d\mathbf{x}dt = \int_{I_T}\int_U \theta^\varepsilon \partial_t(\rho^\varepsilon \mathbf{v}^\varepsilon)\cdot \psi d\mathbf{x}dt = \tag{A.1.1}$$
$$-\int_{I_T}\int_U \theta^\varepsilon(\rho^\varepsilon \mathbf{v}^\varepsilon)\cdot \partial_t \psi d\mathbf{x}dt - \int_{I_T}\int_U (\rho^\varepsilon \mathbf{v}^\varepsilon \cdot \psi)\partial_t \theta^\varepsilon d\mathbf{x}dt -$$
$$\int_U \rho^1 \theta_0 \mathbf{v}_0 \cdot \psi(0,\mathbf{x}) d\mathbf{x}dt.$$

$$\int_{I_T}\int_{V^\varepsilon} \partial_j(\rho^\varepsilon v_i^\varepsilon v_j^\varepsilon)\psi_i d\mathbf{x}dt = \int_{I_T}\int_U \theta^\varepsilon \partial_j(\rho^\varepsilon v_i^\varepsilon v_j^\varepsilon)\psi_i d\mathbf{x}dt = \tag{A.1.2}$$
$$-\int_{I_T}\int_U \theta^\varepsilon \rho^\varepsilon v_i^\varepsilon v_j^\varepsilon \partial_j \psi_i d\mathbf{x}dt - \int_{I_T}\int_U (\rho^\varepsilon \mathbf{v}^\varepsilon \cdot \psi)(\mathbf{v}^\varepsilon \cdot \nabla \theta^\varepsilon) d\mathbf{x}dt.$$

Combining (A.1.1) and (A.1.2) we obtain

$$\int_{I_T}\int_{V^\varepsilon} [\partial_t(\rho^\varepsilon \mathbf{v}^\varepsilon) + \operatorname{div}(\rho^\varepsilon \mathbf{v}\otimes\mathbf{v})]\cdot \psi d\mathbf{x}dt = \tag{A.1.3}$$
$$-\int_U \rho^1 \theta_0 \mathbf{v}_0 \cdot \psi(0,\mathbf{x}) d\mathbf{x}dt - \int_{I_T}\int_U \theta^\varepsilon(\rho^\varepsilon \mathbf{v}^\varepsilon)\cdot \partial_t \psi d\mathbf{x}dt - \int_{I_T}\int_U \theta^\varepsilon(\rho^\varepsilon \mathbf{v}^\varepsilon)\cdot \partial_t \psi d\mathbf{x}dt$$
$$\int_{I_T}\int_U (\rho^\varepsilon \mathbf{v}^\varepsilon \cdot \psi)(\partial_t \theta^\varepsilon + \mathbf{v}^\varepsilon \cdot \nabla \theta^\varepsilon) d\mathbf{x}dt.$$

When θ^ε satisfies the interface evolution equation, the last term in the right hand side is zero. If the interface were frozen, then this term would be present in the weak formulation of the momentum equation. Since θ_0 is piecewise constant, the weak formulation would contain a non-physical term supported on the interface. To avoid such non-physical terms, one needs to use θ^ε in the inertial terms of the momentum equation.

2. Constitutive equation. Moving interface assumption combined with Hook's law would lead to a non-physical dissipation of the elastic energy. Indeed, let the elastic part of the stress be written as

$$\theta_0^\varepsilon(\mathbf{x})\mathbf{A}^1 e(\mathbf{u}^\varepsilon) + (1-\theta_0^\varepsilon(\mathbf{x}))\mathbf{A}^2 e(\mathbf{u}^\varepsilon). \tag{A.1.4}$$

The important condition here is that θ_0 is independent of t. The stiffness tensors $\mathbf{A}^1$, $\mathbf{A}^2$ of the phases are supposed to be constant. Formally multiplying (A.1.4) by $\mathbf{v}^\varepsilon$ and integrating by parts we obtain

$$\int_0^T \int_U (\theta_0 \mathbf{A}^1 + (1-\theta_0)\mathbf{A}^2) e(\mathbf{u}^\varepsilon) \cdot e(\partial_t \mathbf{u}^\varepsilon) d\mathbf{x} dt \tag{A.1.5}$$

$$= \frac{1}{2}\int_0^T \int_U \partial_t \left[(\theta_0 \mathbf{A}^1 + (1-\theta_0)\mathbf{A}^2) e(\mathbf{u}^\varepsilon) \cdot e(\mathbf{u}^\varepsilon)\right] d\mathbf{x} dt$$

$$= \frac{1}{2}\int_U (\theta_0 \mathbf{A}^1 + (1-\theta_0)\mathbf{A}^2) e(\mathbf{u}^\varepsilon) \cdot e(\mathbf{u}^\varepsilon) d\mathbf{x} \,\Big|_0^T . \tag{A.1.6}$$

This expresses the fact that the elastic energy changes by the amount of work done by elastic forces, with no dissipation. If one were to use $\theta^\varepsilon(t,\mathbf{x})$ in (A.1.5), differentiation in time would not commute with multiplication by $\theta^\varepsilon \mathbf{A}^1 + (1-\theta^\varepsilon)\mathbf{A}^2$, and (A.1.6) could not be obtained.

A.2 EXISTENCE OF WEAK SOLUTIONS, OUTLINE OF THE PROOF

In this section we outline the proof of existence of global weak solutions for the system (9.4.13), (9.4.14) for each fixed $\varepsilon > 0$. Since ε is fixed, we drop the superscript to simplify the notation. We follow closely [273], Sect. 2.3, 2.4.

1. The initial conditions. The initial conditions satisfy (9.4.15), (9.4.16).

2. Formal a priori estimates (9.4.20)–(9.4.25) are obtained as explained in Sect. 9.4.3. In particular, renormalization as in [273], Sect 2.3 is used to get $|\{\mathbf{x} \in U : \alpha \leq \rho(t,\mathbf{x}) \leq \beta\}|$ for each $0 \leq \alpha \leq \beta < \infty$, where $|\cdot|$ denotes Lebesgue measure. This implies (9.4.20).

3. Compactness results. Since we wish to approximate exact solutions, a compactness result is needed. Suppose that we have two sequences $\rho^n, \mathbf{v}^n$ satisfying the conditions $0 \leq \rho^n \leq C$ div $\mathbf{v}^n = 0$, a. e. on $I_T \times U$, $\| \mathbf{v}^n \|_{L^2(I_T, H_0^1(U))} \leq C$, $\mathbf{v}^n \rightharpoonup \mathbf{v}$ weakly in $L^2(I_T, H_0^1(U))$. Moreover, assume that

$$\partial_t \rho^n + \operatorname{div}(\rho^n \mathbf{v}^n) = 0$$

in $\mathscr{D}'(I_T \times U)$, $\rho^n |\mathbf{v}^n|^2$ is bounded in $L^\infty(I_T, L^1(U)$, and we have

$$|\langle \partial_t(\rho^n \mathbf{v}^n), \phi \rangle| \leq C \| \phi \|_{L^q(I_T, W^{m,q}(U))}$$

for all $\phi \in L^q(I_T, W^{m,q}(U))$ such that div $\phi = 0$.

Finally suppose that the initial conditions for the density ρ_0^n satisfy

$$\rho_0^n \to \rho_0$$

in $L^1(U)$.

Then, by the compactness Theorem 2.4 in [273] (it applies without any change), we have

$$\rho^n \to \rho \quad \text{in } C([0,T], L^p(U)), \text{ for all } 1 \leq p < \infty,$$

$$\sqrt{\rho^n}v_i^n \to \sqrt{\rho}v_i \quad \text{in } L^p(I_T, L^r(U)), \text{ for } 2 < p < \infty, 1 \le r \le \frac{6p}{3p-4},$$
$$v_i^n \to v_i \quad \text{in } L^\theta(I_T, L^{3\theta}(U)), \text{ for } 1 \le \theta < 2 \quad \text{on the set } \{\rho > 0\}.$$

4. Construction of smooth approximate solutions. As in [273], Sect. 2.4, construct solutions $(\rho, \mathbf{v})$ of the approximate system

$$\partial_t \rho + \operatorname{div}(\rho \mathbf{v}_\delta) = 0 \tag{A.2.1}$$
$$\partial_t(\rho \mathbf{v}) + \operatorname{div}(\rho \mathbf{v}_\delta \otimes \mathbf{v}) - \operatorname{div}(\mathbf{A}_\delta e(\mathbf{u}) + B_\delta e(\mathbf{v})) + \nabla P_\delta = 0 \text{ in } \mathscr{D}' \tag{A.2.2}$$
$$\operatorname{div} \mathbf{v} = 0 \text{ in } \mathscr{D}',$$

where $\mathbf{v}_\delta$, $\mathbf{A}_\delta$, B_δ, are smooth regularizations of the respective quantities. The initial conditions are regularizations of the original ones. Then, using a fixed point argument as in Theorem 2.6, we can prove existence of smooth solutions to (A.2.1), (A.2.2). Existence of a fixed point follows from the a priori estimates. The only issue that needs to be explained here is bootstrap regularity of the constructed solutions. The following proposition replaces Proposition 2.1 in [273].

Proposition A.1. *Consider the system*

$$c\partial_t v_i + b \cdot \nabla v_i - a\Delta v_i - m\Delta u_i + \partial_i P = 0, \tag{A.2.3}$$

$\operatorname{div} \mathbf{v} = 0$ *in* $I_T \times U$, $i = 1,2,3$, *with the initial conditions*

$$\mathbf{v}(0,\cdot) = \mathbf{v}_0, \quad \mathbf{u}(0,\cdot) = 0. \tag{A.2.4}$$

Suppose that $c \in L^\infty(I_T \times U)$, $a, m \in L^\infty(I_T, W^{1,\infty}(U))$, $b \in L^2(I_T, L^\infty(U))$, $c \ge k$, $a \ge k$, $m \ge k$ *a.e. on* $I_T \times U$ *for some* $k > 0$; $\mathbf{v}_0 \in H_0^1(U)$. *Also, assume that* a, m *are independent of* t.

Then the system (A.2.3), (A.2.4) has a unique solution $(\mathbf{v}, P)$ *such that* $\mathbf{v} \in L^2(I_T, H^2(U)) \cap C([0,T], H_0^1(U))$, $\partial_t \mathbf{v} \in L^2(I_T \times U)$, $\nabla P \in L^2(I_T \times U)$.

Outline of the proof. Compared to the proof of Proposition 2.1 in [273], we have one new term $m\Delta u_i$. Multiplying this term by $\partial_t v_i$ and integrating by parts we have

$$-\int_U m\Delta u_i \partial_t v_i = d_t \int_U \nabla u_i \cdot m\nabla v_i - \int_U m|\nabla v_i|^2 + \int_U \nabla u_i \cdot \nabla m \partial_t v_i$$

Multiplying (A.2.3) by $\partial_t v_i$ using the above identity, integrating by parts and summing over i we find

$$\int_U c|\partial_t \mathbf{v}|^2 + \frac{1}{2} d_t \int_U a|\nabla \mathbf{v}|^2 + d_t \int_U \nabla \mathbf{u} \cdot m\nabla \mathbf{v} - \int_U m|\nabla \mathbf{v}|^2 = \int_U \mathbf{f},$$

where $\mathbf{f} = b \cdot \nabla \mathbf{v} \partial_t \mathbf{v} - \nabla \mathbf{u} \cdot (\nabla m \otimes \partial_t \mathbf{v})$. Integrating this identity with respect to t we obtain

$$k\int_0^t \int_U |\partial_t \mathbf{v}|^2 + k\frac{1}{2}\int_U |\nabla \mathbf{v}|^2(t)$$

$$\leq \int_U m|\nabla \mathbf{u}||\nabla \mathbf{v}|(t) + \int_0^t \int_U m|\nabla \mathbf{v}|^2 + \int_0^t \int_U \mathbf{f} + \frac{1}{2}\int_U a|\nabla \mathbf{v}|(0).$$

Next we write

$$\int_U m|\nabla \mathbf{u}||\nabla \mathbf{v}|(t) \leq \| m \|_{L^\infty(U)} \left(\frac{1}{2}\nu \int_U |\nabla \mathbf{v}|^2(t) + \frac{1}{2\nu}\int_U |\nabla \mathbf{u}|^2(t) \right),$$

where we choose $\nu = k/2$. The term containing $\mathbf{f}$ is handled similarly, putting ν in front of $\int_0^t \int_U |\partial_t \mathbf{v}|^2$. Combining the previous two inequalities with the standard a priori bounds on $\nabla \mathbf{u}, \nabla \mathbf{v}$ we have

$$\frac{k}{4}\int_0^t \int_U |\partial_t \mathbf{v}|^2 + \frac{k}{4}\int_U |\nabla \mathbf{v}|^2(t) \leq C_k,$$

where C_k depends only on k and the data. This yields a priori estimates on $\mathbf{v}$ in $L^\infty(I_T, H_0^1(U))$ and on $\partial_t \mathbf{v}$ in $L^2(I_T \times U)$.

Now we can write (A.2.3) as

$$\Delta(a\mathbf{v} + m\mathbf{u}) - \nabla P = h, \quad \text{div}\, \mathbf{v} = 0, \quad \text{in } U, \tag{A.2.5}$$

$\mathbf{v} \in H_0^1(U), \mathbf{u} \in H_0^1(U)$ for almost all $t \in I_T$. Also, h is bounded in $L^2(I_t \times U)$ in terms of the data. Next, estimating pressure P exactly as in [273], Prop. 2.1, we conclude that $a\mathbf{v} + m\mathbf{u} \in L^2(I_T, H^2(U))$. Since a, m are smooth and positive, this implies

$$\partial_t \mathbf{u} + \frac{m}{a}\mathbf{u} = \mathbf{g}, \tag{A.2.6}$$

where $\mathbf{g} \in L^2(I_T, H^2(U))$. Now, formally, $\mathbf{u}(t,\cdot) = \int_0^t e^{-\frac{m}{a}(t-\tau)}\mathbf{g}(\tau,\cdot)d\tau$, and hence

$$\left|\partial^2_{x_i x_j}\mathbf{u}\right|^2(t) \leq \left(\int_0^t \left|\partial^2_{x_i x_j}\mathbf{g}\right|(\tau)d\tau\right)^2 \leq t\int_0^t \left|\partial^2_{x_i x_j}\mathbf{g}\right|^2(\tau)d\tau \leq T\int_0^t \left|\partial^2_{x_i x_j}\mathbf{g}\right|^2(\tau)d\tau.$$

Integrating over U we deduce that $\mathbf{u}$ is bounded in $L^\infty(I_T, H^2(U))$ in terms of the data. Then using $\mathbf{v} = -\frac{m}{a}\mathbf{u} + \mathbf{g}$ we obtain that $\mathbf{v}$ is bounded in $L^2(I_T, H^2(U))$ in terms of the data. The formal calculations can be easily justified by an approximation argument. ■

5. Passage to the limit. This is done using compactness from step 3 exactly as in [273].

Bibliography

1. W. Abendschein and G. Hyatt: *Ultrasonics and selected physical properties of bone*, Clin. Orthop. Rel. Res. **69** (1970), pp. 294–301.
2. P. Arbenz, G.H. van Lenthe, U. Mennel, R. Müller, and M. Sala: *A scalable multi-level preconditioner for matrix-free μ-finite element analysis of human bone structures*, Int. J. Numer. Meth. Eng. **73**(7) (2008), pp. 927–947.
3. J.F. Agassant, P. Avenas, J. Sergent, P. Carreau: *Polymer Processing, Principles and Modeling*, Hasser, Munich, 1993.
4. S. Agmon: *Lectures on Elliptic Boundary Value Problems*, Van Nostrand Mathematical Studies, Princeton, 1965.
5. S. M. Ahmed and A. M. Abd-Alla: *Electromechanical, wave propagation in a cylindrical, porous bone with cavity*, Applied Maths. and Computation, **133**, (2002), pp. 257-286.
6. G. Allaire: *Homogenization of the Stokes flow in a connected porous medium*, Asymptotic Anal. **2** (1989), pp. 203–222.
7. G. Allaire: *Homogenization of the Navier-Stokes equations with a slip boundary condition*, Comm. Pure and Appl. Maths. **44** (1991), pp. 605–664.
8. G. Allaire: *Homogenization and two-scale convergence*, SIAM J. Math. Anal. **23**(6) (1992), pp. 1482–1518.
9. G. Allaire, A. Damlamian, and U. Hornung: *Two-scale convergence on periodic surfaces and applications*, 1995.
10. G. Allaire, A. Mikelić, and A. Piatnitski: *Homogenization of the linearized transport equations in rigid periodic porous media*, Jour. Mathematical Physics, **51** (2010), pp.123103.
11. A. Ambrosetti and C. Sbordone: *Γ-convergenza per problemi nonlineari di tipo elliptico*, Boll. Un, Mat. Ital. **13A**, (1976), pp. 352–362.
12. C. C. Anderson, K. R. Marutyan, M. Holland, K. A. Wear and J. G. Miller: *Interference between wave modes may contribute to the negative dispersion observed in cancellous bone*, J. Acoust. Soc. Amer. **124** (3) (2008), pp. 1781–1789.
13. R. B. Ashman, S. C. Cowin, W. C. Van Buskirk, and J. C. Rice: *A continuous wave technique for the measurement of elastic properties of cortical bone*, J. Biomech. **17** (1984), pp. 349–361.
14. R. B. Ashman, J. D. Corin, and C. H. Turner: *Elastic properties of cancellous bone: Measurement by ultrasonic technique*, J. Biomech. **10** (1987), pp. 979–989.
15. R. B. Ashman and J. Y. Rho: *Elastic modulus of trabecular bone material*, J. Biomech. **21**(3) (1988), pp. 177–181 .
16. R. B. Ashman, J. Y. Rho, and C.H. Turner: *Anatomical variation of orthotropic elastic moduli of the proximal human tibia*, J. Biomech. **22** (1989), pp. 895–900.
17. B. A. Auld: *Acoustic Fields and Waves in Solids*, Volume I and II, 2nd ed., Krueger Publishing Company, Malabar, Florida, 1990.
18. J.-L. Auriault: *Poroelastic media*, in Homogenization and Porous Media, Interdisciplinary Applied Mathematics, Springer, Berlin, 1997, pp. 163–182.
19. T. Azuma and M. Hasegawa: *Distensibility of the vein, from the architectural point of view*, Biorheology **10**(3) (1973), pp. 469–479.

20. N. Bakhvalov and G. Panasenko: *Homogenization: Averaging Processes in Porous Media*, Kluwer, Dordrecht, 1989.
21. K. J. Bathe: *Finite element procedures*, Prentice-Hall International, New Jersey, 1996.
22. A. Q. Bauer, K. R. Marutyan, M. R. Holland, and J.G. Miller: *Negative dispersion in bone: The role of interference in measurements of the apparent phase velocity of two temporally overlapping signals*, J. Acoust. Soc. Amer. **123**(4) (2008), pp. 2407–2414.
23. H. H. Bauschke, R. D. Luke, H. M. Phan, and X. Wang: *Restricted normal cones and the method of alternating projections: Application*, Set-Valued and Variational Analysis **21** (3) (2013), pp. 475–501.
24. A. Bensoussan, J. L. Lions, and G. Papanicolaou: *Asymptotic Analysis for Periodic Structures*, North-Holland, Amsterdam, 1978.
25. L. Berlyand and A. Panchenko: *Strong and weak blow up of the viscous dissipation rates for concentrated suspensions*, J. Fluid Mech. **578** (2007), pp. 1–34.
26. A. Yu. Beliaev and S. M. Kozlov: *Darcy Equation for Random Porous Media*, Commun. Pur. Appl. Math. **69** (1996), pp. 1–34.
27. S. Bike and C. Prieve: *Electrohydrodynamics of thin double layers: A model for the streaming potential profile*, J. Colloid Interf. Sci. **154**(1) (1992), pp. 87–96.
28. M. A. Biot: *Theory of propagation of elastic waves in fluid-saturated porous solid. I. Low-frequency range*, J. Acoust. Soc. Amer. **28**(2), (1956), pp. 168–178.
29. M. A. Biot: *Theory of propagation of elastic waves in fluid-saturated porous solid. II. Higher frequency range*, J. Acoust. Soc. Amer. **28**(2) (1956), pp. 179–191.
30. M. A. Biot: *Mechanics of deformation and acoustic propagation in porous media*, J. Appl. Phys. **33** (1962), pp. 482–498.
31. M. A. Biot. *Generalized theory of acoustic propagation in porous dissipative media*, J. Acoust. Soc. Am. **34** (1962), pp. 1254–1264.
32. R. B. Bird, R. C. Armstrong, and O. Hassager: *Dynamics of Polymeric Liquid: Fluid Mechanics*. Wiley, New York, 1987.
33. J. Black and G. Hastings: *Handbook of Biomaterial Properties*, Springer Science & Business Media, Berlin, 2013.
34. D. Blackstock: *Fundamentals of physical acoustics*, Wiley-Interscience, New York, 2000.
35. E. Bossy, M. Talmant, and P. Laugier: *2D simulation of the axial transmission technique on a cortical bone plate*, Acoust. Imag. **26** (2002), pp 69–76.
36. E. Bossy, M. Talmant, and P. Laugier: *Effect of bone cortical thickness on velocity measurements using ultrasonic axial transmission: 2D simulation study*, J. Acoust. Soc. Am. **112** (2002), pp. 297–307.
37. E. Bossy, M. Talmant, and P. Laugier: *Three-dimensional simulations of ultrasonic axial transmission velocity measurement on cortical bone models*, J. Acoust. Soc. Am. **115**(5) (2004), pp. 2314–2324.
38. E. Bossy, F. Padilla, F. Peyrin, and P. Laugier: *Three-dimensional simulation of ultrasound propagation through trabecular, bone structures measured by synchrotron microtomography*, Physics in Mechanics and Biology **50**(23) (2005), pp. 5545–5556.
39. A. Bourgeat and A. Badea: *Homogenization of two phase flow through a randomly heterogeneous porous media*, in *Mathematical Modelling of flow through porous media*, A. Bourgeat, C. Carasso, S. Luckhaus, A. Mikelić, eds., World Scientific, Singapore, 1996.
40. G. deBotton, T. Hariton, and E.A. Socolsky: *Neo-Hookean fibre reinforced composites in finite elasticity*, J. Mech. Phys. Solids **54**(3) (2006), pp. 533–559.
41. A. Bourgeat and A. Piatnitski: *Approximations of effective coefficients in stochastic homogenization*, Ann. Inst. H. Poincaré **40** (2004), pp. 153–165.

42. A. Bourgeat, A. Mikelić, and A. Piatnitski: *On the double porosity model of single phase flow in random media*, Asymptotic Anal. **34** (2003), pp. 311–332.
43. A. Bourgeat, A. Mikelić, and S. Wright: *On the stochastic two-scale convergence in the mean and applications*, J. Reine Angew. Math. **456** (1994), pp. 19–51.
44. A. Bourgeat and A. Mikelić: *Homogenization of a polymer flow through a porous medium*, Nonlinear Analysis **26** (1996), pp. 1221–1253.
45. A. Bourgeat, O. Gipouloux and E. Marusć-Paloka: *Filtration law for polymer flow through a porous media*, Multiscale. Mod. and Simulation **1** (2003), pp. 432–457.
46. A. Braides and A. Defrancheschi: *Homogenization of multiple integrals*, Clarendon Press, Oxford, 1998.
47. P. I. Brooker: *Two-dimensional simulation by turning bands*, Mathematical Geology **17** (1), (1985), pp. 81–90.
48. J. L. Buchanan, R. P. Gilbert, and K. Khashanah: *Recovery of the poroelastic parameters of cancellous bone using low frequency acoustic interrogation*, in *Acoustic, Mechanics, and the Related Topics of Mathematical Analysis*, A. Wirgin ed., World Scientific, Singapore, 2002, pp. 41–47.
49. J. L. Buchanan and R. P. Gilbert: *Measuring Osteoporosis Using Ultrasound*, in *Advances in Scattering and Biomedical Engineering*, D. I. Fotiadis and C. V. Massalas eds., World Scientific, Singapore, 2004, pp. 484–494.
50. J. L. Buchanan, R. P. Gilbert and K. Khashanah: *Determination of the parameters of cancellous bone using low frequency acoustic measurements*, J. Comput. Acoust. **12**(2) (2004), pp. 99–126.
51. J. Buchanan and R. P. Gilbert: *Determination of the parameters of cancellous bone using high frequency acoustic measurements* Math. Comput. Model. **45** (2007), pp. 281–308.
52. J. Buchanan and R. P. Gilbert: *Determination of the parameters of cancellous bone using high frequency acoustic measurements II: Inverse problems*, Comput. Acoust. **15** (2) (2007), pp. 199–220.
53. J. L. Buchanan, R. P. Gilbert, A. Wirgin, and Y. Xu: *Transient reflection and transmission of ultrasonic waves in cancellous bones*, Math. Comput. Model. **142** (2003), pp. 561–573.
54. J. L. Buchanan, R. P. Gilbert, and K. Khashanah: *Determination of the parameters of cancellous bone using low frequency acoustic measurements*, Comput. Acoust. (2004), pp. 99–126.
55. J. L. Buchanan, R. P. Gilbert, and Y. Xu: *Marine Acoustics: Inverse and Direct Problems*, SIAM, Philadelphia, 2004.
56. J. L. Buchanan, R. P. Gilbert, and Y. M-J. Ou: *Transfer functions for a one-dimensional model of a muscle-bone system subject to an ultrasonic pulse*, Nonlinear Anal. Real World Appl. **13** (2012), pp. 1030–1043.
57. J. L. Buchanan, R. P. Gilbert, and Y. M-J. Ou: *Wavelet decomposition of transmitted ultrasound wave through a 1-D muscle-bone system*, J. Biomech. **44** (2011), pp. 352–358.
58. J. L. Buchanan, R. P. Gilbert, and Y. M-J. Ou: *Recovery of the parameters of cancellous bone by inversion of effective velocities, and transmission and reflection coefficients*, Inverse Probl. **27** (2012), p. 125006.
59. M. J. Buckingham: *Theory of acoustic attenuation, dispersion, and pulse propagation in unconsolidated granular materials, including marine sediments*, J. Acoust. Soc. Am. **102** (1997), pp. 2579–2596.
60. M. J. Buckingham: *Theory of compressional and shear waves in fluid-like marine sediments*, J. Acoust. Soc. Am. **103** (1998), pp. 288–299.

61. M. J. Buckingham: *Wave propagation, stress relaxation, and grain-to-grain shearing in saturated, unconsolidated marine sediments*, J. Acoust. Soc. Am. **108** (6) (2000), pp. 2796–2815.
62. B. Budianski: *On the elastic moduli of some heterogeneous materials*, J. Mech. Phys. Solids. **13** (1965), pp. 223–227.
63. R. Burridge and J.B. Keller: *Poroelasticity equations derived from microstructure*, J. Acoust. Soc. Am. **70** (1981), pp. 1140–1146.
64. A. Bur: *Measurements of the dynamic piezoelectric properties of bone as a function of temperature and humidity*, J. Biomech. **9** (1976), pp. 495–507.
65. H. Callen: *Thermodynamics and an Introduction to Thermodynamics*, Wiley, New York, 1985.
66. J. M. Carcione: *Wave propagation in anisotropic, saturated porous media: Plane-wave theory and numerical simulations*, J. Acoust. Soc. Am. **99** (1996), pp. 2655–2666.
67. J. M. Carcione: *Wave Fields in Real Media: Wave Propagation in Anisotropic, Anelastic, and Porous Media*, Elsevier, Oxford, UK, 2001.
68. J. M. Carcione: *Wave fields in real media: Wave propagation in anisotropic, anelastic, porous and electromagnetic media*, Elsevier, Amsterdam, Netherlands, 2007.
69. J. M. Carcione , C. Morency, and J. E. Santos: *Computational poroelasticity: a review*, Geophys. **75** (2010), pp. 75A229–75A243.
70. P. P. Castañeda: *The effective mechanical properties of nonlinear isotropic composites*, J. Mech. Phys. Solids. **39**(1) (1991), pp. 45–71.
71. P. P. Castañeda: *New variational principles in plasticity and their application to composite materials*, J. Mech. Phys. Solids. **40**(8) (1992), pp. 1757–1788.
72. P. P. Castañeda: *Second-order homogenization estimates for nonlinear composites incorporating field fluctuations I. Theory*, J. Mech. Phys. Solids. **50** (2002), pp.737–757.
73. S. Chaffai, F. Peyrin, S. Nuzzo, R. Porcher, G. Berger, and P. Languier: *Ultrasonic characterization of human cancellous bone using transition and backscatter measurements: Relationships to density and microstructure*, Bone **20** (1) (2002), pp. 229–237.
74. S. Chaffai, F. Padilla, G. Berger, and P. Laugier: *In vitro measurement of the frequency dependent attenuation in cancellous bone between 0.2 and 2 MHz*, J. Acoust. Soc. Am. **108** (2000), pp. 1281-1289.
75. S. Chaffai, F. Padilla, B. Berger. and P. Laugier: *In vitro measurement of the frequency dependent attenuation in cancellous bone between 0.2-2 MHz*, J. Acoust. Soc. Am. **108** (2000), pp. 1281–1289.
76. G. Chaffai, G. Berger, and P. Laugier: *Frequency variation of ultrasonic attenuation coefficient of cancellous bone between 0.2 and 2.0 MHz*, Proc. Ultrasonic Symp. (1998), pp.1397–1400.
77. S. Chaffai, V. Roberjot, F. Peyrin, G. Berger, and P. Laugier: *Frequency dependence of ultrasonic backscattering in cancellous bone: Autocorrelation model and experimental results*, J. Acoust. Soc. Am. **108** (2000), pp. 2403–2411.
78. S. Chaffai, F. Peyrin, S. Nuzzo, R. Porcher, R. Berger, and P. Laugier: *Ultrasonic characterization of human cancellous bone using transmission and backscatter measurements: Relationships to density and microstructure*, Bone **30** (1) (2002), pp. 229–237.
79. J. Charmley: *The Closed Treatment of Common Fractures*, 3rd ed., E. & S Livingston, 1968.
80. A. Chaterjee: *An introduction to the proper orthogonal decomposition*, Current Science, **78**(7) (2000), pp. 808–817.

81. V. Chiadó Piat and A. Defracheschi: *Homogenization of monotone operators*, Nonlinear Analysis **14** (1990), pp. 717–732.
82. H. Chen, R. P. Gilbert, and P. Guyenne: *A Biot model for the determination of material parameters of cancellous bone from acoustic measurements*, Inverse Problems **34**(8) (2018), pp. 085009.
83. A. H.-D. Cheng: *Poroelasticity*, Springer, Berlin, 2016.
84. E. Cherkaev and C. Bonifasi-Lista: *Characterization of structure and properties of bone by spectral measure method*, J. Biomech. **44**, 2011, pp. 345–351.
85. R. M. Christensen: *Theory of Viscoelasticity, An Introduction*, 2nd ed., Academic Press, 1982.
86. P. G. Ciarlet: *Mathematical Elasticity, Vol. 1: Three-Dimensional Elasticity*, Studies in Mathematics and its Applications, **20** North-Holland, Amsterdam, 1988.
87. P. G. Ciarlet: *Mathematical Elasticity, Volume II: Theory of Plates*, North-Holland, Amsterdam, (1997).
88. P. G. Ciarlet and P. Destuynder: *A justification of the of the two-dimensional plate model*, J. Mécanique **18**(1979), pp. 315–344.
89. D. Cioranescu and P. Donato: *An Introduction to Homogenization*, Oxford Lecture Series in Mathematics, Oxford, 1999.
90. G. V. B. Cochran, D. G. Dell, V. R. Palmieri, M. W. Johnson, M. W. Otter, and M. P. Kadaba: *An improved design of electrodes for measurement of streaming potentials on wet bone in vitro and in vivo*, J. Biomech. **22** (1989), pp. 745–750.
91. Th. Clopeau, J. L. Ferrín, R. P. Gilbert, and A. Mikelić: *Homogenizing the acoustic properties of the seabed, Part II*, Modeling in Mathematics and Computation, **33** (2001), pp. 821–841.
92. F. Collino and C. Tsogka: *Application of the perfectly matched absorbing layer model to the linear elastodynamic problem in anisotropic heterogeneous media*, Geophysics **66** (1) (2001), pp. 294–307.
93. J. Cornelius and J. Liu: *Finite-difference time-domain simulation of spacetime cloak*, Optics Express **22** (10) (2014), pp. 12087–12095.
94. S. C. Cowin: *Bone poroelasticity*, J. Biomech. **32** (1999), pp. 217–238.
95. S. C. Cowin: *Bone Mechanics Handbook*, CRC Press, 2001.
96. S. C. Cowin and D. H. Hegedus: *Bone remodeling I: Theory of adaptive elasticity*, J. Elasticity **6** (3) (1976), pp. 313–326.
97. S. C. Cowin and D. H. Hegedus: *Bone remodeling II: Small strain adaptive elasticity*, J. Elasticity **6**(4) (1976), pp. 337–352.
98. S. C. Cowin and R. R. Nachlinger: *Bone remodeling- III: Uniqueness and stability in adaptive elasticity*, J. Elasticity **8** (1978), pp. 285–295.
99. S. C. Cowin and W. C. Van Buskirk: *Surface bone remodeling induced by a medullary pin*, J. Biomech. **12** (4) (1979), pp. 269–276.
100. R. L. Creuss and T. Sakai: *The effect of cortisone upon synthesis rates of some components of rat bone matrix*, Clinical Orthoped. **86** (1972), pp. 253–259.
101. S. R. Cummings, D. Bates, and D. M. Black: *Clinical use of bone densitometry – scientific review*, JAMA **288**(15) (2002), pp. 1889–1897.
102. G. Dal Maso: *An Introduction to* Γ*-convergence*, Burkhäuser, Berlin, 1993.
103. R. Dautray and J.-L. Lions: *Mathematical Analysis and Numerical Methods for Science and Technology*, Evolution Problems I, **5**, Springer, Berlin, 1992.
104. P. D. Delmas, Z. Li, and C. Cooper: *Relationship between changes in bone mineral density and fracture risk reduction with antiresorptive drugs: Some issues with meta-analyses*, J. Bone Miner. Res. **19** (2004), pp. 330–337.

105. L. DeRyck, Z. Fellah, R. P. Gilbert, J.-P. Groby, E. Ogam, N. Sebaa, J.-Y. Chapelon, C. Depollier, Th. Scotti, A. Wirgin, and Y. Xu: *Recovery of the mechanical parameters of long bones from their vibroacoustic impulse response*, SAPEM, 2005.
106. P. Droin, G. Berger, and P. Laugier: *Velocity dispersion of acoustic waves in cancellous bone*, IEEE Trans. Ultrason. Ferroelectr. Freq. Control. **45** (1998), pp. 581–592.
107. P. Droin, P. Laugier, and G. Berger: *Ultrasonic attenuation and dispersion of cancellous bone in the frequency range 200 kHz-600 kHz*, Acoust. Imag. **23** (1997), pp. 157–162.
108. G. Duvaut, J. L. Lions: *Inequalities in Mechanics and Physics*, Springer-Verlag, Berlin, 1972.
109. L. Eldén: *Algorithms for the regularization of ill conditioned least square problems*, BIT **17**(2) (1977), pp. 134–145.
110. I. A. Ene and J. Saint Jean Paulin: *Homogenization and two-scale convergence for a Stokes or Navier-Stokes flow in an elastic thin porous medium*, Math. Models Methods Appl. Sci. **6**(7) (1996), pp. 941–955.
111. A. Erdélyi: *Tables of Integral Transforms*, 2nd ed., McGraw-Hill, New York, 1954.
112. L. C. Evans: *Partial Differential Equations*, AMS, 1998.
113. L. C. Evans: *The perturbed test function method for viscosity solutions of nonlinear PDE*, P. Roy. Soc. Edinb. A. **111** (1989), pp. 359–375.
114. L. C. Evans: *Periodic homogenization of certain fully nonlinear partial differential equations*, P. Roy. Soc. Edinb. A. **120** (1992), pp. 245–265.
115. M. Fang, R. P. Gilbert, A. Panchenko, and A. Vasilic: *Homogenizing the time harmonic acoustics of bone: The monophasic case*, Math. Comput. Model. **46** (3-4) (2007), pp. 331–340.
116. M. Fang, R. P. Gilbert, P. Guyenne, and A. Vasilic: *Numerical homogenizing the time harmonic acoustics of bone: The monophasic case*, Int. Jour. Multiscale Computation **5** (6) (2007), pp. 461–471.
117. M. Fang, R. P. Gilbert, A. Panachenko, and A. Vasilic: *Homogenizing the time harmonic acoustics of bone*, Math. Comput. Model. **46** (2007), pp. 331–340.
118. M. Fang, R. P. Gilbert, and X. Xie: *Deriving the effective ultrasound equations for soft tissue interrogation*, Computers & Mathematics with Applications, **49** (7-8) (2005), pp. 1069–1080.
119. M. Fang, R. P. Gilbert, P. Guyenne, P. Rowe, and A. Vasilic: *Homogenization of the time harmonic acoustics of bone: Biphasic case*, Inter. Journal of Evolution Equations **9** (1) (2014), pp. 71–98.
120. A. Fasano, A. Mikelić, and M. Primicerio: *Homogenization of flows through porous media with permeable grains*, Adv. Math. Sci. Appl. **8** (1998), pp. 1–31.
121. Z. E. A. Fellah, Y. Chapelon, S. Berger, W. Lauriks, and C. Depollier: *Ultrasonic wave propagation in human cancellous bone: Application of Biot theory*, J. Acoust. Soc. Am. **116** (2004), pp. 61–73.
122. M. Fellah, Z. Fellah, F. Mitri, E. Ogam, and C. Depollier: *Transient ultrasound propagation in porous media using Biot theory and fractional Calculus: Application to human cancellous bone*, J. Acoust. Soc. Am. **133** (2013), pp. 1867–1881.
123. F. Feyel and J. L. Chaboche: *FE^2 multiscale approach for modelling the elastoviscoplastic behaviour of long fibre SiC/Ti composite materials*, Comput. Methods Appl. Mech. Eng. **183** (2000), pp. 309–330,.
124. F. Feyel: *A multilevel finite element method (FE^2) to describe the response of highly-nonlinear structures using generalized continua*, Comput. Methods Appl. Mech. Eng. **192** (2003), pp. 3233–3244.

125. I. Figueiredo and L. Trabucho: *Asymptotic model of a nonlinear adaptive elastic rod*, Math. Mech. Solids **4** (2004), pp. 331–353.
126. J. Fish and S. Kuznetsov: *Computational continua*, Int. J. Numer. Meth. Eng. **84**(7) (2010), pp. 774–802.
127. F. J. Fry and J. E. Barger: *Acoustical properties of the human skull*, Jour. Acoustic Soc. Amer. **63** (1978), pp. 1576–1590.
128. Y. C. Fung: *Stress-strain history relations of soft tissue in simple elongation* in *Biomechanics: Its Foundations and Objectives*, Y. C. Fung, N. Perrone, and M. Anliker, eds., Pretice Hall, Englewood Cliffs, 1972.
129. Y. C. Fung: *Biomechanics: Mechanical Properties of Living Tissues*, Springer, New York, 1981.
130. L. J. Gibson: *The mechanical behaviour of cancellous bone*, J. Biomech. **18** (5) (1985), pp. 317–328.
131. F. J. H. Gijsen, E. Allanic, F. N. van de Vosse, and J. D. Janssen: *The influence of the non-Newtonian properties of blood on the flow in large arteries: In steady flow in 90° curved tube*, J. Biomech. **32** (1999), pp. 705–713.
132. F. J. H. Gijsen, E. Allanic, F. N Van de Vosse, and J. D. Janssen: *The influence of the non-Newtonian properties of blood on the flow in large arteries: Unsteady flow in a 90 curved tube*, J. Biomech. **32**(7) (1999), pp. 705–713.
133. D. Gilbarg and N. S. Trudinger: *Elliptic partial differential equations of second order*, Springer, New York, 1983.
134. R. P. Gilbert, S. Gnelecoumbaga, and G. Panasenko: *Wave propagation in a system: Porous media with Dirichlet's condition on the boundary-continuous media*, in *Proc. Third Int. Congr. Acoustics*, G. Cohen, ed., INRIA and SIAM Publ., New York, 1995, pp. 449–455.
135. R. P. Gilbert and A. Mikelić: *Homogenizing the acoustic properties of the seabed: Part I*, Nonlinear Analysis **40** (2000), pp. 185–212.
136. R. P. Gilbert, M. Harik, and A. Panchenko: *Vibration of two bonded composites: Effects of the interface and distinct periodic structures*, Int. Journal of Solids and Structures **40** (2003), pp. 3177–3193.
137. R. P. Gilbert and A. Panchenko: *Acoustics of a stratified poroelastic composite*, J. Anal. Appl. **8**(4) (1999), pp. 977–1001.
138. R. P. Gilbert and A. Panchenko: *Effective acoustic equations for a two-phase medium with microstructure*, Math. Comput. Model. **39**(13) (2004), pp. 1431–1448.
139. R. P. Gilbert and R. J. Ronkese: *An asymptotic model of a nonlinear Kelvin-Voigt viscoelastic plate*, International Journal of Evolution Equations, **2** (3) (2007), pp. 235–253.
140. R. P. Gilbert and R. J. Ronkese: *An asymptotic model of a nonlinear adaptive isotropic viscoelastic rod*, in *Discrete and Computational Mathematics*, F. Liu, G. N'Guerekata, D. Pokrajac, X. Shi, J. Sun, and X. Xia, eds., Nova Science Publishers, 2008, pp. 17–30.
141. R. P. Gilbert and R. Ronkese: *An asymptotic model of a nonlinear adaptive elastic plate*, Math. Comp. Modeling, **48**(3-4) (2008), pp. 505–517.
142. R. P. Gilbert, A. Panchenko, and X. Xie: *A prototype homogenization model for acoustics of granular materials*, Int. Jour. Multiscale Comp. Engin. **4** (2006), pp. 585–600.
143. R. P. Gilbert, A. Panchenko, and X. Xie: *Homogenization of viscoelastic matrix in linear frictional contact*, Math. Methods in the Applied Sciences **28** (2005) pp. 309-328.
144. R. P. Gilbert, Y. Xu, and S. Zhang: *Computing porosity of cancellous bone using ultrasound waves*, in *More Progress in Analysis*, H. G. Begehr and F. Nicolosi, eds., World Scientific, Singapore, 2009, pp. 1393–1405.

145. R. P. Gilbert, Y. Xu, and S. Zhang: *Mathematical model for evaluation of osteoporosis*, Advances in Applied Analysis (2007), pp. 57–67.
146. R. P. Gilbert, D.-S. Lee and M. Y. Ou: *Lamb waves in a poroelastic plate*, J. Comput. Acoust. **21**(2) (2013), p. 1350001.
147. R. P. Gilbert, P. Guyenne, and G. C. Hsiao: *Determination of cancellous bone density using low frequency acoustic measurements*, Appl. Anal. **87** (2008), pp. 1213–1225.
148. R. P. Gilbert, G. C. Hsiao, and L. Xu: *On the variational formulation of a transmission problem for the Biot equations*, Appl. Anal. **89** (2010), pp. 745–755.
149. R. P. Gilbert, A. Panchenko, and A. Vasilic: *Acoustic propagation in a random saturated medium: The monophasic case* , Math. Meth. Appl. Sci. **33**(18) (2010), pp. 2206-2214.
150. R. P. Gilbert, A. Panchenko, and A. Vasilic: *Acoustic propagation in a random saturated medium: The biphasic case*, Appl. Anal. **93** , (4), (2014), pp. 676–697.
151. R. P. Gilbert, P. Guyenne, and M. Y. Ou : *A quantitative ultrasound model of bone with blood as the interstitial fluid*, Math. Comput. Model. **55** (2012), pp. 2029–2039.
152. R. P. Gilbert, P. Guyenne, and J. Li : *Simulation of a mixture model for ultrasound propagation through cancellous bone*, Comput. Acoust. **21**(1) (2013), p. 1250017.
153. R. P. Gilbert, P. Guyenne, and J. Li: *A viscoelastic model for random ultrasound propagation in cancellous bone*, Comput. Math. Appl. **66**(6) (2013), pp. 943–964
154. R. P. Gilbert, P. Guyenne, and J. Li: *Numerical investigation of ultrasonic attenuation through 2D trabecular, bone structures reconstructed from CT scans and random realizations*, Comput. Biol. Med. **45** (2013), pp. 143–156.
155. R. P. Gilbert, P. Guyenne, and M. Shoushani: *Recovery of parameters of cancellous bone by acoustic interrogation*, Inverse Probl. Sci. Eng. **24**(2) (2016), pp. 284–316.
156. R. P. Gilbert and M. Shoushani: *The Biot Model for Anisotropic Poro-elastic Media: The Viscoelastic Fluid Case* J. Comput. Acoust. **25**(3) (2017), p. 1750012.
157. R. P. Gilbert, G. Hsiao, and M. Shoushani: *The Biot model for anisotropic poro-elastic media*, Z. Angew. Math. Mech. **97**(2) (2017), pp. 127–246.
158. R. P. Gilbert, A. Panchenko, and A. Vasilic: *Homogenizing the Acoustics of Cancellous Bone with an Interstitial Non-Newtonian Fluid*, Nonlinear Analysis **74** (2011), pp. 1005–1018.
159. R. P. Gilbert, A. Panchenko, and A. Vasilic: *Acoustic Propagation in a Random Saturated Medium: The Biphasic Case*, Appl. Anal. **93**(4) (2014), pp. 676–697.
160. R. P. Gilbert, A. Panchenko, A. Vasilic, and Y. Xu: *Homogenizing the Ultrasonic Response of Wet Cortical Bone* Poromechanics V, (2013), pp. 1097–1106.
161. R. P. Gilbert and A. Mikelić: *Homogenizing the acoustic properties of the seabed: Part I*, Nonlinear Analysis **40** (1999), pp. 185–212.
162. C. C. Glüer: *Quantitative ultrasound techniques for the assessment of of osteoporosis: Expert agreement on current state*, J. Bone Miner. Res. **12** (1997), pp. 1280–1288.
163. J. D. Goddard: *A dissipative anisotropic fluid model for non-colloidal particle dispersions*, J. Fluid Mech. **568**, (2006), pp. 1–17.
164. M. A. Goodman and S. C. Cowin: *A continuum theory for granular materials*, Arch. Rat. Mech. Analysis. **44** (1971), pp. 249–266.
165. R. W. Graves: *Simulating seismic wave propagation in 3D elastic media using staggered-grid finite differences*, Bull. Seismological Soc. Amer. **86**(4) (1996), pp. 1091–1106.
166. J.-P. Groby, E. Ogam, A. Wirgin, Z. E. A. Fellah, W. Lauriks, J.-Y. Chapelon, C. Depollier, L. DeRyck, N. Sebaa, R. P. Gilbert, and Y. Xu: *2D mode excitation in a porous slab saturated with air in the high frequency approximation*, SAPEM, 2005.

167. L. F. Greengard and V. Rokhlin: *A fast algorithm for particle simulations*, J. Comput. Phys. **73** (1987), pp. 325–348.
168. R. Guillemin, L. Philippe, J.-P. Sessarego, and A. Wirgin: *Inversion of synthetic and experimental acoustical scattering data for the comparison of two reconstruction methods employing the Born approximation*, Ultrasonics **39**(2) (2001), pp. 121–131.
169. N. Guzelsu and H. Demiray: *Electromechanical Properties and Related Models of Bone Tissue: A Review*, Int. J. Eng. Sci. Recent Advances, **17** (1979), pp. 813–851 .
170. N. Guzelsu and S. Saha: *Electro-mechanical wave propagation in long bones*, J. Biomech. **14** (1981), pp. 19–33.
171. N. Guzelsu and S. Saha: *Electro Mechanical Behavior of Wet Bone - Part I Theory*, J. Biomech. Eng. **106**(3) (1984), pp. 249–261.
172. N. Guzelsu and S. Saha: *Electro Mechanical Behavior of Wet Bone - Part II, Wave Propagation*, J. Biomech. Eng. **106**(3) (1984), pp. 262–271. .
173. N. Guzelsu: *Electromechanical properties and electromagnetic stimulation of bone*, in *Biomechanics of Normal and Pathological Human Articulating Joint*, N. Berme, A. Engin, and Da Silva, eds., Martinuz Nijhoff, 1985, pp. 267–284.
174. N. Guzelsu and W. R. Walsh: *Electrical Stimulation of Bones and Peripheral Nerves*, in *Spinal Cord Injury Medical Engineering*, D. N. Ghista, C. Thomas, and H. L. Frankel, eds., Pub. Springfield, Illinois 1986, pp. 511–529.
175. K. Hackl: *Asymptotic methods in underwater acoustics*, in *Generalized Analytic Functions*, H. Florian, ed., Kluwer Academic, 1998, pp. 229–240.
176. G. Haïat, F. Padilla, F. Peyrin, and P. Laugier: *Variation of ultrasonic parameters with microstructure and material properties of trabecular bone: A 3D model simulation.*, J. Bone Miner. Res. **22**(5) (2007), pp. 665–674.
177. G. Haïat, F. Padilla, F. Peyrin and P. Laugier: *Fast wave ultrasonic propagation in trabecular bone: Numerical study of the influence of porosity and structural anisotropy*, J. Acoust. Soc. Am.**123**(3) (2008), pp. 1694–1705.
178. G. Haïat, A. Lhémery, F. Renaud, F. Padilla, P. Laugier, and S. Naili: *Velocity dispersion in trabecular bone: Influence of multiple scattering and of absorption*, J. Acoust. Soc. Am. **124**(6) (2008), pp. 4047–4058.
179. T.J. Haire and C. M. Langton: *Biot theory: A review of its application to ultrasound propagation through cancellous bone*, Bone **24** (1999), pp. 291–295.
180. D. E. Hall: *Basic Acoustics*, Krieger Publishing Company, Malabar Florida, 1993.
181. D. Harris: *Characteristic relations for a model for the flow of granular materials*, Proc. Roy. Soc. A. **457** (2001), pp. 349–370.
182. Z. Hashin and S. Shtrikman: *On some variational principles in anisotropic and nonhomogeneous elasticity*, J. Mech. Phys. Solids, **10** (1962), pp. 335–342.
183. Z. Hashin and S. Shtrikman: *A variational approach to the theory of the elastic behaviour of polycrystals*, J. Mech. Phys. Solids, **10** (1962), pp. 343–352.
184. Z. Hashin and S. Shtrikman: *A variational approach to the theory of the elastic behaviour of multiphase materials*, J. Mech. Phys. Solids **11** (1963), pp. 127–140.
185. S. Hazanov and M. Amieur: *On overall properties of elastic heterogeneous bodies smaller than the representative volume*, Int. J. Eng. Sci. **33**(9) (1995), pp. 1289–1301.
186. D. M. Hegedus and S. C. Cowin: *Bone remodeling- II: Small-strain adaptive elasticity*, J. Elasticity **6** (1976), pp. 337–352.
187. T. Hildebrand and P. Rüegsegger: *A new method for the model independent assessment of thickness in three-dimensional images*, J. of Microsc. **185** (1997), pp. 67–75.

188. T. Hildebrand and P. Rüegsegger: *Quantification of bone architecture with structure model index* CMBBE, **1** (1997), pp. 5–23.
189. J. M. Hill: *Similarity "hot-spot" solutions for a hypoplastic granular material*, Proc. Roy. Soc. A. **456** (2000), pp. 2653–2671.
190. R. Hill: *The elastic behaviour of a crystalline aggregate*, Proc. Phys. Soc. A, **65** (1952), pp. 349–354.
191. R. Hill: *Elastic properties of reinforced solids: Some theoretical principles*, J. Mech. Phys. Solids, **11** (1963), pp. 357–372.
192. D. Hoff: *Discontinuous solutions of the Navier-Stokes equations for multidimensional flows of heat-conducting fluid*, Arch. Rat. Mech. Anal. **139**(4) (1997), pp. 303–354.
193. B. K. Hofmeister, S. A. Whitten, and J. Y. Rao: *Low megahertz ultrasonic properties of bovine cancellous bone* , Bone **26**(9) (2000), pp. 635–642.
194. T. Hildebrand and P. Rüegsegger: *A new method for the model-independent assessment of thickness in three-dimensional images*, J. Microscopy **185** (1996), pp. 67–75.
195. T. Hildebrand and P. Rüegsegger: *Qualification of bone microarchitecture with the structure model index*, Comp. Meth. Biomech. Biomed. Eng. **1** (1997), pp.15–23.
196. A. Laib, T. Hildebrand, H. J. Hauselmann, and P. Rüegsegger: *Ridge Number Density: A New Parameter for In Vivo Bone Structure Analysis*, Bone **21**(6) (1997), pp. 541–546.
197. R. Hill: *On Constitutive Macro-Variables for Heterogeneous Solids at Finite strain*, Proc. R. Soc. Lond. A. **326** (1972), pp. 131–147.
198. R. Hill: *Elastic properties of reinforced solids: Some theoretical principles*, J. Mech. Phys. Solids **11** (1963), pp. 357–372.
199. M. C. Hobatho, J. Y. Rho and R. B. Ashman: *Atlas of mechanical properties of human cortical and cancellous bone*, in *In Vivo Assessment of Bone Quality by Vibration and Wave Propagation Techniques, Pt. II*, G. Van der Perre, G. Lowet. and A. Borgwardt-Christensen (eds.), Leuven, 1990, pp. 7–38.
200. R. Hodgskinson and J. D. Currey: *Young's modulus, density and material properties in cancellous bone over a large density range*, J. Mater. Sci. Mater. Med. **3** (1992), pp. 377–381.
201. R. Hodgskinson, C. F. Njeh, J. D. Curey, and C. M. Langton: *The ability of ultrasound velocity to predict the stiffness of cancellous bone in vitro*, Bone **21**(2) (1997), pp. 183–190.
202. C. E. Hoffler, K. E. Moore, K. Kozloff, P. K. Zysset, M. B. Brown, and S. A. Goldstein: *Heterogeneity of Bone Lamellar-Level Elastic Moduli.* Bone, **26** (2000), pp. 603–609.
203. A. Hosokawa and T. Otani: *Ultrasonic wave propagation in bovine cancellous bone*, Jour. Acoust. Soc. Am. **101** (1997), pp. 558–562.
204. A. Hosokawa and T. Otani: *Ultrasonic wave propagation in bovine cancellous bone*, J. Acoust. Soc. Am. **101** (1997), pp. 558–562.
205. A. Hosokawa and T. Otani: *Acoustic anisotropy in bovine cancellous bone*, J. Acoust. Soc. Am. **103** (1998), pp. 2718–2722.
206. A. Hosokawa: *Ultrasonic pulse waves in cancellous bone analyzed by finite-difference time-domain methods*, Ultrasonics **44** (2006), pp. 227–231.
207. C. Huet: *Universal conditions for assimilation of a heterogeneous material to an effective medium*, Mech. Res. Commun. **9**(3) (1982), pp. 165–170.
208. C. Huet: *On the definition and experimental determination of effective constitutive equations for assimilating heterogeneous materials*, Mech. Res. Commun. **11**(3) (1984), pp. 195–200.

209. C. Huet: *Application of variational concepts to size effects in elastic heterogeneous bodies*, J. Mech. Phys. Solids **38**(6), (1990), pp. 813–841.

210. E. R. Hughes, T. G. Leighton, G. W. Petley and P. R. White: *Ultrasonic propagation in cancellous bone: A new stratified model*, Ultrasound Med. Biol. **25** (1999), pp. 811–821.

211. G. C. Hsiao, R. E. Kleinman, and G. F. Roach: *Weak solutions of fluid-solid interaction problem*, Math. Nachrichten **218**(1) (2000), pp. 139–163.

212. G. C. Hsiao, F.-J. Sayas and R. J. Weinacht: *Time-dependent fluid-structure interaction*, Math. Meth. Appl. Sci. **40** (2017), pp. 486–500.

213. G. C. Hsiao, T. S'anchez-Vizuet, and F.-J. Sayas: *Boundary and coupled boundary-finite element methods for transient wave-structure interaction*, IMA J. Numer. Anal. **37** (2016), pp. 237–265.

214. G. C. Hsiao, T. S'anchez-Vizuet, F.-J. Sayas and R. J. Weinacht: *A time-dependent fluid-thermoelastic solid interaction*, IMA J. Numer. Anal. **39**(2) (2019), pp. 924–956.

215. G. C. Hsiao, O. Steinbach, and W. L. Wendland: *Boundary element methods: Foundation and error analysis*, in *Encyclopedia of Computational Mechanics Second Edition*, E. Stein et al., eds., John Wiley & Sons, Chichester, UK, 2017, pp. 1–62.

216. G. C. Hsiao, and W. L. Wendland: *Boundary Integral Equations*, Applied Mathematical Sciences **164**, Springer-Verlag, Berlin, 2008.

217. D. L. Johnson, J. Koplik, and R. Dashen: *Theory of dynamic permeability and tortuousity in fluid-saturated porous media*, J. Fluid. Mech. **176** (1987), pp. 379–402.

218. S. Ilic, K. Hackl, and R. Gilbert: *Effective parameters of cancellous bone*, Proc. Appl. Math. Mech. **8** (2008), pp. 10175-10176.

219. S. Ilic, K. Hackl, and R. Gilbert: *Multiscale modeling for cancellous bone by using shell elements*, in *Shell Structures: Theory and Applications*, W. Pietrasckiewicz and I. Kreja, eds., Taylor & Francis Group CRC Press, 2009.

220. S. Ilic, K. Hackl, and R.P. Gilbert: *Application of the multiscale FEM to the modeling of cancellous bone*, Biomechan. Model. Mechanobiol. **9**(1) (2010), pp. 87–102.

221. S. Ilic, K. Hackl, and R. Gilbert: *Application of biphasic representative volume element to the simulation of wave propagation through cancellous bone I*, J. Comput. Acoust. **19**(2) (2011), pp. 111–138.

222. S. Ilic, K. Hackl, and R. P. Gilbert: *Application of biphasic representative volume element to the simulation of wave propagation through cancellous bone. I*, Theoretical and Computational Acoustics, **19** (2) (2011), pp. 111–138.

223. S. Ilic and K. Hackl: *Application of the multiscale FEM to the modeling of nonlinear multiphase materials*, J. Theor. Appl. Mech. **47** (2009), pp. 537-551.

224. M. Isakson, S. McJunkin, and N. P. Chotiros: *A comparison of inversions for elastic and poro-elastic models for high-frequency reflection from a smooth water-sediment interface*, 16th ASCE Engineering Mechanics Conference, Seattle, (2003).

225. V. Isakov: *Inverse Problems for Partial Differential Equations*, Applied Mathematical Sciences **127**, Springer, New York, 1998.

226. C. R. Jacobs: *The mechanobiology of cancellous bone structure adaption*, J. Rehabilitation Res. Devel. **37** 2000, pp. 209–216.

227. D. L. Johnson, J. Koplik, and R. Dashen: *Theory of dynamic permeability and tortuosity in fluid-saturated porous media*, J. Fluid Mech. **176** (1987), pp. 379–402.

228. J. T. Jenkins and S. B. Savage: *A theory for the rapid flow of identical, smooth, nearly elastic, spherical particles*, J. Fluid Mech. **130** (1983), pp. 187–202.

229. V. V. Jikov, S. M. Kozlov, and O.A. Oleinik: *Homogenization of Differential Operators and Integral Functionals*, Springer, Berlin, 1994.

230. M. Kaczmarek, M. Pakula, and J. Kubik: *Multiphase nature and structure of biomaterials studied by ultrasound*, Ultrasonics, **38** (2000), pp. 703–707.

231. J. A. Kanis, P. Delmas, P. Burckhardt, C. Cooper, and D. Torgerson, *Guidelines for diagnosis and management of osteoporosis*, Osteoporos. Int. **7** (1997), pp. 390–406.

232. J. Kastelic, A. Galeski, and E. Baer *The multicomposite ultrastructure of tendon*, Connective Tissue Research, **6** (1978), pp. 11–23.

233. Y. Kantor and D. J. Bergman: *Improved rigorous bounds on the effective elastic moduli of a composite material*, J. Mech. Phys. Solids **32**(1) (1984), pp. 41–62.

234. T. S. Keller: *Predicting the compressive mechanical behavior of bone*, J. Biomech. **27** (1994), pp. 1159–1168.

235. N. Kikuchi and J. T. Oden: *Contact problems in elasticity*, SIAM, Philadelphia, 1988.

236. N. P. Kirchner: *Thermodynamically consistent modelling of abrasive granular materials. I. Non-equilibrium theory*, Proc. Roy. Soc. A. **458** (2002), pp. 2153–2176.

237. S. Klinge, R. P. Gilbert and G. A. Holzapfel: *Multiscale FEM-simulations of cross-linked actin network embedded in cytosol with the focus on the filament orientation*, Int. J. Num. Meth. Biomed. Eng. **34**(7) (2018), e2993.

238. S. Klinge and K. Hackl: *Application of the multiscale FEM to the determination of macroscopic deformations caused by dissolution-precipitation creep*, Int. J. Mult. Comp. Eng. **14**(2) (2016), pp. 95–111.

239. S. Klinge, A. Bartels, and P. Steinmann: *The multiscale approach to the curing of polymers incorporating viscous and shrinkage effects*, Int. J. Solid. Struct. **49** (2012), pp. 3883–3900.

240. S. Klinge: *Determination of the geometry of the RVE for cancellous bone by using the effective complex shear modulus*, Biomechan. Model. Mechanobiol. **12**(2) (2013), pp. 401–412.

241. S. Klinge, K. Hackl, and R.P. Gilbert: *Investigation of the Influence of Reflection on the Attenuation of Cancellous Bone*, Biomechan. Model. Mechanobiol. **12**(1) (2013) pp. 185–199.

242. S. Klinge: *Parameter Identification for Two-Phase Nonlinear Composites*, Comput. Struct. **108-109** (2012), pp. 118–124.

243. S. Klinge and K. Hackl: *Application of the multiscale FEM to the modeling of nonlinear composites with a random microstructure*, Int. J. Multiscale Comp. Eng. **10**(3) (2012), pp. 213–227.

244. S. Klinge: *Inverse analysis for multiphase nonlinear composites with random structure*, Int. J. Multiscale Comp. Eng. **10**(4) (2012), pp. 361–373.

245. D. Kolymbas: *A novel constitutive law for soils.* in *Proc. 2nd Int. Conf. Constitutive laws for Engineering Materials*, C. S. Desai, ed., 1987, pp. 319–326.

246. E. Krner: *Elastic moduli of perfectly disordered composite materials*, J. Mech. Phys. Solids, **15** (1967), pp. 319–329.

247. K. L. Kuttler and M. Shillor: *Set-valued pseudomonotone maps and degenerate evolution inclusions*, Comm. Contemp. Math. **1** (1999), pp. 87–123.

248. T. Kundu: *Ultrasonic Nondestructive Evaluation*, CRC Press, Boca Raton, 2004.

249. V. D. Kupradze: *Three-dimensional Problems of the Mathematical Theory of Elasticity and Thermoelasticity*, North-Holland, Amsterdam, 1979.

250. K. Kuratowski: *Topology*, v. 1, Academic Press, New York, 1966.

251. R. S. Lakes, H. S. Yoon, and J. L. Katz: *Slow compressional wave propagation in wet human cortical bone*, Science, **220** (1992), pp. 513–515.

252. R. S. Lakes, H. S. Yoon, and J.L. Katz.: *Ultrasonic wave propagation and attenuation in wet bone*, J .Biomed. Eng. **8** (1986), pp. 143–148.
253. R. Lakes: *Viscoelastic Properties of Cortical Bone*, in *Bone Mechanics Handbook*, S. C. Cowin, ed., CRC, Boca Raton, 2001.
254. A. R. Laliena and F.-J. Sayas: *Theoretical aspects of the application of convolution quadrature to scattering of acoustic waves*, Numer. Math. **112** (2009), pp. 637–678.
255. L. Landau and E. Lifchitz: *Mécanique des Fluides*, Mir, Moscow, 1971.
256. C. M. Langton, S. B. Palmer, and R. W. Porter: *The measurement of broadband ultrasonic attenuation in cancellous bone*, Eng. Medicine **13** (1984), pp. 89–91.
257. C. M. Langton, C. F. Njeh, R. Hodgskinson, and J. D. Curey: *Prediction of mechanical properties of human cancellous by broadbeam ultrasonic attenuation* Bone, **18** (1996), pp. 495–503.
258. C. M. Langton and C. F. Njeh: *The Physical Measurement of Bone*, IOP, Bristol, 2004.
259. W. Lauriks, J. Thoen, I. Van Ashbroek, G. Lowt, and G. Vanderperre: *Propagation of ultrasonic pulses through trabecular bone*, J. Phys. Colloq. **4** (1994), pp. 1255–1258.
260. P. Y. Le Bas, F. Luppe, J. M. Conoir, and H. Franklin: *N-shell cluster in water: Multiple scattering and splitting of resonances*, J. Acoust. Soc. Am. **115**(4) (2004), pp. 1460–1467.
261. R. A. Lebensohn: *N-site modeling of a 3D viscoplastic polycrystal using fast Fourier Transfrom*. Acta materialia **49** (2001), pp. 2723–2737.
262. R. Leis: *Initial Boundary Value Problems*, John Wiley, New York, 1986.
263. S. G. Lekhnitskii: *Theory of Elasticity of an Anisotropic Elastic Body*, Holden Day, San Francisco, 1963.
264. T. Lemaire, E. Capiez-Lernout, J. Kaiser, S. Naili, E. Rohan and V. Sansalone: *A multiscale theoretical investigation of electric measurements in living bone: Piezoelectricity and Electrokinetics*, Bull. Math. Biol. **73** (2011), pp. 2649–2677.
265. J. Lemaitre, J.L. Chaboche: *Mechanics of Solid Materials*, Cambridge, 1990.
266. G. L. Lemoine, M. Y. Ou, and R. J. Leveque: *High-resolution finite volume modeling of wave propagation in orthotropic poroelastic media*, SIAM J. Sci Comput. **35**(1) (2013), pp. B176–B206.
267. T. Levy. *Acoustic phenomena in elastic porous media*, Mech. Res. Comm. **4**(4) (1977), pp. 253–257.
268. T. Levy and E. Sanchez-Palencia: *Equations and interface conditions for acoustic phenomena in porous media*, J. Math. Anal. Appl. **6** (1977), pp. 830-834.
269. T. Levy: *Fluids in porous media and suspensions*, in *Homogenization Techniques for Composite Media, Lecture Notes in Physics*, **272**, Springer, Berlin, 1987, pp. 64–119.
270. T. Levy. *Propagation waves in a fluid-saturated porous elastic solid*, Int. J. Eng. Sci. **17** (1979), pp. 1005–1014.
271. J. Li: *Staggered Grid FDTD for Ultrasound Propagation Through Cancellous Bones*, Ph.D. diss., University of Delaware, 2013.
272. J. L. Lions and E. Magenes: *Non-Homogenenous Boundary Value Problems and Applications*, v. 1, Springer, New York, 1972.
273. P.- L. Lions: *Mathematical Topics in Fluid Mechanics*, v.1, Clarendon Press, Oxford, 1996.
274. P.- L. Lions: *Mathematical Topics in Fluid Mechanics*, v.2, Clarendon Press, Oxford, 1998.
275. Y. J. Liu and L. Shen: *A dual BIE approach for large-scale modelling of 3-D electrostatic problems with the fast multipole boundary element method*, Int. J. Numer. Meth. Eng. **71** (2007), pp. 837–855.

276. J. Liu, M. Brio, and J. V. Moloney: *Overlapping Yee FDTD Method on Nonorthogonal Grids*, J. Sci. Comput. **39**(1) (2009), pp. 129–143.
277. A. S. Lewis, D. R. Luke, and J. Malick: *Local, convergence for alternating and averaged nonconvex projections*, Found. Comput. Math. **9**(4) (2008), pp. 485–513.
278. J. Liu, M. Brio, Y. Zeng, A. Zakharian, W. Hoyer, S. W. Koch, and J. V. Moloney: *Generalization of the FDTD algorithm for simulations of hydrodynamic nonlinear Drude model*, J. Comput. Phys., **229**(17) (2010), pp. 5921–5932.
279. J. Liu, M. Brio, and J. V. Moloney: *Subpixel smoothing finite-difference time-domain method for material interface between dielectric and dispersive media*, Opt. Lett. **37** (2012), pp. 4802–4804.
280. J. Liu, M. Brio, and J. V. Moloney: *Transformation optics based local mesh refinement for solving Maxwell's Equations*, J. Comput. Phys. **258** (2014), pp. 359–370.
281. J. Liu, M. Brio, and J. V. Moloney: *An overlapping Yee FDTD method for material interfaces between anisotropic dielectrics and general dispersive or PEC media*, International Journal of Numerical Modelling: Electronic Networks, Devices and Fields, **27** (2014), pp. 22–33.
282. O. Lopez-Pamies and P. P. Castañeda: *Second-order estimated for the macroscopic response and loss of ellipticity in porous rubbers at large deformations*, J. Elasticity, **76** (2005), pp. 247–287.
283. Ch. Lubich: *On the multistep time discretization of linear initial-boundary value problems and their boundary integral equations*, Numer. Math. **67** (1994), pp. 365–389.
284. Ch. Lubich and R. Schneider: *Time discretization of parabolic boundary integral equations*, Numer. Math. **63**(4), (1992), pp. 455–481.
285. G. L. Lukacs, P. Haggie, O. Seksek, D. Lechardeur, N. Freedman, and A. S. Verkman: *Size-dependent DNA Mobility in Cytoplasm and Nucleus*, J. Biol. Chem. **275**(3) (2000), pp. 1625–1629.
286. F. Luppé, J.M. Conoir and H. Franklin: *Scattering by fluid cylinder in a porous medium: Application to trabecular bone*, J. Acoust. Soc. Am. **111**(6) (2002), pp. 2573–2582.
287. J. Lyklema: *Fundamentals of colloid and Interface Surfaces*, Academic Press, London, 1993.
288. A. Mantoglou and J. L. Wilson: *The turning bands method for simulation of random fields using line generation by a spectral method*, Water Resources Research **18** (1982), pp. 1379–1394.
289. R. B. Martin: *Porosity and specific surface of bone*, Crit. Rev. Biomed. Eng. **10** (1984), pp. 97–222.
290. D. Marshall, O. Johnell, and H. Wedel: *Meta-analysis of how well measures of bone mineral density predict occurrence of osteoporotic fractures*, BMJ, **312** (1996), pp. 1254–1259.
291. J. E. Marsden and T. J. R. Hughes: *Mathematical Foundations of Elasticity*, Dover Publications, New York, 1983.
292. W. Maurel, Y. Wu, N. M. Thalmann, and D. Thalmann: *Biomechanical Models for Soft Tissue Simulation*, Basic Research Series, Springer, Berlin, 1991.
293. R. W. McCalden , J. A. McGeough, and C. W. Court-Brown: *Age-related changes in the compressive strength of cancellous bone. The relative importance of changes in density and trabecular architecture*, J. Bone Jt. Surg. **79** (1997), pp. 421–427.
294. M. L. McKelvie and S. B. Palmer: *The interaction of ultrasound with cancellous bone*, Phys. Med. Biol. **10** (1991), pp. 1331–1340.

295. W. Maurel, Y. Wu, N.M. Thalmann, and D. Thalmann: *Biomechanical Models for Soft Tissue Simulation*, Springer, Berlin, 1998.
296. M. D. McKee and Jaro Sodek: *Bone matrix proteins*, in THE OSTEOPOROSIS PRIMER, J. E. Hendersen and D. Goltzman, eds., Cambridge, 2000.
297. M. C. van der Meulen, K. J. Jepsen, and B. Mikic: *Understanding bone strength: Size isn't everything*, Bone, **29** (2001), pp. 101–104.
298. J. C. Michel and P. Suquet: *Nonuniform transformation field analysis*, Int. J. Solids Struct. **40** (2003), pp. 6937–6955.
299. J. C. Michel and P. Suquet: *Computational analysis of nonlinear composite structures using the nonuniform transformation fields analysis*, Comput. Methods Appl. Mech. Eng. **193**(48-51) (2004), pp. 5477–5502.
300. C. Miehe, J. Schrder, and C. Bayreuther: *On the homogenisation analysis of composite materials based on discretized fluctuations on the microstructure*, Acta Mechanica, **155**, (2002), pp. 1–16.
301. C. Miehe, J. Schotte, and M. Lambrecht: *Homogenisation of inelastic solid materials at finite strains based on incremental minimization principles*, J. Mech. Phys. Solids, **50**, (2002), pp. 2123–2167.
302. A. Mikelić: *Mathematical derivation of the Darcy-type law with memory effects, governing transient flow through porous medium*, Glasnik Matematički **29**(49) (1994), pp. 57–77.
303. C. Miranda: *Partial Differential Equations of Elliptic Type*, Springer, Berlin, 1955.
304. J. Moré: *The Levenberg-Marquardt algorithm. Implementation and theory*, in *Numerical analysis. Proc. Biennial Conf. Dundee 1977*, G. Watson, ed., Springer series in Lecture Notes in Mathematics, **30**, 1978, pp. 105–116.
305. T. Mori and K. Tanaka: *Average stress in matrix and average elastic energy of materials with misfitting inclusions*, Acta Metallurgica, **21** (1973), pp. 571–574.
306. J. Monnier and L. Trabucho: *An existence and uniqueness result in bone remodeling theory*, Computer Meth. Applied Mech. Eng. **151** (1998), pp. 539–544.
307. E. F. Morgan and T. M. Keaveny: *Dependence of yield strain of human trabecular bone on anatomic site*, J. Biomech., **34**, (2001), pp. 569–577.
308. R. Müller, H. Van Campenhout, B. Van Damme, G. Van Der Perre, J. Dequeker, T. Hildebrand, and P. Rüegseggers: *Morphometric Analysis of Human Bone Biopsies: A Quantitative Structural Comparison of Histological Sections and Micro-Computed Tomography*, Bone **23** (1) (1998), pp. 59–66.
309. C. Moyne and M. A. Murad: *Electro-chemo-mechanical coupling in swelling clays derived from a micro/macro-homogenization procedure*, Solids and Structures, **39** (2002), pp. 6159–6190.
310. H. Moulinec and P. Suquet: *Intraphase strain heterogeneity in nonlinear composites: A computational approach*, Europ. J. of Mech. A. Solids, **22** (2003), pp 751–770.
311. E. F. Morgon, O. C. Yeh, W. C. Chang, and T. M. Keaveny: *Nonlinear behavior of trabecular bone at small strains*, J. Biomech. Eng. **123**(1) (2001), pp. 1–9.
312. P. M. Morse and H. Feshbach: *Methods of Mathematical Physics* volumes I & II, McGraw-Hill, New York, 1953.
313. F. Murat and L. Tartar: *H-convergence*, in *Topics in the mathematical modeling of composite materials*, A. Charkaev, R. Kohn, and Burkhauser, eds., 1997, pp. 21–43.
314. M. Neuss-Radu: *Some extensions of two-scale convergence*, C.R. Acad. Sci. Paris t. 322 Serie I (1996), pp. 899–904.

315. G. Nguetseng: *A general convergence result for a functional related to the theory of homogenization*, SIAM J. Math. Anal. **20** (1989), pp. 608–623.
316. G. Nguetseng: *Asymptotic analysis for a stiff variational problem arising in mechanics*, SIAM J. Math. Anal. **21**(6) (1990), pp. 1394-1414.
317. P. H. F. Nicholson, R. Müller, X. G. Cheng, P. Rüegsegge, G. Van Der Perre, J. Dequeker, and S. Boonen: *Quantitative Ultrasound and Trabecular Architecture in the Human Calcaneus*, J. Bone and Mineral Research, **16**(10) (2001), pp. 1186–1892.
318. P. H. F. Nicholson, G. Lowet, C. M. Langton, J. Dequeker, and G. Van der Perre: *A comparison of time-domain and frequency-domain approaches to ultrasonic velocity measurement in trabecular bone*, Phys. Med. Biol. **41** (1996), pp. 2421–2435.
319. P. H. F. Nicholson and M. L. Bouxsein: *Bone marrow influences quantitative ultrasound measurements in human cancellous bone*, Ultrasound in Med. & Biology bf 28 (3) (2002), pp. 369–375.
320. C. F. Njeh, C. M. Langton: *The effect of cortical endplates on ultrasound velocity through the calcaneus: An in vitro study*, Br. J. Radiol. **70** (1997), pp. 504–510.
321. C.F. Njeh, D. Hans, T. Fuerst, C. C. Gluer, and H. K. Genant: *Quantitative Ultrasound. Assessment of Osteoporosis and Bone Status*, London: Martin Duniz, 1999, pp. 391–399.
322. J. Nocedal and S. J. Wright: *Numerical Optimization*, Springer-Verlag, Wien, New York, 1999.
323. J. T. Oden and J. A. C. Martins: *Model and computational methods for dynamic friction phenomena*, Comput. Meth. Appl. Mech. Eng. **52** (1985), pp. 527–634.
324. J. T. Oden and T. I. Zohdi: *Analysis and adaptive modeling of highly heterogeneous elastic structures*, Comp. Met. Appl. Mech. Eng. **148** (1997) pp. 367–391.
325. E. Ogam, A. Wirgin, Z. E. A. Fellah, J.-P. Groby, W. Lauriks, J.-Y. Chapelon, C. Depollier, L. DeRyck, N. Sebaa, R.P. Gilbert, and Y. Xu: *Recovery of the mechanical parameters of long bones from their vibroacoustic impulse response,* SAPEM, Lyon, France, 2005.
326. E. Ogam, C. Masson, S. Erard, A. Wirgin, Z. E. A. Fellah, J.-P Groby, and Y. Xu: *On the vibratory response of a human tibia: Comparison of 1D Timoshenko model, 3d FEM and experiment*, Proceedings SFA2006, Paris, 2006.
327. E. Ogam and A. Wirgin: *Recovery of the ARMA model parameters from the vibration response of a finite length elastic cylinder using neural networks*, in *Proceedings of ISMA2004 - Noise and Vibration Engineering*, P. Sas and M. De Munck, eds., Leuven, 2004, pp. 2437–2446.
328. E. Ogam: *Caractéisation ultrasonore et vibroacoustique de la santé mécanique des os humains*, Ph.D. diss., Université de provence aix Marseille I, Marseille, France, 2007.
329. E. Ogam, A. Wirgin, Z. Fellah, C. Masson, P. Guillemain, F. Gabrielli, J.-P. Groby, and R. Gilbert: *Vibration spectroscopy and guided wave propagation data as indicators of structural and mechanical degradation of human bones*, J. Acoust. Soc. Am. **123** (2008), p. 3786.
330. O. A. Oleinik, A. S. Shamaev, and G. A. Yosifian: *Mathematical Problems in Elasticity and Homogenization*, North Holland, Amsterdam, 1992.
331. C. Oskay and J. Fish: *On calibration and validation of eigendeformation-based-multiscale models for failure analysis of heterogeneous systems*, Comput. Meth. **42** (2008), pp. 181–195.
332. C. Oskay and J. Fish: *Eigendeformation-based reduced order homogenization*, Comp. Meth. Appl. Mech. Engng. **196**(7) (2007), pp. 1216–1243.
333. M. Ostoja-Starzewski: *Material spatial randomness: From statistical to representative volume element*, Probabilistic engineering mechanics, **21**(2) (2006), pp. 112–132.

334. R. Othman and G. Gary: *Dispersion identification using the Fourier analysis of resonances in elastic and viscoelastic rods*, in *Acoustics, Mechanics, and the Related Topics of Mathematical Analysis*, A. Wirgin, ed., World Scientific, 2002, pp. 265–272.
335. A. Panchenko, L. Barannyk, and R. P. Gilbert: *Closure method for spatially averaged dynamics of particle chains*, Nonlinear Analysis: Real World Applications **12** (2011), pp. 1681–1697.
336. F. Padila, F. Jenson, V. Bousson, F. Peyrin, and P. Laugier: *Relationships of trabecular bone structure with quantitative ultrasound parameters: In vitro study on human proximal femur using transmission and backscatter measurements*, Bone **42** (2008), pp. 1193–1202.
337. F. Padilla and P. Laugier: *Phase and group velocities of fast and slow compressional waves in trabecular bone*, J. Acoust. Soc. Amer. **108** (2000), pp. 1949–1952.
338. F. Padilla, F. Peyrin and P. Laugier: *Prediction of backscatter coefficient in trabecular bones using a numerical model of three-dimensional microstructure*, J. Acoust. Soc. Am. **113** (2003), pp. 1122–1129.
339. D.H. Pahr and P.K. Zysset: *Influence of boundary conditions on computed apparent elastic properties of cancellous bone*, Biomech. Model Mechanobiol. **7** (2008), pp. 4963–4976.
340. A. Panchenko: *G-convergence and homogenization of viscoelastic flows* ArXiv preprint http://xxx.lanl.gov/abs/0706.1088.
341. A. Pankov: *G-convergence and Homogenization of Nonlinear Partial Differential Operators*, Kluwer, Dordrecht, 1997.
342. A. M. Parfitt, C. H. E. Mathews, A. R. Villaneuva, M. Kleerekopper, B. Frame, and D. D. Rao: *Relationships between surface and volume, and thickness of iliac trabecular bone in aging and osteoporosis*, Calcified Tissue Int. **72** (1983), pp. 396–409.
343. C. Picart and P. Carpentier: *Human blood shear yield stress and its hematocrit dependence*, J. Rheol. **42**(1) (1998), pp. 1–12.
344. W. H. Press, S. A. Teukolsky, W. T. Vetterling, and B.P. Flannery: *Numerical Recipes in Fortran*, Cambridge University Press, 1992.
345. S. R. Pride: *Governing equations for the coupled electromagnetics and acoustics of porous media*, Phys. Review B **50**(21) (1994), pp. 678–696.
346. S. R. Pride and M. W. Haartsen: *Electroseismic wave properties*, J. Acoust. Soc. Am. **100**(3) (1996), pp. 1301–1315.
347. M. Reed and B. Simon: *Methods of Modern Mathematical Physics V.1, Functional Analysis*, Academic Press, New York, 1972.
348. D. T. Reilly and A. H. Burstein: *The elastic and ultimate properties of compact bone tissue.*, J. Biomech. **8** (1975), pp. 393–405.
349. G. Reinish and A. Nowick: *Piezoelectric properties of bone as functions of moisture content.* Nature, **253** (1975), pp. 626–627.
350. M. Renardy and R. C. Rogers: *An Introduction to Partial Differential Equations*, Springer, Texts in Applied Mathematics, Berlin, 2002.
351. J. Y. Rho, R. B. Ashman, and C. H. Turner: *Young's modulus of trabecular and cortical bone material: Ultrasonic and microtensile measurements*, J. Biomech. **26** (1993), pp. 111–119.
352. J.-Y. Rho: *An ultrasonic method for measuring the elastic properties of human tibial cortical and cancellous bone*, Ultrasonics, **34** (1996), pp. 777–783.
353. C. Rich, E. Klinik, R. Smith, and B. Graham: *Measurement of bone mass from ultrasonic transmission time*, Proc. Soc. Exper. Biol. Med. **123** (1966), pp. 282–285.

354. B. van Rietbergen, R. Huiskes: *Elastic constants of canellous bone*, in *Bone Mechanics Handbook*, S.C. Cowin,ed., CRC, Boca Raton, 2001, pp. 15.1–15.24.
355. R. T. Rockafellar and R. J.-B. Wets: *Variational Analysis*, Springer, New York, 2004.
356. D. Royer and E. Dieulesaint: *Elastic Waves in Solids, Free and Guided Propagation, vol. 1*, Advanced Texts in Physics, Springer, Berlin, 2000.
357. M. R. Rubin, D. W. Dempster, T. Kohler, M. Stauber, H. Zou, E. Shane, T. L. Nickolas, E. Stein, J. Sliney Jr, S.J. Silverberg, J.P. Bilezikian and R. Müller: *Three dimensional cancellous bone structure in hypoparathyroidism*, Bone **46**(1) (2010), pp. 190–195.
358. W. Rudin: *Functional Analysis*, McGraw-Hill, New York, 1973.
359. S. Saha and R. S. Lakes: *A noninvasive technique for detecting stress waves in bone using the piezoelectric effect*, Biomedical Eng. **24** (1977), pp. 508–512.
360. E. Sarkar, B. H. Mitlak, M. Wong, J. L. Stock, and K. D. Harper: *Relationships between bone mineral density and incident vertebral fracture risk with raloxifene therapy*, Bone Miner. Res. **17** (2002), pp. 1–10.
361. A. Saied, K. Raum, L. Leguerney, and P. Laugier: *Spatial distribution of anisotropic acoustic impedance as assessed by time-resolved 50-mHz scanning acoustic microscopy and its relation porosity in human cortical bone*, Bone **43** (2008), pp. 187–194.
362. J. Sanchez-Hubert: *Asymptotic study of the macroscopic behavior of a solid-liquid mixture*, Math. Methods Appl. Sci. **2** (1980), pp. 1–18.
363. E. Sanchez-Palencia. *Non-Homogeneous Media and Vibration Theory*, Springer Lecture Notes in Physics **129**, 1980, pp. 158–190.
364. D.S. Sankara and K. Hemalathab: *Non-Newtonian fluid flow model for blood flow through a catheterized artery: Steady flow* Appl. Math. Model. **31**(9) (2007), pp. 1847-1864.
365. V. Sasidhar and E. Ruckstein: *Electroyte osmosis through capillaries*, J. Colloid. Interf. Sci. **82**(2) (1981), pp. 439–457.
366. V. Sasidhar and E. Ruckstein: *Anomolous effects during electrolyte osmosis across charged porous membranes*, **85**(2) (1982), pp. 332–362.
367. M. Sasso, G. Haiat, Talmant, P. Laugier, and S. Naili: *Singular value decomposition-based wave extraction in axial transmission; Application to cortical bone ultrasonic characterization*, IEEE Trans. on Ultrasonics, Ferroelectrics and Freq. Control **55** (2008), pp. 1328–1342.
368. F.-J. Sayas: *Retarded potentials and time domain boundary integral equations: A roadmap*, Comput. Math. **50** Springer Series, 2016.
369. D. G. Schaeffer: *A mathematical model for localization in granular flow*, Proc. Roy. Soc. A. **436** (1992), pp. 217–250.
370. M. Schanz: *Wave Propagation in Viscoelastic and Poroelastic Continua*, Lecture Notes in Applied Mechanics **1**, Springer, Berlin, 2001.
371. S. C. Schuit, M. van der Klift , A. E. Weel, C. H. de Laet, H. Burger, E. Seeman, A. Hofman, A. G. Uitterlinden, J.P. van Leeuwen, and H. P. Pols: *Fracture incidence and association with bone mineral density in elderly mean and women*, Bone, **34** (2004), pp. 195–202.
372. A. Sierou and J. F. Brady: *Rheology and microstructure in concentrated noncolloidal suspensions*, J. Rheology, **46** (2002), pp. 1031–1056.
373. T. H. Smit, J. M. Huyghe and S. C. Cowin: *Estimation of poroelastic parameters of cortical bone*, J. Biomech. **35** (2002), pp. 829–835.
374. S. Spagnolo: *Sul limite della soluzione di problemi di Cauchy relativi all'equazione del calore*, Ann. Scu. Norm. Super. Piza, Cl. Sci, **21** (1967), pp. 657–699.

375. S. Spagnolo: *Sul convergenza di soluzione di equazioni paraboliche e ellipticihe*, Ann. Scu. Norm. Super. Piza, Cl. Sci, **22** (1968), pp. 577–598.
376. R. Strelitzki and J. A. Evans: *On the measurement of the velocity of ultrasound in the os calcis using short pulses*, Eur. J. Ultrasound, **4**(3) (1996), pp. 205–213.
377. P. Suquet: *Effective properties of nonlinear composites.* in Continuum Micromechanics, P. Suquet, eds. , CISM, **377**, Springer, Wien New York, 1997.
378. D. R .S. Talbot and J. R. Willis: *Variational principles for inhomogeneous non-linear media*, IMA-Journal of Applied Mathematics, **35** (1985), pp. 39–54.
379. A. Taflove and S. Hagness: *Computational Electrodynamics: The Finite-Difference Time-Domain Method*, Artech House, Norwood, MA, 3rd edition, 2005.
380. L. Tartar: *Cours Peccot au Collége de France*, preprint (1977).
381. R. L. Taylor: *Finite element analysis of linear shell problems*, in *The Mathematics of finite element and Applications, VI*, In J.R Whiteman, ed., 1998.
382. L. Tartar: *Appendix* in *Non-Homogeneous, Media and Vibration Theory*, Springer, Berlin, 1980.
383. R. Temam: *Naiver-Stokes Equations*, 3rd ed., North-Holland, New York, 1984.
384. K. Terada and N. Kikuchi: *A class of general algorithms for multi-scale analysis of heterogeneous media*, Comput. Methods Appl. Mech. Eng. **192** (2001), pp. 5427–5464.
385. G.B. Thurston: *Viscoelasticity of Human Blood*, Biophysical Journal **12** (1972), pp. 1205–1217.
386. A. Tompson, R. Ababou, and L. Gelhar: *Implementation of the three-dimensional turning bands random field generator*, Water Resources Research **25** (1989), pp. 2227–2243.
387. C. A. Truesdell: *A first Course in Rational Continuum Mechanics*, Academic Press, New York, 1997.
388. L. Trabucho and J. M. Viano: *Mathematical modelling of rods*, in Handbook of Numerical Analysis, P.G. Cialet and J.L. Lions, eds., North-Holland, Amsterdam, 1996, pp. 487–974.
389. L. Trabucho: *Non-linear bone remodeling: An existence and uniqueness result*, Math. Meth. Appl. Sci. **23** (2000), pp. 1331–1346.
390. L. N. Trefethen: *Spectral Methods in MATLAB*, SIAM, Philadelphia, 2000.
391. E. Vanmarcke: *Random Fields: Analysis and Synthesis*, World Scientific Publishing Company, 2010.
392. C. R. Vogel: *Computational methods for inverse problems*, SIAM, Philadelphia, 2002.
393. L. Wang, S. P. Fritton, S. C. Cowin, and S. Weinbaum: *Fluid pressure on depends upon osteonal microstructure: Modeling an oscillatory bending experiment*, J. Biomech. **32** (1999), pp. 663–672.
394. K. A. Wear: *Ultrasonic attenuation in human calcaneus from 0.2 to 1.7 MHz*, IEEE Trans. Ultrason. Ferroelectr. Freq. Control **48**(2) (2001), pp. 602–608.
395. K. A. Wear: *Frequency dependence of ultrasonic backscatter from human trabecular bone: Theory and experiment*, J. Acoust. Soc. Am. **106** (1999), pp. 3659–3664.
396. K. A. Wear: *Measurements of phase velocity and group velocity in human calcaneus.* Ultrasound Med. Bio. **26**(4) (2000), pp. 641–646.
397. K. A. Wear: *Ultrasonic attention in human calacaneus from 0.2 to 1.7 MHz*, IEEE Trans. Ultrasin. Ferroelectr. Freq. Control **48** (2001), pp. 602–608.
398. G. B. Whitham: *Linear and Nonlinear Waves*, Pure and Applied Mathematics Series, Wiley, New York (1999).

399. J. L. Williams and W. J. H. Johnson: *Elastic constants of composites formed from PMMA bone cement and anisotropic bovine tibial cancellous bone*, J. Biomech. **22**(6/7) (1989), pp. 673–682.
400. J. L. Williams: *Ultrasonic wave propagation in cancellous and cortical bone: Prediction of some experimental results by Biot's theory*, J. Acoust. Soc. Am. **91** (1992), pp.1106–1112.
401. K. L. Williams, D. R. Jackson, E. I. Thorsos, D. Tang, and S. G. Schock: *Comparison of sound speed and attenuation measured in a sandy sediment to predictions based on the Biot theory of porous media*, IEEE J. Oceanic Eng. **27**(3) (2002), pp. 413–428.
402. J. R. Willis: *Bounds and self-consistent estimates for the overall properties of anisotropic composites*, J. Mech. Phys. Solids **25** (1977), pp. 185–202.
403. A. Wirgin: *Ill-Posedness and Accuracy in Connection with the Recovery of a Single Parameter from a Single Measurement*, Inverse Problems in Engineering **10**(2) (2002), pp. 105-115. DOI: 10.1080/10682760290031186
404. Y. Xu: *Transmission of ultrasonic wave in cancellous bone and evaluation of osteoporosis*, in *Acoustics, Mechanic, and the Related Topics of Mathematical Analysis*, A. Wirgin, ed., World Scientific, 2003, pp. 265–271.
405. G. Yang, J. Kabel, B. van Rietbergen, A. Odgaard, R. Huiskes, and S. C. Cowin: *The anisotropic Hooke's law for cancellous bone and wood.*, J. Elast. **53** (1999), pp. 125–146.
406. K. S. Yee: *Numerical solution of initial boundary value problems involving Maxwell's equations in isotropic media*, IEEE Trans. Antennas Propag. **14** (1966), pp. 302–307.
407. H. S. Yoon and J. L. Katz: *Ultrasonic wave propagation in human cortical bone III. Piezoelectric contribution*, J. Biomech. **9** (1976), pp. 537–540.
408. L. You, S. C. Cowin, M. B. Schaeffer, and S. Weinbaum: *A model for strain amplification in the actin cytoskeleton of osteocytes due to fluid drag on pericellular matrix*, J. Biomech. **34** (2001), pp. 1375–1386.
409. J. Yvonet and Q.C. He: *The reduced model multiscale method (R3M) for the non-linear homogenization of hyperelastic media at finite strains*, J. Comput. Phys. **223** (2007), pp. 341–368.
410. Y. Zhao and J. Liu: *FDTD for Hydrodynamic Electron Fluid Maxwell Equations*, Photonics, **2**(2) (2015), pp. 459–467.
411. Y. Zeng, W. Hoyer, J. Liu, S. W. Koch, and J. V. Moloney: *Classical theory for second-harmonic generation from metallic nanoparticles*, Phys. Rev. B, **79** (2009), p. 235109.
412. V. V. Zhikov: *On two-scale convergence*, J. Math. Sci. **120**(3) (2004), pp. 1328–1352.
413. V. V. Zhikov, S. M. Kozlov, and O. A. Oleinik: *G-convergence of Parabolic Operators*, Uspehkhi Matematicheskih Nauk, **36**, (1), (1981), pp. 11–58.
414. V. V. Zhikov, S. M. Kozlov, and O. A. Oleinik: *Homogenization and G-convergence of differential operators*, Russian Math Surveys **43** (1979), pp. 65–147.
415. O. C. Zienkiewicz and R. L. Taylor: *The Finite Element Method*, Butterworth-Heinemann, 2000.
416. T. I. Zohdi, J. T. Oden, and G. J. Rodin: *Hierarchical modeling of heterogeneous bodies*, Comp. Met. Appl. Mech. Eng. **138** (1996), pp. 273–298.
417. T. I. Zohdi and P. Wriggers: *A domain decomposition method for bodies with heterogeneous microstructure based on the material regularization*, Int. J. Sol. Struct. **36** (1999), pp. 2507–2525.
418. T. I. Zohdi, P. Wriggers, and C. Huet: *A method of substructuring large-scale computational micromechanical problems*, Comp. Met. Appl. Mech. Eng. **190**(13) (2001), pp. 5639–5656.

419. T. I. Zohdand and P. Wriggers: *Introduction to computational micromechanics*, in *Lecture Notes in Applied and Computational Mechanics*, **20**, Springer, 2005.
420. P. K. Zysset: *A review of morphology-elasticity relationships in human trabecular bone: Theories and experiments*, J. Biomech. **36** (2003), pp. 1469–1485.

Index

Milton Keynes UK
Ingram Content Group UK Ltd.
UKHW040446071024
449327UK00020B/1037

9 780367 713720